Dancing with Python

Learn to code with Python and Quantum Computing

Robert S. Sutor

BIRMINGHAM - MUMBAI

Dancing with Python

Producer: Tushar Gupta
Acquisitions Editor – Peer Reviews: Saby D'silva
Project Editor: Parvathy Nair
Content Development Editor: Bhavesh Amin
Copy Editor: Safis Editing
Technical Editor: Aditya Sawant
Proof Reader: Safis Editing
Presentation Designer: Pranit Padwal

First published: August 2021

Production reference: 3281221

Published by Packt Publishing Ltd.
Livery Place
35 Livery Street
Birmingham B3 2PB, UK.

ISBN 978-1-80107-785-9

www.packt.com

For my mother and father,
Anita and Benjamin Sutor.

Contributors

About the author

Robert S. Sutor has been a technical leader and executive in the IT industry for over 35 years. More than two decades of that have been spent in IBM Research in New York. During his time there, he worked on and led efforts in symbolic mathematical computation, mathematical programming languages, optimization, AI, blockchain, and quantum computing. He is the author of *Dancing with Qubits: How quantum computing works and how it can change the world,* also with Packt. He is the published co-author of several research papers and the book *Axiom: The Scientific Computation System* with the late Richard D. Jenks.

Sutor was an IBM executive on the software side of the business in areas including Java web application servers, emerging industry standards, software on Linux, mobile, and open source. He's a theoretical mathematician by training, has a Ph.D. from Princeton University, and an undergraduate degree from Harvard College. He started coding when he was 15 and has used most of the programming languages that have come along.

I want to thank my wife, Judith Hunter, and children, Katie and William, for their love, patience, and humor while this book was being written.

I also want to thank the following for their conversations, insight, and inspiration regarding coding, Python, Computer Science, and Quantum Computing: Abe Asfaw, Andrew Wack, Aparna Prabhakar, Barry Trager, Blake Johnson, Chris Nay, Christine Vu, Christopher Schnabel, David Bryant, Fran Cabrera, Hanhee Paik, Heather Higgins, Ismael Faro, James Weaver, James Wooten, Jay Gambetta, Jeanette Garcia, Jenn Glick, Jerry Chow, Joseph Broz, Julianna Murphy, Julien Gacon, Katie Pizzolato, Luciano Bello, Matthew B. Treinish, Mark Mattingley-Scott, Matthias Steffen, Michael Houston, Paul Nation, Rajeev Malik, Robert Loredo, Ryan Mandelbaum, Samantha Davis, Sarah Sheldon, Sean Dague, Stefan Woerner, and Zaira Nazario.

About the reviewer

Martin Renou is a Scientific Software Engineer working at QuantStack (Paris, France). Before joining QuantStack, Martin also worked as a Software Developer at Enthought (Cambridge, UK). He studied at the French Aerospace Engineering School ISAE-Supaero, with a major in autonomous systems and programming. As an open source developer, Martin has worked on a variety of projects, mostly around the Jupyter project ecosystem.

Table of Contents

List of Figures

Preface

Skill is the unified force of experience, intellect and
passion in their operation.

—John Ruskin

Coding is the art and engineering of creating software. *Code* is the collection of written instructions and functionality in one or more programming languages that provides directions for how computing hardware should operate. A *coder* creates code.

Coders go by other names as well. They are often called *software developers* or just *developers*. More traditionally, they have been called *programmers*.

The range of hardware devices that need code to tell them what to do is astounding. Cars have many computer processors in them to control how they operate and how to entertain you. As you can imagine, a vehicle with any degree of self-driving capability contains a lot of code. It's not simple programming either: artificial intelligence (AI) software makes many operating decisions.

Your mobile phone is both a computing and a communication device. Low-level code controls how your phone connects to Wi-Fi or cellular networks. Someone wrote that code, but as an app developer, you don't need to redo it; you call *functions* that access the Internet. Similarly, someone wrote the low-level graphics routines that put the color dots on the screen in the right places. While you may want to do that in some cases as an app developer, you mostly call higher-level functions that draw lines or shapes, show photos, or play videos.

Even at this level, several kinds of hardware get involved within your phone. There is the communications chip, general processor, arithmetic processor, floating-point processor, and the graphics processing unit (GPU). These are what we call *classical computers* or *classical processors*. Their architecture is descended from computers of the 1940s. While there is a range of ways of programming them, it is all called *classical coding*.

However, there is another kind of computer that has only been available on the cloud for general users since 2016. That is the *quantum computer*. Coding a quantum computer is radically different from classical device programming at the level close to the hardware. However, if you know or can learn Python, a programming language estimated to be used by over 8 million software developers globally and taught in many universities, you have a tremendous advantage in that you can do both classical and quantum computing together.

Classical hardware and software have proven themselves over the last seven decades, while quantum computing is new. It promises to help solve some kinds of problems that would take too much time, too much processing power, or too much memory, even for a classical supercomputer. Experts expect quantum computing to be useful in the future in areas including financial services, logistics, chemistry, materials science, drug discovery, scientific simulation of physical systems, optimization, and artificial intelligence.

If you plan to be a professional software developer or someone who needs high-performance computing for research, you should learn about quantum computing systems and how to code for them.

Why did I write this book?

How do you learn to code in this new world that involves both classical and quantum hardware?

One way to do it is to learn classical computing by itself. This is the traditional way of doing it, using a language such as C, C++, JavaScript, Java, Go, or Python. Along the way, you would learn how to use extra functionality in libraries of code along with the programming tools or from a third-party provider. Examples of these are the C++ Standard Library; the Java Platform, Enterprise Edition; the Python Standard Library; or the thousands of Python packages listed in the Python Package Index. [PYPI]

Once you have the philosophy, syntax, structure, and idioms of the classical programming language understood, you then learn quantum computing on top of that. For example, you could use the Qiskit open source quantum computing software development kit (SDK) along with Python. [QIS] These mesh together and operate exceptionally well. Thousands of people are already Qiskit coders. If you know Python, this is a great approach.

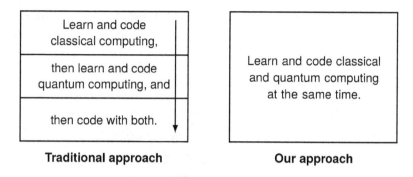

Figure 0.1: Our learning strategy

But what if you are learning to code or have only a small amount of experience? What if I could offer you the chance to learn classical and quantum computing in a unified manner? Would it be useful if I could help you understand the concepts of both so that you don't see them as different disciplines? That's what I do in this book.

For whom did I write this book?

I believe this book will be useful and engaging if one or more of these descriptions apply to you:

- You are learning to code as part of a class or course.
- Through self-study, you want to learn to think like someone who writes software.
- You want to understand and use the fundamentals of modern Python programming.
- You recognize that quantum computing will be one of the most important technologies of this century, and you want to learn the basics of how to code for it.
- You think that solving a problem through software means you use all the tools available to you, and you want to employ both classical and the newer quantum coding techniques.

Assumed reader prerequisites

I do not expect you to have any experience in coding or using Python. Some of the discussion and implementation in *Chapter 11, Searching for the Quantum Improvement*, presumes comfort with some mathematics, but nothing that isn't typically covered before a calculus course. This includes algebra, geometry, trigonometry, and logarithms.

This book does not cover the detailed mathematics and theory of quantum computing. For that, I direct you to my book *Dancing with Qubits* if you wish to learn more. [DWQ]

What does this book cover?

Given the wide use of Python and the wide variety of learning and reference materials, I have chosen to structure this book into three main parts. I give you the information you need as you need it.

Before jumping into those, however, we together explore what coders do, how they think about using programming languages, and what they expect from the tools they use. That chapter,

1. Doing the Things That Coders Do

is not specific to Python and is occasionally philosophical about the art and engineering of writing code.

After that introduction, the rest of the book proceeds in the following way.

Part I. Getting to Know Python

Being a full-featured programming language, Python implements the features described in the first chapter mentioned above. In this part, we learn how to write basic expressions including numbers and textual strings, collect objects together using data structures such as lists, and explore Python's core and extended mathematical facilities.

We then jump into defining functions to organize and make our code reusable, introduce object-oriented coding through classes, and finally interact with information within the computing environment via files.

2. Working with Expressions

3. Collecting Things Together

4. Stringing You Along

5. Computing and Calculating

6. Defining and Using Functions

7. Organizing Objects into Classes

8. Working with Files

The Python modules we introduce in this part include **abc**, **cmath**, **collections**, **datetime**, **enum**, **fractions**, **functools**, **glob**, **json**, **math**, **os**, **pickle**, **random**, **shutil**, **sympy**, and **time**.

<u>Part II. Algorithms and Circuits</u>

Now that we understand Python's core features, we're ready to explore how to make it useful to solve problems. Although many books only speak about functions and classes, we enlarge our discussion to include gates and circuits for classical and quantum computing. It's then a good time to see how we can test our code and make it run faster.

We then look at traditional problems and see how we can attack them classically. Quantum computing's reason for existence and development is that it might solve some of those problems significantly faster. We explore the how and why of that, and I point you to further reading on the topic.

<u>9. Understanding Gates and Circuits</u>

<u>10. Optimizing and Testing Your Code</u>

<u>11. Searching for the Quantum Improvement</u>

The Python modules we introduce in this part include **coverage**, **pytest**, **qiskit**, **time**, **timeit**, and **wrapt**.

<u>Part III. Advanced Features and Libraries</u>

In the final part, we address some heavy-duty but frequent applications of Python. Though we worked with text as strings earlier in the book, we revisit it with more sophisticated tools such as regular expressions and natural language processing (NLP).

The final three chapters focus on data: how to bring it into an application and manipulate it, how to visualize what it represents, and how to gain insights from it through machine learning. Machine learning itself is worth a book or two (or three or ten), so I introduce the key concepts and tools, and you can then jump off into more sophisticated Python and AI applications.

<u>12. Searching and Changing Text</u>

<u>13. Creating Plots and Charts</u>

<u>14. Analyzing Data</u>

<u>15. Learning, Briefly</u>

The Python modules we introduce in this part include **flashtext**, **matplotlib**, **nltk**, **pandas**, **pillow**, **pytorch**, **re**, **scikit-learn**, **spacy**, **string**, and **textblob**.

What conventions do I use in this book?

This book uses version 3.9 of Python or later. Most of the features are available in earlier versions of Python 3, but we do not show nor further mention the now-unsupported Python 2.

Executable Python code and its produced results are shown in a monospace font and are marked off in the text in the following way:

```
2**50
```

```
1125899906842624
```

The second expression shown is indented and is the result of running the code.

Code can also span several lines as in this example where we create and display a set of numbers that contains no duplicates:

```
print({1, 2, 3, 2, 4,
       1, 5, 3, 6, 7,
       1, 3, 8, 2})
```

```
{1, 2, 3, 4, 5, 6, 7, 8}
```

For long function, method, and class definitions, I number the lines so I can refer to them more easily in the text. The numbers are not part of the Python input, and you should not enter them if you type in the code.

```
1  def display_string(the_string, put_in_uppercase=False):
2      if put_in_uppercase:
3          print(the_string.upper())
4      else:
5          print(the_string)
```

When I refer to Python function, method, and property names in text, they appear like this: *print*, *__add__*, and *left*. Example Python type and class names in the text are **int**, **Fraction**, and **Guitar**. Python module and package names appear like **math**, **os.path**, and **qiskit**.

This book has exercises throughout the text. Work them out as you encounter them before proceeding.

Exercise 0.1

Is this a sample exercise?

The exercises are numbered within chapters.

Exercise 0.2

Is this another sample exercise?

Due to some typographical restrictions, square roots in mathematical expressions within sentences may not have lines over them. For example, an expression like $\sqrt{(x + y)}$ in a sentence is the same as

$$\sqrt{x + y}$$

when it appears within a standalone centered formula.

Occasionally, you may see something like [DWQ] or [DWQ, Section 3.5]. This is a reference to a book, article, or web content. The <u>References</u> section provides details about the works cited.

Dancing with Qubits covers the mathematics for and of quantum computing in great detail. That book is not a prerequisite to this, but I point you to sections in *Qubits* that pertain to any related material here for your convenience. For example:

I cover real numbers, including floating-point numbers, in section 3.5 of *Dancing with Qubits* section 3.5. [DWQ]

Download the example code files

The code bundle for the book is hosted on GitHub at `https://github.com/PacktPublishing/Dancing-with-Python`. We also have other code bundles from our rich catalog of books and videos available at `https://github.com/PacktPublishing/`. Check them out!

Download the color images

We also provide a PDF file that has color images of the screenshots/diagrams used in this book. You can download it here: `https://static.packt-cdn.com/downloads/9781801077859_ColorImages.pdf`.

Get in touch

Feedback from my readers is always welcome.

General feedback: If you have questions about any aspect of this book, mention the book title in the subject of your message and email us at `customercare@packtpub.com`.

Errata: Although we have taken every care to ensure our content's accuracy, mistakes do happen. If you have found an error in this book, we would be grateful if you report this to us. Please visit `http://www.packt.com/submit-errata`, selecting your book, clicking on the Errata Submission Form link, and entering the details.

Piracy: If you come across any illegal copies of our works in any form on the Internet, we would be grateful if you would provide us with the location address or website name. Please contact us at `copyright@packt.com` with a link to the material.

If you are interested in becoming an author: If there is a topic that you have expertise in, and you are interested in either writing or contributing to a book, please visit `http://authors.packtpub.com`.

Share your thoughts

Once you've read *Dancing with Python*, we'd love to hear your thoughts! Scan the QR code below to go straight to the Amazon review page for this book and share your feedback.

https://packt.link/r/1-801-07785-1

Your review is important to us and the tech community and will help us make sure we're delivering excellent quality content.

1

Doing the Things
That Coders Do

Talk is cheap. Show me the code.

—Linus Torvalds

There is not just one way to code, and there is not only one programming language in which to do it. A recent survey by the analyst firm RedMonk listed the top one hundred languages in widespread use in 2021. [PLR]

Nevertheless, there are common ideas about processing and computing with information across programming languages, even if their expressions and implementations vary. In this chapter, I introduce those ideas and data structures. I don't focus on how we write everything in any particular language but instead discuss the concepts. I do include some examples from other languages to show you the variation of expression.

Topics covered in this chapter

1.1 Data

If you think of all the digital data and information stored in the cloud, on the web, on your phone, on your laptop, in hard and solid-state drives, and in every computer transaction, consideration comes down to just two things: **0** and **1**.

These are *bits*, and with bits we can represent everything else in what we call "classical computers." These systems date back to ideas from the 1940s. There is an additional concept for quantum computers, that of a *qubit* or "quantum bit." The qubit extends the bit and is manipulated in quantum circuits and gates. Before we get to qubits, however, let's consider the bit more closely.

First, we can interpret **0** as *false* and **1** as *true*. Thinking in terms of data, what would you say if I asked you the question, "Do you like Punk Rock?" We can store your response as **0** or **1** and then use it in further processing, such as making music recommendations to you. When we talk about *true* and *false*, we call them *Booleans* instead of bits.

Second, we can treat **0** and **1** as the numbers 0 and 1. While that's nice, if the maximum number we can talk about is 1, we can't do any significant computation. So, we string together more bits to make larger numbers. The *binary* numbers 00, 01, 10, and 11 are the same as 0, 1, 2, and 3 when we use a decimal representation. Using even more bits, we represent 72 decimal as 1001000 binary and 83,694 decimal as 10100011011101110 binary.

$$\text{decimal: } 72 = 7 \times 10^1 + 2 \times 10^0$$
$$\text{binary: } 1001000 = 1 \times 2^6 + 0 \times 2^5 + 0 \times 2^4 + 1 \times 2^3 + 0 \times 2^2 + 0 \times 2^1 + 0 \times 2^0$$

Exercise 1.1

How would you represent 245 decimal in binary? What is the decimal representation of 1111 binary?

With slightly more sophisticated representations, we can store and use negative numbers. We can also create numbers with decimal points, also known as floating-point numbers. [FPA] We use floating-point numbers to represent or approximate real numbers. Whatever programming language we use, we must have a convenient way to work with all these kinds of numbers.

When we think of information more generally, we don't just consider numbers. There are words, sentences, names, and other textual data. In the same way that we can encode numbers using bits, we create characters for text. Using the Extended ASCII standard, for example, we can create my nickname, "Bob": [EAS]

$$01000010 \rightarrow B$$
$$01101111 \rightarrow o$$
$$01100010 \rightarrow b$$

Each run of zeros and ones on the left-hand side has 8 bits. This is a *byte*. One thousand (10^3) bytes is a *kilobyte*, one million (10^6) is a *megabyte*, and one billion (10^9) is a *gigabyte*.

If we limit ourselves to a byte to represent characters, we can only represent $256 = 2^8$ of them. If you look at the symbols on a keyboard, then imagine other alphabets, letters with accents and umlauts and other marks, mathematical symbols, and emojis, you will count well more than 256. In programming, we use the Unicode standard to represent many sets of characters and ways of encoding them using multiple bytes. [UNI]

Exercise 1.2

How can you create 256 different characters if you use 8 bits? How many could you form if you used 7 or 10?

When we put characters next to each other, we get *strings*. Thus "abcd" is a string of length four. I've used the double quotes to delimit the beginning and end of the string. They are not part of the string itself. Some languages treat characters as special objects unto themselves, while others consider them to be strings of length one.

This is a good start for our programming needs: we have bits and Booleans, numbers, characters, and strings of multiple characters to form text. Now we need to do something with these kinds of data.

1.2 Expressions

An *expression* is a written combination of data with operations to perform on that data. That sounds quite formal, so imagine a mathematical formula like:

$$15 \times 6^2 + 3 \times 6 - 4 \,.$$

This expression contains six pieces of data: 15, 6, 2, 3, 6, and 4. There are five operations: multiplication, exponentiation, addition, multiplication, and subtraction. If I rewrite this using symbols often used in programming languages, it is:

```
15 * 6**2 + 3 * 6 - 4
```

See that repeated 6? Suppose I want to consider different numbers in its place. I can write the formula:

$$15 \times x^2 + 3 \times x - 4$$

and the corresponding code expression:

```
15 * x**2 + 3 * x - 4
```

If I give x the value 6, I get the original expression. If I give it the value 11, I calculate:

```
15 * 11**2 + 3 * 11 - 4
```

We call x a *variable* and the process of giving it a value *assignment*. The expression

```
x = 11
```

means "assign the value 11 to x and wherever you see x, substitute in 11." There is nothing special about x. I could have used y or something descriptive like *kilograms*.

An expression can contain multiple variables.

Exercise 1.3

In the expression

$$a * x**2 + b * x + c,$$

what would you assign to *a*, *b*, *c*, and *x* to get

$$7 * 3**2 + 2 * 3 + 1?$$

What assignments would you do for

$$(-1)**2 + 9 * (-1) - 1?$$

The way we write data, operations, variables, names, and words together with the rules for combining them is called a programming language's *syntax*. The syntax must be unambiguous and allow you to express your intentions elegantly. In this chapter, we do not focus on syntax but more on the ideas and the meaning, the *semantics*, of what programming languages can do. In *Chapter 2, Working with Expressions*, we begin to explore the syntax and features of Python.

We're not limited to arithmetic in programming. Given two numbers *x* and *y*, if I saw **maximum**(*x*, *y*), I would expect it to return the larger value. Here "**maximum**" is the name of a *function*. We can write our own or use functions created, tested, optimized, and stored in libraries by others.

1.3 Functions

The notion of "expression" includes functions and all that goes into writing and using them. The syntax of their definition varies significantly among programming languages, unlike simple arithmetic expressions. In words, I can say

> The function `maximum(x, y)` returns x
> if it is larger than y. Otherwise, it returns y.

Let's consider this informal definition.

- It has a descriptive name: *maximum*.
- There are two variables within parentheses: x and y. These are the *parameters* of the function.
- A function may have zero, one, two, or more parameters. To remain readable, it shouldn't have too many.

- When we employ the function as in `maximum(3, -2)` to get the larger value, we *call* the function *maximum* on the *arguments* 3 and -2.

- To repeat: x and y are parameters, and 3 and -2 are arguments. However, it's not unusual for coders to call them all arguments.

- The *body* of the function is what does the work. The statement involving if-then-otherwise is called a conditional. Though "if" is often used in programming languages, the "then" and "otherwise" may be implicit in the syntax used to write the expression. "else" or a variation may be present instead of "otherwise."

- The test "x larger than y" is the inequality x > y. This returns a Boolean *true* or *false*. Some languages use "predicate" to refer to an expression that returns a Boolean.

- *maximum* returns a value that you can use in further computation. Coders create some functions for their actions, like updating a database. They might not return anything.

For comparison, this is one way of writing **maximum** in the C programming language:

```
int maximum(int x, int y) {
    if (x > y) {
        return x;
    }
    return y;
}
```

This is a Python version with a similar definition:

```
def maximum(x, y):
    if x > y:
        return x
    return y
```

There are several variations within each language for accomplishing the same result. Since they look so similar in different languages, you should always ask, "Am I doing this the best way in this programming language?". If not, why are you using this language instead of another?

Expressions can take up several lines, particularly function definitions. C uses braces "{ }" to group parts together, while Python uses indentation.

1.4 Libraries

While many books are now digital and available online, physical libraries are often still present in towns and cities. You can use a library to avoid doing something yourself, namely, buying a book. You can borrow it and read it. If it is a cookbook, you can use the recipes to make food.

A similar idea exists for programming languages and their environments. I can package together reusable data and functions, and then place them in a library. Via download or other sharing, coders can use my library to save themselves time. For example, if I built a library for math, it could contain functions like *maximum, minimum, absolute_value,* and *is_prime.* It could also include approximations to special values like π.

Once you have access to a library by installing it on your system, you need to tell your programming environment that you want to take advantage of it. Languages use different terminology for what they call the contents of libraries and how to make them available.

- Python imports modules and packages.
- Go imports packages.
- JavaScript imports bindings from modules.
- Ruby requires gems.
- C and C++ include source header files that correspond to compiled libraries for runtime.

In languages like Python and Java, you can import specific constructs from a package, such as a class, or everything in it. With all these examples, the intent is to allow your environment to have a rich and expandable set of pre-built features that you can use within your code. Some libraries come with the core environment. You can also optionally install third-party libraries. Part III of this book looks at a broad range of frequently used Python libraries.

What happens if you import two libraries and they each have a function with the same name? If this happens, we have a *naming collision* and ambiguity. The solution to avoid this is to embellish the function's name with something else, such as the library's name. That way, you can explicitly say, "use this function from here and that other function from over there." For example, `math.sin` in Python means "use the `sin` function from the `math` module." This is the concept of a *namespace.*

1.5 Collections

Returning to the book analogy we started when discussing libraries, imagine that you own the seven volumes in the *Chronicles of Narnia* series by C. S. Lewis.

- *The Lion, the Witch and the Wardrobe*
- *Prince Caspian: The Return to Narnia*
- *The Voyage of the Dawn Treader*
- *The Silver Chair*
- *The Horse and His Boy*
- *The Magician's Nephew*
- *The Last Battle*

The order I listed them is the order in which Lewis published the volumes from 1950 through 1956. In the list, the first book is the first published, the second is the second published, and so on. Had someone been maintaining the list in the 1950s, they would have appended a new title when Lewis released a new book in the series.

A programming language needs a structure like this so that you can store and insert objects as above. It shouldn't surprise you that this is called a *list*. A list is a *collection* of objects where the order matters.

If we think of physical books, you can see that lists can contain duplicates. I might have two paperback copies of *The Lion, the Witch and the Wardrobe,* and three of *The Horse and His Boy.* The list order might be based on when I obtained the books.

Another simple kind of collection is a *set*. Sets contain no duplicates, and the order is not specified nor guaranteed. We can do the usual mathematical unions and intersections on our sets. If I try to insert an object into a set, it becomes part of the set only if it is not already present.

The third basic form of collection is called a *dictionary* or *hash table*. In a dictionary, we connect a *key* with a *value*. In our example, the key could be the title and the value might be the book itself. You see this for eBooks on a tablet or phone. You locate a title within your library, press the text on the screen with your finger, and the app displays the text of the book for you to read. A key must be *unique*: there can be one and only one key with a given name.

Authors of computer science textbooks typically include collections among data structures. With these three collections mentioned above, you can do a surprising amount of coding. Sometimes you need specialized functionality such as an ordered list that contains no duplicates. Most programming libraries give you access to a broad range of such structures and allow you to create your own.

C. S. Lewis wrote many other books and novels. Another dictionary we could create might use decades like "1930s", "1940s", "1950s", and "1960s" as the keys. The values could be the list of titles in chronological order published within each ten-year span.

Collections can be parts of other collections. For example, I can represent a table of company earnings as a list of lists of numbers and strings. However, some languages require the contained objects all to be of the same type. For example, maybe a list needs to be all integers or all strings. Other languages, including Python, allow you to store different kinds of objects.

Exercise 1.4

In the section on functions, we looked at **maximum** with two parameters. Think about and describe in words how a similar function that took a single parameter, a list of numbers, might work.

An object which we can change is called *mutable*. The list of books is mutable while an author is writing and publishing. Once the author passes away, we might lock the collection and declare it *immutable*. No one can make any further changes. For dictionaries, the keys usually need to be immutable and are frequently strings of text or numbers.

Exercise 1.5

Is a string a collection? If so, should it be mutable or immutable?

1.6 Conditional processing

When you write code, you create instructions for a computer that state what steps the computer must take to accomplish some task. In that sense, you are specifying a recipe. Depending on some conditions, like if a bank balance is positive or a car's speed exceeds a stated limit, the sequence of steps may vary. We call how your code moves from one step to another its *flow*.

Let's make some bread to show where a basic recipe has points where we should make choices.

Figure 1.1: Loaves of bread. Photo by the author.

Ingredients

- 1 ½ tablespoons sugar
- ¾ cup warm water (slightly warmer than body temperature)
- 2 ¼ teaspoons active dry yeast (1 package)
- 1 teaspoon salt
- ½ cup milk
- 1 tablespoon unsalted butter
- 2 cups all-purpose flour

Straight-through recipe

1. Combine sugar, water, and yeast in a large bowl. Let sit for 10 minutes.
2. Mix in the salt, milk, butter, and flour. Stir with a spatula or wooden spoon for 5 minutes.
3. Turn out the dough onto a floured counter or cutting board. Knead for 10 minutes.
4. Butter a large bowl, place the dough in it, turn the dough over to cover with butter.

5. Cover the bowl and let it sit in a warm place for 1 hour.

6. Punch down the dough and knead for 5 minutes. Place in a buttered baking pan for 30 minutes.

7. Bake in a pre-heated 375° F (190° C, gas mark 5) oven for 45 minutes.

8. Turn out the loaf of bread and let it cool.

Problems

I call that a "straight-through" recipe because you do one step after another from the beginning to the end. The flow proceeds in a straight line with no detours. However, the recipe is too simple because it does not include conditions that must be met before moving to later steps.

- In step 1, your bread will not rise if the yeast is dead. We must ensure that the mixture is active and foamy.

- In step 2, we need to add more flour in little portions until the stirred dough pulls away from the bowl.

- In step 3, if the dough gets too sticky, we need to add even more flour.

- In step 5, an hour is only an estimate for the time it will take for the dough to double in size. It may take more or less time.

- In step 6, we want the dough to double in size in the loaf pan.

- In step 7, 45 minutes is an estimate for the baking time for the top of the bread to turn golden brown.

Conditional recipe

1. Combine sugar, water, and yeast in a large bowl. Let sit for 10 minutes. **If the mixture is foamy, then** continue to the next step. **Otherwise**, discard the mixture, and combine sugar, water, and fresh yeast in a large bowl. Let sit for 10 minutes.

2. Mix in the salt, milk, butter, and flour. Stir with a spatula or wooden spoon for 5 minutes. **If** the mixture does not pull aside from the sides of the bowl, **then** add 1 tablespoon of flour and stir for 1 more minute. **Otherwise**, continue to the next step.

3. Turn out the dough onto a floured counter or cutting board. Knead for 10 minutes. **If** the mixture is sticky, **then** add 1 tablespoon of flour and stir for 1 more minute. **Otherwise**, continue to the next step.

4. Butter a large bowl, place the dough in it, turn the dough over to cover with butter.

5. Cover the bowl and let it sit in a warm place for 1 hour. **If** the dough has not doubled in size, **then** let it sit covered in the bowl for 15 more minutes. **Otherwise**, continue to the next step.

6. Punch down the dough and knead for 5 minutes. Place in a buttered baking pan for 30 minutes. **If** the dough has not doubled in size, **then** let it sit covered in the baking pan for 15 more minutes. **Otherwise**, continue to the next step.

7. Bake in a pre-heated 375° F (190° C, gas mark 5) oven for 45 minutes. **If** the top of the bread is not golden brown, **then** bake for 5 more minutes. **Otherwise**, continue to the next step.

8. Turn out the loaf of bread and let cool.

Figure 1.2 is a *flowchart*, and it shows what is happening in the first step. The rectangles are things to do, and the diamond is a decision to be made that includes a condition.

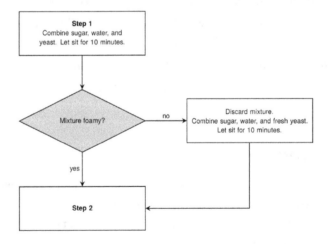

Figure 1.2: A flowchart for the first step of bread making

You may find it useful to use such a chart to map out the flow of your code.

These instructions are better. We test several conditions to determine the next course of action. We still are not checking whether a condition is met after repeating a step multiple times.

1.7 Loops

If I ask you to close your eyes, count to 10, then open them, the steps look like this:

1. Close your eyes
2. Set count to 1
3. While count is not equal 10, increment count by 1
4. Open your eyes

Steps 2 and 3 together constitute a *loop*. In this loop, we repeatedly do something while a condition is met. We do not move to step 4 from step 3 while the test returns *true*.

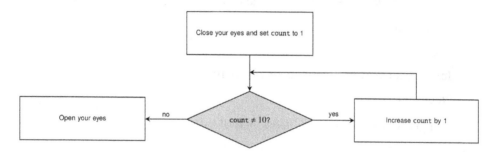

Figure 1.3: A while-loop flowchart

Compare the simplicity of Figure 1.3 to

1. Close your eyes
2. Set count to 1
3. Increment count to 2
4. Increment count to 3
5. ...
6. Increment count to 10
7. Open your eyes

If that doesn't convince you, imagine if I asked you to count to 200. Here is how we do it in C++:

```cpp
int n = 1;

while( n < 201 ) {
    n++;
}
```

 Exercise 1.6

Create a similar while-loop flowchart for counting backward from 100 to 1.

That loop was a *while-loop*, but this is an *until-loop*:

1.   Close your eyes

2.   Set count to 1

3.   Increment count by 1 until count equals 10

4.   Open your eyes

Many languages do not have *until-loops*, but VBA does:

```vba
n = 1
Do Until n>200
    n = n + 1
Loop
```

We saw earlier that the code within a function's definition is its *body*. The repeated code in a loop is also called its body.

Exercise 1.7

What is different between while-loops and until-loops regarding when you test the condition? Compare this until-loop flowchart with the previous while-loop example.

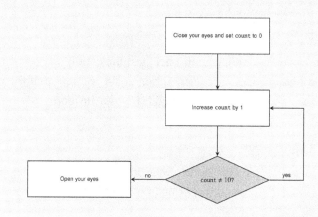

Figure 1.4: An until-loop flowchart

Our next example is a *for-loop*, so named because of the keyword that many programming languages use. A for-loop is very useful when you want to repeat something a specified number of times. This example uses the Go language:

```
sum := 0
for n := 1; n <= 50; n++ {
    sum += n
}
```

It adds all the numbers between 1 and 50, storing the result in sum. Here, := is an assignment, n++ means "replace the value of n by its previous value plus 1," and sum += n means "replace the value of sum by its previous value plus the value of n."

There are four parts to this particular syntax for the for-loop:

- the initialization: n := 1
- the condition: n <= 50
- the post-body processing code: n++
- the body: sum += n

The sequence is: do the initialization once; test the condition and, if *true*, execute the body; execute the post-body code; test the condition again and repeat. If the condition is ever *false*, the loop stops, and we move to whatever code follows the loop.

Exercise 1.8

What are the initialization, condition, post-body processing code, and body of the "count to 10" example rewritten with a for-loop?

Exercise 1.9

Draw a flowchart of a for-loop, including the initialization, condition, post-body processing code, and the body. Use the template in Figure 1.5.

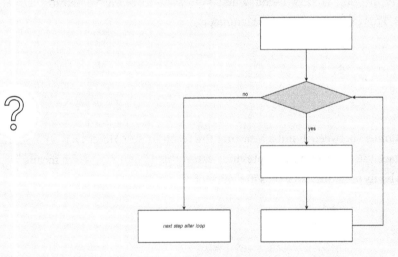

Figure 1.5: A for-loop flowchart

1.8 Exceptions

In many science fiction stories, novels, movies, and television series, characters use "teleportation" to move people and objects from one place to another. For example, the protagonist could teleport from Toronto to Tokyo instantly, without going through the time-consuming hassle of air flight.

It is also used in such plots to allow characters to escape from dangerous and unexpected conditions. Perhaps a monster might have the team cornered in the back of a cave, and there is no way to fight their way to safety. A hurried call by the leader on a communications device might demand "get us out of here!" or "beam us up now!" The STAR TREK® series and movies often featured the latter exclamation.

So, characters use teleportation for regular transportation, but also for exceptional situations that are usually life-threatening. Many programming languages also have the concept and mechanism of *exceptions*.

In a loop or a well-written function, you can see how your code's execution moves from one step to another. With an exception, you say, "try this, and if it doesn't work out, come back here." Or, you might be thinking, "I'm going about my business, but something may happen infrequently, and I need to handle it if it does." The place in your code where you handle the exception is called its *catch point*.

Suppose you want to delete a file. You have its name, and you call a function to remove the file. The only problem is that the file does not exist. How do you handle this? One way is to use a condition: if the file exists, then delete it. Another way is to use an exception: try to delete the file, and if something goes wrong (for example, it doesn't exist), *raise* an exception, and go back and decide what you should do about it.

Another good example of why you might raise an exception is division by zero. Something is seriously wrong in your code if it tries to do this, and the solution may not be obvious to you. Using an exception can help you find processing or code design errors.

An exception usually includes information that explains why it was raised. Perhaps we could not delete the file because it wasn't present. Alternatively, maybe we didn't have permission to delete it. When we *catch* the exception, we can see what happened and proceed accordingly.

An exception can be raised and caught within a single function. Or, it might originate hundreds of steps away from the catch point. Functions often call other functions, which call yet other functions. The exception could be raised from one of these deeply nested function executions.

The philosophy and accepted conventions for using exceptions vary from programming language to programming language. Like several topics in software development, exceptions can be the focus of fervor and heated debate.

1.9 Records

Consider this one-sentence paragraph with some text formatted in bold and italics:

Some words are **bold**, some are *italic*, and some are ***both***.

How could you represent this as data in a word processor, spreadsheet, or presentation application?

Let's begin with the text itself for the paragraph. Maybe we just use a string.

```
"Some words are bold, some are italic, and some are both."
```

The words are there, but we lost the formatting. Let's break the sentence into a list of the formatted and unformatted parts. I use square brackets "[ ]" around the list and commas to separate the parts.

```
[ "Some words are ",
  "bold",
  ",some are ",
  "italic",
  ", and some are ",
  "both",
  "." ]
```

Now we can focus on how to embellish each part with formatting. There are four options:

1. the text has no formatting,
2. the text is only bold,
3. the text is only italic, and
4. the text is both bold and italic.

Each part needs three pieces of information: the string holding the text, a Boolean indicating whether it is bold, and another Boolean indicating whether it is italic.

```
[ {text: "Some words are ", bold: false, italic: false},
  {text: "bold", bold: true, italic: false},
  {text: ",some are ", bold: false, italic: false},
  {text: "italic", bold: false, italic: true},
  {text: ", and some are ", bold: false, italic: false},
  {text: "both", bold: true, italic: true},
  {text: ".", bold: false, italic: false} ]
```

The object that I've shown inside braces "{ }" is a *record*. It has named members that I can access independently.

Languages often use a period "." to access the value for a member. If I assign the first record in the list to the variable "part_1," you might see something like `part_1.text` or `part_1.italic` in the code. The values are "Some words are " and *false*.

Exercise 1.10

Word processors allow much more formatting than shown here. How would you extend the example to include fonts? Make sure you take into account font names, sizes, and colors.

In C, a record is called a *struct*, and most languages have some way of representing records. The syntax I used in the example is **not** Python.

While records allow us to create objects with constituent parts, they are not much better than lists that contain items of different types, such as integers and strings. We need to raise the level of abstraction and make new kinds of objects that have functions that work just on them. We also want to hide the representation and workings of these new compound objects. That way, we can alter and improve them without breaking dependent code and applications.

1.10 Objects and classes

Let's return to cooking. I recently ran out of bay leaves in our kitchen. Bay leaves are used to flavor many foods, most notably soups, stews, and other braised meat dishes. Okay, I thought, I'll just order some bay leaves online. I found four varieties:

- Turkish or Mediterranean bay leaves, from the Bay Laurel tree,
- Indian bay leaves,
- California bay leaves, and
- Caribbean bay leaves.

I eventually discovered that what I wanted was Turkish bay leaves. I ordered six ounces. When they arrived, I realized that I had enough to last the rest of my life.

Let's get back to the varieties. While numbers and strings are built into most programming languages, bay leaves most certainly are not. Nevertheless, I want to create a structure to represent bay leaves in a first-class way. Perhaps I'm building up a collection of ingredient objects to use in digital recipes.

Before we consider bay leaves in particular, let's think about leaves in general. Without worrying about syntax, we define a class called **Leaf**. It has two properties: is_edible and latin_name.

Once I create a **Leaf** object, you can ask if it is edible and what its Latin name is. Depending on the programming language, you can either examine the property directly or use a function on an object of class **Leaf** to get the information.

A function that operates on an object of a class is called a *method*.

For most users of the class, it is best to call a method to get object information. I do not want to publish to the world how I store and manipulate that data for objects. If I do this, I, as the class designer, am free to change the internal workings. For example, maybe I initially store the Latin names as strings, but I later decide to use a database. If a user of **Leaf** assumed I always had a string present, their code would break when I changed to using the database.

This hiding of the internal workings of classes is called *encapsulation*.

Now we define a class called **BayLeaf**, and we make it a *child class* of **Leaf**. Every **BayLeaf** object is also a **Leaf**. All the methods and properties of **Leaf** are also those of **BayLeaf**. **Leaf** is the *parent class* of **BayLeaf**.

Moreover, I can redefine the methods in **BayLeaf** that I inherit from **Leaf** to make them correct or more specific. For example, by default, I might define **Leaf** so that an object of that type always returns *false* when asked if it is edible. I *override* that method in **BayLeaf** so that it returns *true*.

I can add new properties and methods to child classes as well. Perhaps I need methods that list culinary uses, taste, and calories per serving. These are appropriate to objects of **BayLeaf** because they are edible, but not to all objects in class **Leaf**. Every object that is a bay leaf should be able to provide this information.

Continuing in this way, I define four child classes of **BayLeaf**: **TurkishBayLeaf**, **IndianBayLeaf**, **CaliforniaBayLeaf**, and **CaribbeanBayLeaf**. These can provide specific information about taste, say, and this can vary among the varieties.

In many languages, a class can have at most one parent. In some, multiple inheritance is allowed, but the languages provide special rules for properties and methods. That's especially important when there are name collisions; two methods from different parents may have the same name but different behavior. Which method is called? For an analogy, consider the rules of genetics that determine the eye color of a child.

If we define the **Herb** class, then we can draw this class hierarchy. The arrows point from the class parents to their children.

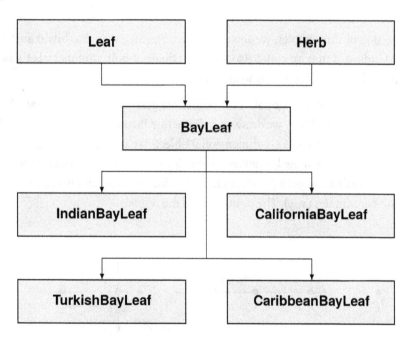

Figure 1.6: Class hierarchy for Leaf and BayLeaf

Exercise 1.11

My brother likes bay rum cologne, which is made from Caribbean bay leaf. In which classes would you place a method that tells you whether you can make cologne from a variety of bay leaf? Where would you put the default definition? Where would you override it?

In this example involving bay leaves, the methods returned information. If I create a class for polynomials that all use the same variable, then I would likely define methods for addition, subtraction, negation, multiplication, quotient and remainder, degree, leading coefficient, and so forth.

In this section, I discussed what we call object-oriented programming (OOP) as it is implemented in languages like Python, C++, Java, and Swift. The terms *superclass* and *subclass* often replace parent class and child class, respectively. Once you are familiar with Python's style of OOP, I encourage you to look at alternative approaches, such as how JavaScript does it.

1.11 Qubits

In the first section of this chapter, we looked at data. Starting with the bits **0** and **1**, we built integers, and I indicated that we could also represent floating-point numbers. Let's switch from the arithmetic and algebra of numbers to geometry.

A single point is 0-dimensional. Two points are just two 0-dimensional objects. Let's label these points 0 and 1. When we draw an infinite line through the two points and consider all the points on the line, we get a 1-dimensional object. We can represent each such point as (x), or just x, where x is a real number. If we have another line drawn vertically and perpendicular to the first, we get a plane. A plane is 2-dimensional, and we can represent every point in it with coordinates (x, y). The point $(0, 0)$ is the *origin*.

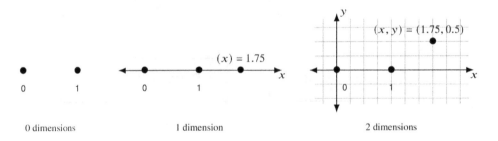

Figure 1.7: Points in 0, 1, and 2 dimensions

This is the Cartesian geometry you learned in school. As we move from zero dimensions to two in Figure 1.7, we can represent more information. In an imprecise sense, we have more room in which to work. We can even link together the geometry of the plane with *complex numbers*, which extend the real numbers. We see those in section 5.5.

Now I want you to think about taking the standard 2-dimensional plane and wrapping it completely around a sphere. If that request strikes you as bizarre, consider this: start with a sphere and balance the plane on top. Make the origin of the plane sit right on the north pole of the sphere. Now start uniformly bending the plane down over the sphere, compressing it together as you move out from the origin.

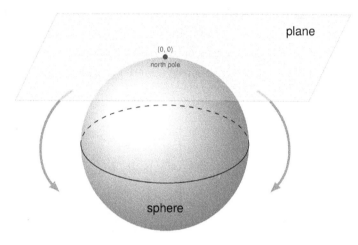

Figure 1.8: A plane balanced on a sphere

Carry this to its completion, and the "infinity" ∞ of the plane sits at the south pole. In mathematical terms, this is a stereographic projection of the plane onto the sphere.

That was mind-warping as well as plane-warping! The sphere sits in three dimensions, but we can think about its surface as two-dimensional.

Exercise 1.12

We use longitude and latitude measured in degrees to locate any location on our planet. Those are the two coordinates for the "dimensions" of the surface of the Earth. Look at an image of the planet with longitude lines drawn and see how they get closer as you approach the poles. This is very different from the usual xy-plane, where the lines stay equally spaced.

With that geometric discussion as the warm-up, I am now ready to introduce the *qubit*, which is short for "quantum bit." While a bit can only be **0** or **1**, a qubit can exist in more states. Qubits are surprising, fascinating, and powerful. They follow strange rules which may not initially seem natural to you. According to physics, these rules may be how nature itself works at the level of electrons and photons.

A qubit starts in an initial state. We use the notation $|0\rangle$ and $|1\rangle$ when we are talking about a qubit instead of the **0** and **1** for a bit. For a bit, the only non-trivial operation you can perform is switching **0** to **1** and vice versa. We can move a qubit's state to any point on the sphere shown in the center of Figure 1.9. We can represent more information and have more room in which to work.

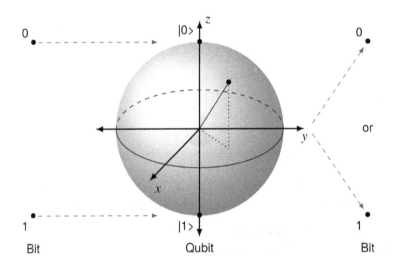

Figure 1.9: A quantum Bloch sphere

This sphere is called the *Bloch sphere*, named after physicist Felix Bloch.

Things get even better when we have multiple qubits. One qubit holds two pieces of information, and two qubits hold four. That's not surprising, but if we add a third qubit, we can represent eight pieces of information. Every time we add a qubit, we double its capacity. For 10 qubits, that's 1,024. For 100 qubits, we can represent 1,267,650,600,228,229,401,496,703,205,376 pieces of information. This illustrates exponential behavior since we are looking at $2^{\text{the number of qubits}}$.

You're going to see how to use bits, numbers, qubits, strings, and many other useful objects and structures as we progress through this book. Before we leave this section, let me point out several weird yet intriguing qubit features.

- While we can perform operations and change the state of a qubit, the moment we look at the qubit, the state *collapses* to **0** or **1**. We call the operation that moves the qubit state to one of the two bit states "measurement."

- Just as we saw that bits have meaning when they are parts of numbers and strings, presumably the measured qubit values **0** and **1** have meaning as data.

- Probability is involved in determining whether we get **0** or **1** at the end.

- We use qubits in algorithms to take advantage of their exponential scaling and other underlying mathematics. With these, we hope eventually to solve some significant but currently intractable problems. These are problems for which classical systems alone will never have enough processing power, memory, or accuracy.

Scientists and developers are now creating quantum algorithms for use cases in financial services and artificial intelligence (AI). They are also looking at precisely simulating physical systems involving chemistry. These may have future applications in materials science, agriculture, energy, and healthcare.

1.12 Circuits

Consider the expression maximum(2, 1 + round(x)), where *maximum* is the function we saw earlier in this chapter, and *round* rounds a number to the nearest integer. For example, round(1.3) equals 1. Figure 1.10 shows how processing flows as we evaluate the expression.

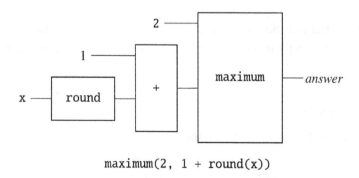

maximum(2, 1 + round(x))

Figure 1.10: A function application circuit

To fix terminology, we can call this a function application *circuit*. When we draw it like this, we call **maximum**, **round**, and "+" functions, operations, or *gates*.

The **maximum** and "+" gates each take two inputs and have one output. The **round** gate has one input and one output. In general, **round** is not reversible: you cannot go from the output answer it produced to its input.

Going to their lowest level, classical computer processors manipulate bits using logic operators. These are also called logic *gates*.

The simplest logic gate is **not**. It turns **0** into **1** and **1** into **0**. If you think of Booleans instead of bits, **not** interchanges *false* and *true*. **not** is a reversible gate. **not**(*x*) means "**not** applied to the Boolean *x*."

Exercise 1.13

What is the value of **not**(**not**(*x*)) when *x* equals each of **0** and **1**?

The **and** gate takes two bits and returns **1** if they are both the same and **0** otherwise.

Exercise 1.14

What are the values for each of the following?

$$\text{and}(0, 0)$$
$$\text{and}(0, 1)$$
$$\text{and}(1, 0)$$
$$\text{and}(1, 1)$$

The **or** gate takes two bits and returns **1** if either is **1** and **0** otherwise. The **xor** gate takes two bits and returns **1** if one and only one bit is **1** and **0** otherwise. **xor** is short for "exclusive-or."

Exercise 1.15

What are the values for each of the following?

$$\text{xor}(0, 0)$$
$$\text{xor}(0, 1)$$
$$\text{xor}(1, 0)$$
$$\text{xor}(1, 1)$$

Figure 1.11 is a diagram of a logic circuit with two logic gates that does simple addition.

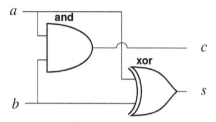

Figure 1.11: A logical addition circuit

Try it out with different bit values for *a* and *b*. Confirm that the sum *s* and the carry bit *c* are correct.

By analogy to the function application and logic cases, we also have quantum gates and circuits. These operate on and include one or more qubits.

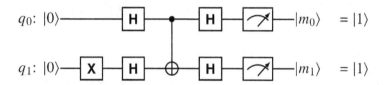

Figure 1.12: A quantum reverse CNOT circuit

This is a 2-qubit circuit containing one quantum **X** gate and four quantum **H** gates. Both **X** and **H** are reversible gates. We cover their definitions and uses later, but note that **X** interchanges the north and south poles in our qubit sphere model. The **H** gate moves the poles to locations on the equator of the sphere.

Coders who have done classical programming know that arithmetic and algebra are involved. When we bring in quantum computing, geometry also becomes useful!

The gates that look like dials on the right of the quantum circuit perform quantum measurements. These produce the final bit values.

There is one more gate in the circuit, and that is **CNOT**. It is an essential and core gate in quantum computing. It looks like a • on the horizontal line from qubit q_0, a ⊕ on the q_1 line, and a line segment between them. It has two inputs and two outputs. Unlike the logic gates **and**, **or**, and **xor**, it is a reversible gate.

Exercise 1.16

Written in functional form, **CNOT** has the following behavior.

$$\text{CNOT}(|0\rangle, |0\rangle) = |0\rangle, |0\rangle$$
$$\text{CNOT}(|0\rangle, |1\rangle) = |0\rangle, |1\rangle$$
$$\text{CNOT}(|1\rangle, |0\rangle) = |1\rangle, |1\rangle$$
$$\text{CNOT}(|1\rangle, |1\rangle) = |1\rangle, |0\rangle$$

The gate has two outputs, which I separated with a comma. The first qubit input is called the *control*, and the second is the *target*.

In what way is **CNOT** a reversible xor?

Quantum gates and circuits are the low-level building blocks that will allow us to see significant advantages over classical methods alone. You will see qubits, quantum gates, and quantum circuits as we progress through this book. In *Chapter 11, Searching for the Quantum Improvement*, we focus on quantum algorithms and how they obtain their speed-ups.

1.13 Summary

In my coding career, I have used dozens of programming languages and created several. [AXM] When I have a new problem to solve or am considering a new project, I research how languages have evolved and improved. For each language I investigate, I ask myself,

- "can this language do what I have in mind?",
- "can this language do what I want better than the others?", and
- "what are the special features and idioms within the language that enable me to produce an elegant result?".

This chapter introduced many of the ideas widely used today in languages for coding. It showed that you could think about features and functionality separately from what one specific language provides. In the next chapter, we get concrete and see how Python implements these and other concepts.

PART I

Getting to Know Python

2

Working with Expressions

It is the supreme art of the teacher to awaken joy in
creative expression and knowledge.

—Albert Einstein

At one level, a programming language is a super-calculator with the ability to work with data. You can do arithmetic with numbers or manipulate text represented as strings. More than that, a language allows you to create functions that perform custom operations and even to create new structures for data.

In this chapter, we introduce the basics of Python and how it implements the programming features described in *Chapter 1, Doing the Things That Coders Do*. In subsequent chapters, we go into greater detail and expand upon these ideas.

Via a search of the web, I saw an estimate that Python is known and used by more than 8 million developers. It is an excellent modern language that has been built by dozens of open source coders over the last thirty years. [HPY]

Topics covered in this chapter

2.1 Numbers

The two most frequently used forms of numbers in Python are integers and floating-point. An integer contains no decimal point and can be negative, zero, or positive.

```
-4

-4

0

0

83737540038882653884746329099383700004846673

83737540038882653884746329099383700004846673
```

As you can see, integers can be quite large. Python uses the type **int** for integers. We'll learn more about this later, but it's common to say something like "-17 is an **int**."

A floating-point number, known as a **float**, contains a decimal point but is sometimes limited in how much information the system retains from what you type or what is computed.

```
-3.6836

-3.6836

17.981735993576742996244536277253636

17.98173599357674
```

Here we lost precision when we had too many digits to the right of the decimal point for Python to store. If there are too many on the left, Python uses an exponential form.

```
10000000000000000000000000000000000000000000000000000.1
```

```
1e+52
```

That result means 1×10^{52}. The ".1" was lost.

Python allows you to insert underscores in numbers to make them more readable.

```
1_000_000_000
```

```
1000000000
```

2.2 Strings

We use strings for text. They are a sequence of characters.

```
'Better a witty fool, than a foolish wit.'
```

```
'Better a witty fool, than a foolish wit.'
```

If I am manipulating the text of Shakespeare's *Twelfth Night*, I can represent it this way. I *delimited* the string at its beginning and end with single quotes. Notice that Python displayed single quotes in its output. You may use either single or double quotes as *delimiters*; they are not part of the data representing the string and its contained characters.

```
"Some are born great, others achieve greatness."
```

```
'Some are born great, others achieve greatness.'
```

If you want to create a string that includes a single quote, use double quotes as the delimiters. Reverse this if your string includes a double quote. If I can use either, my convention is to use single quotes for strings of length one (for example, 'Z') and double quotes otherwise (for example, "hola" or "").

```
"I use single quotes around single characters like 'Z'."
```

```
"I use single quotes around single characters like 'Z'."
```

Finally, you can always precede a single quote or double quote within a string with a backslash. You then don't need to worry about which delimiters you use. We call using a backslash this way "escaping a character."

```
'I\'ll use a backslash in front of my single quote.'
```

```
"I'll use a backslash in front of my single quote."
```

The type of a string is **str**. We'll see additional rules for using quotes, both for strings (section 4.1) and the descriptions you add to your code (section 7.8).

2.3 Lists

If we could only write down numbers and strings, Python would not be a practical programming environment. Of course, we want to do things to numbers such as add them, multiply them, and use them to count things. For strings, we'd like to search and construct text we display to our users.

More than that, we need ways to hold collections of data together. If I had three children, I could hold their ages together as a list:

```
[12, 16, 19]
```

```
[12, 16, 19]
```

and use another list for their names:

```
["Robin", "Robert", "Roberta"]
```

```
['Robin', 'Robert', 'Roberta']
```

I could put the names and ages in the same list:

```
["Robin", 12, "Robert", 16, "Roberta", 19]
```

```
['Robin', 12, 'Robert', 16, 'Roberta', 19]
```

Square brackets "[" and "]" delimit the items of the list. Python uses the type **list** for lists.

It's awkward to keep repeating the data like this. Using a variable is a better way of referencing and using the information.

2.4 Variables and assignment

If we wanted to assign a reusable name to the list of ages, what might be suitable? Well, "ages" starts with "a", so we could use "a". However, that's not very descriptive. What if we also had a list of kinds of apples? Or the names of your aunts? A good and descriptive name is ages itself. We make the association between the name and the data using an equal sign "=".

```
ages = [12, 16, 19]
```

This is an *assignment*, and ages is a *variable*. The name of the variable is an example of an *identifier*. We *assign* the value [12, 16, 19] to the variable ages. Python produces no output from an assignment.

If we type the variable name into the Python interpreter, we get the list we assigned to it.

```
ages
```

```
[12, 16, 19]
```

We compute the number of elements in the list using the Python built-in *len* function. [P.Y.B]

```
len(ages)
```

```
3
```

We use square brackets and a number to access items within a list. Here is how we get the first item:

```
ages[0]
```

```
12
```

The 0 is called an *index* into the list. The [0] says, "return the object at index 0." Most computer languages, Python included, start counting indices at 0 instead of 1. **The item in the first position is at index 0.** This syntax works with strings too.

```
"Maria"[1]
```

```
'a'
```

2.4.1 Naming rules

The rules for identifiers and naming variables are simple:

1. You may use an uppercase letter A-Z in any part of the name.
2. You may use a lowercase letter a-z in any part of the name.
3. You may use an underscore "_" in any part of the name.
4. You may use a digit 0-9 in any part of the name, *except* in the first position.

The name `revenue_2021` is valid, but `2021_revenue` is not. Python has conventions for using the variable name "_" and names that start or end with an underscore. We'll see them when we discuss function parameters in section 6.3 and methods in classes in section 7.2.

2.4.2 List unpacking

Let's look at one kind of expression where you take apart a list and assign the pieces to variables. The assignment

```
robins_age, roberts_age, robertas_age = ages
```

gives the first value to `robins_age`

```
robins_age
```

```
12
```

and the others to `roberts_age` and `robertas_age`.

```
roberts_age
```

```
16
```

```
robertas_age
```

```
19
```

This example shows *list unpacking*, and we discuss it in detail in section 3.2.9.

2.4.3 Multiple assignments

We can combine several assignments into a single statement.

```
a = b = 10
a
```

```
10
```

```
b
```

```
10
```

That assigned the same value to the variables a and b. We can also assign multiple values to multiple variables.

```
m, n = 17, 6
```

Exercise 2.1

What are the values of m and n after you run the following?

```
m, n = 17, 6
m, n = n, m
```

This is *simultaneous assignment.*

Exercise 2.2

Does the following work the same as the last example? Why or why not?

```
m, n = 17, 6
m = n
n = m
```

This is *sequential assignment.*

2.5 True and False

In section.1.1, we saw how you could treat the bits **0** and **1** like the Booleans *false* and *true*. In Python, we represent these as False and True.

The not operator inverts them.

```
not True
```

```
False
```

```
not False
```

```
True
```

2.5.1 Equality testing

Recall that we use "==" to compare whether two things are equal. Remember: assignment uses "=" and equality testing uses "==".

```
a = 10
-a == -10
```

```
True
```

```
a == "10"
```

```
False
```

The last expression shows that the *string* "10" is not the same as the *number* 10. We can, however, create the *string representation* of a number.

```
str(a)
```

```
'10'
```

```
str(a) == "10"
```

```
True
```

To test if two things are not equal, use "!=".

```
"London" != "london"
```

```
True
```

2.5.2 "and", "or", and "^"

While not operates on one Boolean, and and or operate on two. If boolean_1 and boolean_2 are Booleans, the expression

boolean_1 and boolean_2

returns True if both boolean_1 and boolean_2 are True, and False otherwise.

Figure 2.1 is the *truth table* for and.

| | boolean_2 | |
and	False	True
boolean_1 False	False	False
boolean_1 True	False	True

Figure 2.1: Boolean truth table for "and"

Pick a value for boolean_1 in the second column, a value for boolean_2 in the second row, and then find the result where the column intersects the row.

For or,

boolean_1 or boolean_2

returns True if either boolean_1 or boolean_2 are True, and False otherwise.

Exercise 2.3

Complete the truth table in Figure 2.2 for or.

| | boolean_2 | |
or	False	True
boolean_1 False		
boolean_1 True		

Figure 2.2: Incomplete Boolean truth table for "or"

Python uses the "^" symbol for "exclusive-or."

$$boolean\_1 \; \text{^} \; boolean\_2$$

returns True if exactly one of boolean_1 and boolean_2 is True, and False otherwise. (On my US English keyboard, "^" is Shift-6.)

Exercise 2.4

Complete the truth table in Figure 2.3 for exclusive-or "^".

	^	boolean_2	
		False	True
boolean_1	False		
	True		

Figure 2.3: Incomplete Boolean truth table for "xor"

Numeric comparisons are a common source of Boolean results used in conditional statements.

```
3 < 4

True

-3 <= -4

False

20 > -6

True

0 >= 0.01

False
```

These examples show "less than," "less than or equal," "greater than," and "greater than or equal."

2.6 Arithmetic

If we are to think of Python as a super-calculator, we must do the basic arithmetic operations. We write addition, negation, and subtraction in the usual ways with "+", "-", and "-", respectively.

```
m, n = 17, 6
```

```
m + n
```

```
23
```

```
-n
```

```
-6
```

```
n - m
```

```
-11
```

There is no multiplication sign "×" on a keyboard, so we use the asterisk "*".

```
m * n
```

```
102
```

Floating-point numbers are contagious. Once you use a decimal point, the remaining arithmetic operations usually proceed as **float**.

```
5 * 3.4
```

```
17.0
```

```
5 * 3.0
```

```
15.0
```

Python tries to do the right thing when comparing **int** and **float** numbers. We test for equality using the "==" operator.

```
a = 100000000000000000000
```

```
b = 100000000000000000000.
```

```
a == b
```

```
True
```

Beware of round-off errors when working with **float**.

```
1000000000000000000000 == 1000000000000000000000.1
```

```
True
```

The Python documentation describes and explains this behavior. [PYD] [PYT] Stick to integers when you can because they can be arbitrarily large. With any programming language, you need to understand the rules of working with floating-point. This is also true in applications like Microsoft Excel®.

2.6.1 Division and exponentiation

Subject to the above, division works as you would expect with two floating-point numbers, or a **float** and an **int**.

```
2.75/4.0
```

```
0.6875
```

```
2.75/4
```

```
0.6875
```

You must take care when dividing two integers. If you use "/", you get a **float**.

```
11/4
```

```
2.75
```

```
-13/5
```

```
-2.6
```

What if you wanted the integer quotient? Since 11 divided by 4 is not an integer, we have a remainder. In particular, 11 divided by 4 is 2 with remainder 3: 2 * 4 + 3. We use the division operator with two slashes "//" for the integer quotient and the percent sign "%" for the remainder.

```
11 // 4
```

```
2
```

```
11 % 4
```

```
3
```

If you need both the quotient and remainder, it's inefficient to compute them separately. Use the *divmod* function to get both.

```
quotient, remainder = divmod(-13, 5)
```

```
quotient
```

```
-3
```

```
remainder
```

```
2
```

```
quotient * 5 + remainder
```

```
-13
```

We use "**" for exponentiation. The code b**n is b^n. b^n is $1/(b^{-n})$ when n is negative.

```
4**3
```

```
64
```

```
4**-3
```

```
0.015625
```

```
1/(4**3)
```

```
0.015625
```

2.6.2 Built-in functions

We define the absolute value $|x|$ of an **int** or **float** x to be the non-negative number

- 0 if x is zero,
- x if x is positive, or
- $-x$ if x is negative.

The *abs* function implements this in Python.

```
abs(78.3)
```

78.3

```
abs(0)
```

0

```
abs(-3)
```

3

The graph of the absolute value is in Figure 2.4.

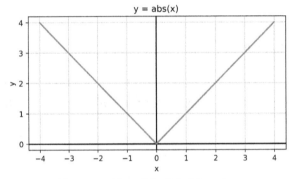

Figure 2.4: Plot of the "abs" function

Most of the mathematical functions are in Python modules, and you must import them into your environment. We discuss modules in section 5.1.

abs and a few other functions are built in, and you can use them directly. The *round* function comes in two forms. In the first, it rounds to the nearest **int**.

```
round(2.7)
```

3

```
round(-7.51)
```

-8

Exercise 2.5

What does *round* do with a value exactly between two integers? What are round(7.5) and round(-67.5)? What are the rules for how *round* works?

In its second form, *round* takes a second parameter, an `int`, which is the number of digits to which it should round.

```
round(2.71828, 2)
```

```
2.72
```

An `int` still rounds to itself when the second parameter is not negative.

```
round(714, 0)
```

```
714
```

```
round(714, 2)
```

```
714
```

It may surprise you that the second parameter can be negative.

```
round(714, -2)
```

```
700
```

```
round(83567356.3846, -5)
```

```
83600000.0
```

Exercise 2.6

Explain what is happening in the last two examples.

If you try using a non-`int` number for the second parameter, Python stops the computation and shows an error message.

```
round(3.73, 2.1)
```

```
Traceback (most recent call last):
  File "<stdin>", line 1, in <module>
TypeError: 'float' object cannot be interpreted as an integer
```

A *traceback* is a sequence of displayed nested function calls to where the error occurred. Python is reading the "standard input," also called "stdin," to see what you typed. Finally, the error message is displayed, and it gives you details of the problem. From this point forward, I will usually only show the error message in a traceback.

Exercise 2.7

What error do you see when you enter 1/0?

In section 1.3, I introduced a **maximum** function that took two numbers. Python implements this as *max*. To get the minimum value, use *min*.

```
max(0, 19.7)
```

```
19.7
```

```
min(0, 19.7)
```

```
0
```

You are not limited to two numbers.

```
max(0, 19.7, -8, 33.2)
```

```
33.2
```

```
min(0, 19.7, -8, 33.2)
```

```
-8
```

You can also use a list as the sole argument to these functions.

```
max([0, 19.7, -8, 33.2])
```

```
33.2
```

```
min([0, 19.7, -8, 33.2])
```

```
-8
```

Exercise 2.8

Show how *sum* works with the same forms of arguments we coded for *max* and *min*. Why do you get an error with max([ ]) but not sum([ ])?

I cover numbers and mathematical functions in detail in *Chapter 5, Computing and Calculating.*

2.6.3 Operation assignments

Python offers a shorthand way of combining assignments with operations like addition. Suppose we initialize count to zero:

```
count = 0
count

0
```

Instead of doing an addition followed by assignment,

```
count = count + 1
count

1
```

you can do them together:

```
count = 0
count += 1
count

1
```

This shorthand works with many Python operators, including the arithmetic ones we've seen so far: "+", "-", "*", "/", "//", and "**". We call short and convenient ways of writing expressions "syntactic sugar."

Exercise 2.9

What is the final value of count in this sequence of expressions? What are the intermediate values after the operation assignments?

```
count = 1
count += 3
count *= 7
count -= 5
count //= 4
count **= 2
```

Show intermediate and final values for count when starting with count = -2.

2.7 String operations

In this section, we work with this line fragment from Shakespeare's *Hamlet*:

```
quote = "perchance to dream"
quote
```

```
'perchance to dream'
```

Since we'll need it several times, let's save the number of characters in the **str**.

```
quote_length = len(quote)
quote_length
```

```
18
```

2.7.1 Accessing characters

The first character is at index 0, and the last is at index `quote_length - 1`.

```
quote[0]
```

```
'p'
```

```
quote[quote_length - 1]
```

```
'm'
```

Python raises an error if you try to access a character beyond this final valid index.

```
quote[quote_length]
```

```
IndexError: string index out of range
```

 Exercise 2.10

What happens if you try to access a character at index 0 in an empty string?

Python does something interesting if you use a negative integer as an index.

```
quote[-1]
```

```
'm'
```

Using the index -1 is shorthand for accessing the last character in the string. It counts backward from the end. This counts backward to the first position:

```
quote[-quote_length]
```

```
'p'
```

You can't go too far, though, or you will be out of the valid index range.

```
quote[-quote_length - 1]
```

```
IndexError: string index out of range
```

Figure 2.5 shows the valid indices of the characters in the quote.

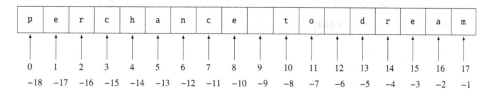

Figure 2.5: Valid string indices in "perchance to dream"

Exercise 2.11

Let `string_index` be a variable holding an index you would like to access in an expression like `quote[string_index]`. What are x and y in

$$x \leq \text{string\_index} \leq y$$

if `string_index` is a valid index?

2.7.2 Searching for substrings

Use *find* to locate specific text within a **str**.

```
quote
```

```
'perchance to dream'
```

```
quote.find("dream")
```

```
13
```

This result says that "dream" is at index 13. We say that "dream" is a substring of "perchance to dream". It also illustrates a form of function application we have not seen before. In this case, *find* is a *method* within the class **str**, and so operates on objects of that class. The format is

$$object.method\_name\,(argument_1, argument_2, ...)$$

You can have zero or more arguments, but you must have the parentheses. In our example,

- *object* is quote,
- *method_name* is find, and
- *argument_1* is "dream".

Since "dream" is present in quote, we say "dream" is a *substring* of quote. If the argument to *find* is not in the string, the method returns -1.

```
quote.find("snooze")

-1
```

The *find* method on strings is defined within Python to have three parameters with the last two having default values.

$$string\_to\_be\_searched.find\,(string\_to\_be\_found, start\_index, end\_index)$$

The default value of *start_index* is 0: if you do not include it, Python assumes it is 0. If you provide an explicit value for *start_index*, you can then also give a value for *end_index*. Its default otherwise is len(*string_to_be_searched*). We search from index *start_index* to but **not** including *end_index*.

```
quote

'perchance to dream'

quote.find('a', 7)

16

quote.find('a', 3, 8)

5

quote.find('a', 0, 4)

-1
```

If you are only interested in knowing whether one **str** is contained within another, use in.

```
"dream" in quote
```

```
True
```

```
"nightmare" in quote
```

```
False
```

To test whether a string is not contained in another, use not in.

```
"Hamlet" not in quote
```

```
True
```

2.7.3 Slicing and extracting substrings

We've just seen how to access individual characters and find the position of substrings in a **str**. We use *slice notation* to extract portions of strings and return them as new strings.

For our quote

```
quote
```

```
'perchance to dream'
```

we extract the substring starting at index 3 and of length 6 by executing

```
quote[3:9]
```

```
'chance'
```

The number before the colon is the starting index, and the number after the colon is the starting index plus the substring length. The expression quote[3:9] means

> Beginning at index 3, collect together all characters in
> quote at that index and the valid indices less than 9.

 Exercise 2.12

What does quote[0:len(quote)] return?

We extract the first five characters in our quote by

```
quote[:5]
```

```
'perch'
```

The value before the colon defaults to 0.

We get the characters in our quote from index 5 to the end of the string by

```
quote[5:]
```

```
'ance to dream'
```

because the value after the colon defaults to the length of the string.

Exercise 2.13

What does `quote[:]` return?

2.7.4 Creating strings

As we saw in section 2.2, you enter strings between quote delimiters.

```
"Finely baked bread"
```

```
'Finely baked bread'
```

You can also build strings by concatenating text together using "+".

```
"Finely baked bread" + " available today!"
```

```
'Finely baked bread available today!'
```

Even though we used "+" for the addition of numbers, it has the application for strings shown here. We have *overloaded* the operation to make it useful for arguments of different types.

Suppose I want to concatenate three copies of a **str** together.

```
text = "la "
text + text + text
```

```
'la la la '
```

By analogy with addition, we use multiplication to repeat the string several times.

```
3 * text
```

```
'la la la '
```

```
10 * text
```

```
'la la la la la la la la la la '
```

```
0 * text
```

```
''
```

Python provides several ways of formatting data and text into a **str**. The newest is called a *formatted string literal* or, more simply, an *f-string*.

To create an f-string, precede the beginning quote delimiter with either **f** or **F**. Within the string, any expression within braces "{" and "}" is evaluated. The result is pasted back into the string at the same place.

```
a = 2
b = 5
f"The sum of a and b is {a + b}"
```

```
'The sum of a and b is 7'
```

Don't use f-strings if you are not evaluating expressions within them as Python wastes time scanning them for braces. I often use f-strings together with the *print* function to format the information I display to users.

I cover strings and their operations in greater detail in *Chapter 4, Stringing You Along.*

2.8 List operations

You can perform many of the same operations on lists as you can on strings, including slicing.

```
years = [2021, 1983, 1976, 1997, 1990]
years
```

```
[2021, 1983, 1976, 1997, 1990]
```

```
len(years)
```

```
5
```

```
years[0]
```

```
2021
```

```
years[-2]
```

```
1997
```

```
years[1:3]
```

```
[1983, 1976]
```

A major difference between strings and lists is that you can change the items in a **list**: lists are *mutable.*

```
years[0] = 2000
years
```

```
[2000, 1983, 1976, 1997, 1990]
```

With strings, you must construct a new and different string with any alterations. Strings are *immutable,* meaning that you cannot change their structure or contents. I can have a variable point to a new string, but I cannot change the string object itself.

```
letters = "abcd"
letters
```

```
'abcd'
```

```
letters[0] = "A"
```

```
TypeError: 'str' object does not support item assignment
```

```
letters = "A" + letters[1:]
letters
```

```
'Abcd'
```

Now that we know we can change lists, we can look at several operations and methods that alter them in place.

```
years = [2021, 1983, 1976, 1997, 1990]
years
```

```
[2021, 1983, 1976, 1997, 1990]
```

The "+" operator concatenates two lists into a new list.

```
years + [1961, 1980]

[2021, 1983, 1976, 1997, 1990, 1961, 1980]

years

[2021, 1983, 1976, 1997, 1990]
```

The *extend* method appends the items in the second list after the items in the first list.

```
years.extend([1961, 1980])
years

[2021, 1983, 1976, 1997, 1990, 1961, 1980]
```

Use *append* to place a new item at the end of the **list**.

```
years.append(2012)
years

[2021, 1983, 1976, 1997, 1990, 1961, 1980, 2012]
```

Coders frequently use this when they are processing information and building a list of data. For example, they might be programmatically scanning revenue reports and making a list of all years where the income was greater than $1 million.

The *insert* method places a new item at a specific index. All items at the previous index and higher are moved one to the right. This example inserts the number 100 in the second position. Remember: the second position has index 1.

```
years.insert(1, 1812)
years

[2021, 1812, 1983, 1976, 1997, 1990, 1961, 1980, 2012]
```

We can take it out again with *pop*.

```
years.pop(1)

1812

years

[2021, 1983, 1976, 1997, 1990, 1961, 1980, 2012]
```

If you use *pop* with no arguments, it removes and returns the last item in the list. In that sense, it is the opposite of *append*.

```
years.pop()
```

```
2012
```

```
years
```

```
[2021, 1983, 1976, 1997, 1990, 1961, 1980]
```

I cover lists and their operations in detail in section 3.2.

2.9 Printing

So far, we have only seen Python displaying the value of objects or evaluated expressions.

```
"This is a sample string"
```

```
'This is a sample string'
```

If you are displaying information to a consumer of your code, you need more sophisticated tools to generate text from data. The *print* function is a versatile way of combining strings and objects into readable forms.

In its simplest form, *print* takes a **str**, or an object that can be converted to a **str**, and displays it on the console. Python does not show the delimiting quotes.

```
print("This is a sample string")
```

```
This is a sample string
```

```
print(17 + 4)
```

```
21
```

You can print several objects at once, and Python separates them with spaces.

```
print(57, 99, -4.3)
```

```
57 99 -4.3
```

2.9.1 Concatenating strings

You can concatenate strings using "+" within *print*. Use *str* on any non-string objects when joining strings in this way.

```
print("Last year's revenue was " + "3" + " million euros")

Last year's revenue was 3 million euros

print("Last year's revenue was " + 3 + " million euros")

TypeError: can only concatenate str (not "int") to str

print("Last year's revenue was " + str(3) + " million euros")

Last year's revenue was 3 million euros
```

2.9.2 f-strings

Even better, use an f-string.

```
print(f"Last year's revenue was {3} million euros")

Last year's revenue was 3 million euros
```

In this case, we could have coded an expression or function call inside the braces.

```
print(f"Last year's revenue was {revenue(2020)} million euros")

Last year's revenue was 3 million euros
```

I have not defined *revenue* here, but you may assume it accesses a database or gets its result from a list or another data structure.

We saw in section 2.2 that we can use a quote within a string that is the same quote type as the delimiter:

```
"Don't use \"odd\" quoted words in your prose."

'Don\'t use "odd" quoted words in your prose.'
```

Python inserted a backslash in the output to escape the single quote.

2.9.3 Newlines

This is *escaping* a character by inserting a backslash before it. Another example is to put a line break within a **str**: use '\n' to insert a newline.

```
"See how I broke\nthis line?"

'See how I broke\nthis line?'
```

The newline is evident when you print the string.

```
print("See how I broke\nthis line?")

See how I broke
this line?
```

Spaces and newlines are examples of *whitespace* within strings.

By default, *print* inserts a newline after it displays its output. You can use the optional *keyword argument* end to change that to something else. The most common values for end are the space ' ' and the empty string "".

```
print("First string")
print("Second string")

First string
Second string

print("First string", end=' ')
print("Second string")

First string Second string
```

The default behavior uses the newline '\n' for end. We will learn more about keyword arguments in section 6.5.

2.10 Conditionals

A conditional statement is one that says, "If this is true, do that thing, otherwise do that other thing." It's up to us to decide what we mean by "this," "that thing," and "that other thing."

Conditional statements change the flow of execution. Instead of evaluating expression after expression in sequence, conditionals allow us to branch and do different things based on whether some condition is true.

When we evaluate one expression after another with no variation, we perform *sequential processing*. When a condition can alter the evaluation flow, we do *dynamic processing*.

2.10.1 "if" and "else"

Let's look at an example of a vending machine that dispenses cookies. Suppose you press a button requesting a shortbread cookie.

```
if number_of_shortbread_cookies == 0:
    print("Sorry, we are out of shortbread cookies.")
else:
    print("Here is your shortbread cookie.")
    number_of_shortbread_cookies -= number_of_shortbread_cookies
```

Here we have one variable used in a cookie transaction with a machine. Because you are hungry or are exceptionally generous to a friend, you asked for a shortbread cookie. If the vending machine is out of that kind, you get a message saying that. Otherwise, you get your cookie, and the internal count of shortbread cookies is reduced by one.

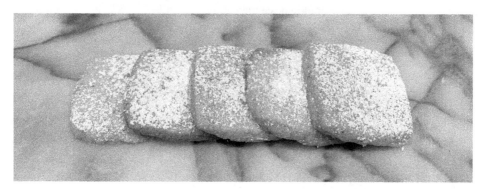

Figure 2.6: Delicious shortbread cookies

The conditional statement begins with the keyword `if`. The `if` is followed by a condition, and then a colon ":". If the condition is true, then the indented lines following the statement are executed. If not, we move to the `else` clause and execute the statements following that. Note how we used the colon ":" again and aligned the indented lines. By convention, we indent four spaces.

The flowchart in Figure 2.7 shows the decisions and actions.

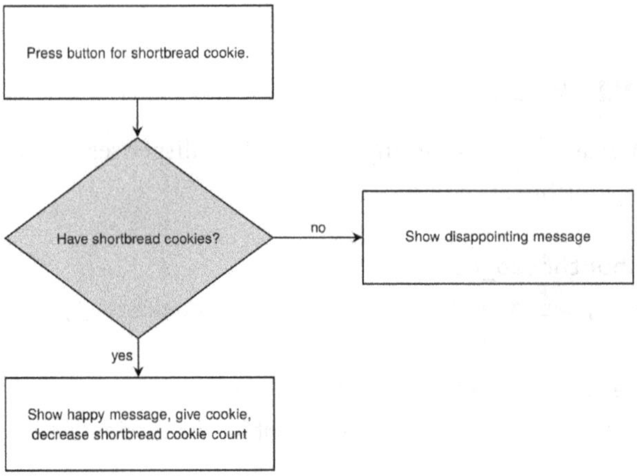

Press button for shortbread cookie.

Have shortbread cookies?

no

Show disappointing message

yes

Show happy message, give cookie, decrease shortbread cookie count

Figure 2.7: Flowchart for shortbread cookies

A vending machine with one kind of cookie isn't interesting enough for us, so let's introduce sugar cookies and the variable number_of_sugar_cookies. If we are out of shortbread, we give you a sugar cookie if we have one.

```python
if number_of_shortbread_cookies == 0:
    print("Sorry, we are out of shortbread cookies.")
    if number_of_sugar_cookies == 0:
        print("Sorry, we are also out of sugar cookies.")
    else:
        print("Here is a sugar cookie instead.")
        number_of_sugar_cookies -= number_of_sugar_cookies
else:
    print("Here is your shortbread cookie.")
    number_of_shortbread_cookies -= number_of_shortbread_cookies
```

Here you can see a conditional nested inside another conditional. The indentation makes the structure of the code easy to follow. Python uses indentation to wrap a sequence of expressions and statements that should be executed sequentially.

Exercise 2.14

Draw a flowchart of this nested conditional example.

While we need the `else` clause in this example, you don't need to include it if you don't need it. You might simply want to do something if a condition is true and then move on.

```
if number_of_shortbread_cookies == 0:
    print("Shortbread cookies are out of stock.")


if number_of_sugar_cookies == 0:
    print("Sugar cookies are out of stock.")
```

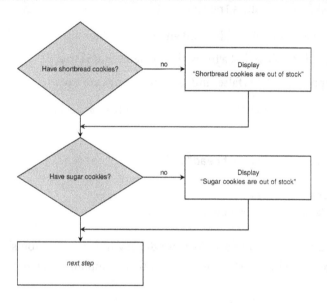

Figure 2.8: Flowchart for shortbread and sugar cookies

2.10.2 "elif"

One or more `elif` clauses are also optional. The `elif` is short for "else if."

```
if number_of_shortbread_cookies > 0:
    print("Here is a shortbread cookie.")
elif number_of_sugar_cookies > 0:
    print("Here is a sugar cookie.")
else:
    print("Sorry, I have no cookies to give you.")
```

Exercise 2.15

Draw a flowchart of this `elif` example.

You may use as many `elif` clauses as you wish, though beyond three or four, there are probably more efficient ways to write your code.

Python is very flexible in what it considers to be true or false in a condition, beyond the literal or computed Boolean values `True` and `False`.

- The number 0 (or 0.0) is false, and all other numbers are true.
- An empty list is false, but a list with any items in it is true.
- The empty string `""` is false, and all other strings are true.

Suppose `cookies_in_stock` is a list of the kinds of cookies we have available for purchase.

```
cookies_in_stock = ["shortbread", "sugar", "oatmeal"]

if not cookies_in_stock:
    print("We are out of cookies.")
```

The condition is not true and Python does not display the message. You should always use a form like this to test whether a list is empty rather than computing its length and comparing it to 0.

2.11 Loops

We use a loop to repeat some action or set of actions while a condition is true.

2.11.1 while

Suppose we want to know the sum of the positive integers less than or equal to 5. This is not much work for you to type and compute directly.

```
1 + 2 + 3 + 4 + 5
```

```
15
```

Now suppose you want the sum of the positive integers less than or equal to 500. That's certainly more than you or I would want to type! We can do it easily with a while loop.

```
sum, count = 0, 1

while count <= 500:
    sum += count
    count += 1

sum
```

```
125250
```

We initialize sum and count and then increment sum by count and count by one while count is less than or equal to 500.

Exercise 2.16

What does Python do if you forget to include count += 1?

Let me make the problem smaller and insert a *print* call so we can see what is happening.

```
sum, count = 0, 1

while count <= 5:
    sum += count
    count += 1
    print(f"{count = } and {sum = }")

sum
```

```
count = 2 and sum = 1
count = 3 and sum = 3
count = 4 and sum = 6

count = 5 and sum = 10
count = 6 and sum = 15
15
```

I used an f-string with a new kind of expression inside the braces. If you place an equal sign "=" after a variable, Python prints out both the variable's name and value. The spaces around "=" are optional, but I included them for readability.

Exercise 2.17

Rewrite the call to *print* to get an equivalent display, but only use {count} and {sum}.

2.11.2 break

The break keyword allows you to leave a loop immediately.

```
sum, count = 0, 1

while True:
    sum += count
    count += 1
    if count > 500:
        break


sum

125250
```

The code "while True" says, "the following code should be repeated indefinitely until something happens to stop the repetition." The break is that "something."

Exercise 2.18

How many addition operations did we perform in this while loop?

I find while loops with the terminating condition and break buried in the code harder to read and debug than the first version above. Your aim should be that anyone who glances at your code can tell what it is doing as quickly as possible.

2.11.3 continue

The `continue` keyword tells Python, "stop what you are doing and return to the top of the loop at the point immediately before testing the condition." This code adds the odd numbers less than or equal to 500:

```
sum, count = 0, 1

while count <= 500:
    if count % 2 == 0:
        count += 1
        continue

    sum += count
    count += 1

sum
```

```
62500
```

Note how I tested whether `count` was even with the remainder % operator.

Exercise 2.19

How would you test for `count` being odd? How would you test for `count` being evenly divisible by 5?

2.11.4 for loops

A `for` loop is often much more concise than a `while` loop if you are iterating over a sequence of values. We often use the *range* function to deliver a sequence of integers starting with its first parameter and ending before and not including its second.

```
sum = 0

for count in range(1, 501):
    sum += count

sum
```

```
125250
```

We initialize sum to zero, and start count with the value one. This is the first parameter to *range*. We repeat the body of the for loop while count is less than 501, the second parameter to *range*.

We can use a third parameter to *range*, and that is the step. If you only provide one parameter, the starting **int** defaults to 0, the step defaults to 1, and the given number is the end. With two parameters, the step defaults to 1.

```
sum = 0

for count in range(1, 501, 2):
    sum += count

sum

62500
```

Here we iterated over the odd integers greater than or equal to 1 and less than 501.

Exercise 2.20

The first parameter to *range* need not be less than the second, and the arguments may be any integers, including negative ones. Write a for loop that prints 5, 4, 3, 2, 1 on a single line. Start by printing the numbers in the stated order and then experiment with *print*.

Use for to iterate over items in a list or characters in a string.

```
for year in [2020, 2022, 2025, 2026]:
    print(year)

2020
2022
2025
2026
```

```
for character in "abc":
    print(character)
```

```
a
b
c
```

The expression range(0, x) is equivalent to range(x). Here we show the substrings of a given string beginning with the first character:

```
string = "Zdrasti"

for index in range(len(string)):
    print(string[0:index + 1])
```

```
Z
Zd
Zdr
Zdra
Zdras
Zdrast
Zdrasti
```

Study this example carefully because it pulls together several very useful features and functions of Python.

Exercise 2.21

Change the example to show the substrings in the string ending with the final character.

One of my favorite parts of Python is how you can use a for loop to build a list. This example is a *list comprehension* and creates a list of the first five squares:

```
[x*x for x in range(1, 6)]
```

```
[1, 4, 9, 16, 25]
```

We return to these in section 3.5.

2.12 Functions

In this section, we combine many of the techniques in this chapter into examples of functions.

2.12.1 A simple function

This Python code is the definition of an *identity* function:

```
def identity(x):
    return x
```

identity takes one parameter x, does nothing to it, and returns it.

A function definition begins with def followed by the function name. After that, we list zero or more named parameters in parentheses. We end the line with a colon. The body of the function starts on the next line and is indented.

Because of the way I wrote *identity*, it doesn't care what the type of x is. It is overloaded to work on anything.

```
identity(7)
```

```
7
```

```
identity("just a string")
```

```
'just a string'
```

```
identity(["one", 2, 3.0])
```

```
['one', 2, 3.0]
```

The definition of *identity* defined the *parameter* x. A parameter is a variable assigned the value of an *argument* when a function is called. In identity(7), the parameter x is assigned the argument 7.

Slightly more interesting is a function that "doubles" its input.

```
def double(x):
    return 2 * x
```

```
double(7)
```

```
14
```

```
double("just a string")
```

```
'just a stringjust a string'
```

```
double(["one", 2, 3.0])
```

```
['one', 2, 3.0, 'one', 2, 3.0]
```

Exercise 2.22

Why does *double* work for an **int**, **str**, or **list**?

2.12.2 An error creeps in

We're less lucky if we write a function to divide its argument in half.

```
def half(x):
    return x // 2
```

```
half(7)
```

```
3
```

```
half("just a string")
```

```
Traceback (most recent call last):
  File "<stdin>", line 1, in <module>
  File "<stdin>", line 2, in half
TypeError: unsupported operand type(s) for //: 'str' and 'int'
```

This traceback has additional information from the ones we have seen before. The third line states that the problem is on line 2 of the definition of *half.* This message helps us locate and fix the problem. Welcome to debugging!

I cover function definitions in detail in *Chapter 6, Defining and Using Functions.*

2.12.3 The selection sort

For the remainder of this section, we examine several variations on a technique for ordering lists of items called a *selection sort.*

To fix ideas, let's start by thinking about a list of numbers. We want to manipulate the list to move the smallest number in the list to index 0. Before we write a function to do this, let's write some test code. Here is the list:

```
numbers = [3, 2, 1, 4]
```

and here is the code to percolate the smallest number to the first position. Note the loop, the conditional, and the simultaneous assignment.

```
for index in range(1, 4):
    if numbers[0] > numbers[index]:
        numbers[0], numbers[index] = numbers[index], numbers[0]

numbers

[1, 3, 2, 4]
```

 Exercise 2.23

What are the 1 and the 4 in the *range* function call?

That worked as expected. Figure 2.9 shows how the code transformed the list. You can see how the smallest number, 1, is moved to the start of the list. Only two swaps were necessary, and I showed these with line segments with circles at their ends,

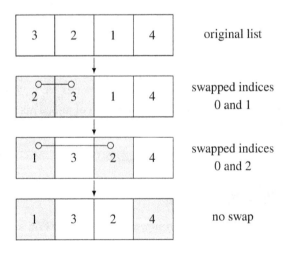

Figure 2.9: Movement of the list items while sorting

Now we wrap this code in a function and allow arguments.

```python
def least_first(items):
    for index in range(1, len(items)):
        if items[0] > items[index]:
            items[0], items[index] = items[index], items[0]
    return items

least_first([3, 2, 1, 4])

[1, 3, 2, 4]

least_first([5, -1, 8, -3, 0])

[-3, 5, 8, -1, 0]
```

This technique worked well, but what if the list has only one item or is empty? Will we get an error?

```python
least_first([10])

[10]

least_first([])

[]
```

These *edge cases* worked because of the way *range* works and because we did not access items[0] for the empty list.

The variables items and index are both *local* to *least_first*. I have not defined them in the Python environment, and they are not known there.

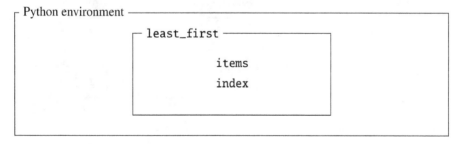

Figure 2.10: Local variables in the "least_first" function

```python
items

NameError: name 'items' is not defined
```

Now that we have the smallest number in the first position, we want the second-lowest number in the second position, and so forth. This repetition sorts the list in ascending order.

```python
def selection_sort(items):
    for i in range(0, len(items) - 1):
        for j in range(i + 1, len(items)):
            if items[i] > items[j]:
                items[i], items[j] = items[j], items[i]
    return items

selection_sort([3, 2, 1, 4])

[1, 2, 3, 4]

selection_sort([5, -1, 8, -3, 0])

[-3, -1, 0, 5, 8]

selection_sort([10])

[10]

selection_sort([])

[]
```

This function worked nicely! Figure 2.11 shows the three passes that *selection_sort* makes when given the list [3, 2, 1, 4]. The gray boxes show the two positions compared, and the line segments with circles at the ends show where *selection_sort* made swaps. This technique is not the most efficient way of sorting, and methods like Quicksort and Merge Sort perform much better. [ALG]

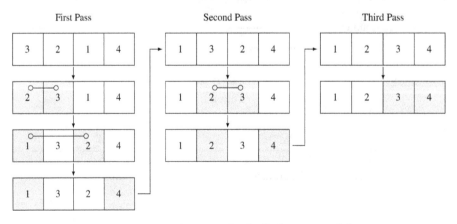

Figure 2.11: List transformation by the "selection_sort" function

Though you should try to use descriptive names for variables, it is common practice for coders to use i, j, and k as indices, as I did in this function definition.

Exercise 2.24

Remember that *range* can return an **int** up to but not including its second parameter. Why was this second parameter len(items) - 1 for i and len(items) for j?

Exercise 2.25

What would you change to sort the list in descending order? Add a Boolean-valued parameter to *selection_sort* to allow the user to select the order of sorting.

Exercise 2.26

For our original *selection_sort*, calculate the number of comparisons and item swaps it does for each of the following lists.

```
[]
[5]
[5, 4]
[5, 4, 3]
[5, 4, 3, 2]
[5, 4, 3, 2, 1]
```

Augment the function with calls to *print* to show the counts. Can you see a pattern?

Why would you consider these to be "worst-case scenarios" for sorting lists of numbers of the given lengths?

2.12.4 Nesting function definitions

You can define one function inside another function. Here I define *test_and_swap* as a nested function within *nested_selection_sort*.

```
1  def nested_selection_sort(items):
2      def test_and_swap(first_index, second_index):
3          if items[first_index] > items[second_index]:
```

```
4              items[first_index], items[second_index] = \
5                  items[second_index], items[first_index]
6
7      for i in range(0, len(items) - 1):
8          for j in range(i + 1, len(items)):
9              test_and_swap(i, j)
10
11     return items
12
13 nested_selection_sort([3, 2, 1, 4])
```

```
[1, 2, 3, 4]
```

I used the *line continuation character* "\" on line 4 to spread the very long assignment expression over two lines.

Our loop-within-a-loop is now much easier to read. We can also look at the definition of *test_and_swap* and directly see what it does.

test_and_swap is not visible by name outside *nested_selection_sort*. What variables can be used where and by what is called *variable scoping*. We look at this in section 6.9.

Exercise 2.27

By inserting a call to *print*, see if `first_index` is usable outside of the definition of *test_and_swap* but within *nested_selection_sort*.

2.12.5 Recursive functions

A *recursive* function is one that calls itself. Here is a recursive definition of selection sort that includes an optional parameter with a default value.

```
def recursive_selection_sort(items, first_index=0):
    if first_index < len(items) - 1:
        for second_index in range(first_index + 1, len(items)):
            if items[first_index] > items[second_index]:
                items[first_index], items[second_index] = \
                    items[second_index], items[first_index]
        recursive_selection_sort(items, first_index + 1)

    return items
```

```
recursive_selection_sort([3, 2, 1, 4])
```

```
[1, 2, 3, 4]
```

If you call *recursive_selection_sort* with only one argument, the list of numbers, Python gives the second parameter `first_index` a value of 0. By convention, we omit the spaces around the equal sign "=". We return to recursion in section 6.12.

Though I have been showing the code of our functions, I should be inserting comments. These remarks allow other coders to understand what the code does, improve it, and maybe fix it. In section 7.7, I discuss the conventions we use to comment Python code. Any line that begins with a hash symbol "#" is a comment and Python does not execute it.

```
def do_nothing():
    # do_nothing does nothing whatsoever
    pass
```

Use the `pass` keyword when syntax requires an expression to be present, but we don't want to evaluate anything. Its presence is usually self-documenting.

Exercise 2.28

Write a new function *nested_recursive_selection_sort*. Include the recursion from *recursive_selection_sort* and the nesting from *nested_selection_sort*.

Comment the function with information about what the function does, what the parameters are, and what each section of code does.

2.13 Summary

In this chapter, we looked at many of the general features of programming languages from *Chapter 1, Doing the Things That Coders Do*, and saw how Python implements them. My coverage here was a survey of some of Python's core functionality. The remaining chapters in this part go into greater detail and breadth. Parts II and III look at many of the advanced features and libraries that give Python its preeminence among coding environments.

3

Collecting Things Together

O! thou hast damnable iteration, and art indeed able to
corrupt a saint.

—William Shakespeare, *Henry IV*

Imagine a wall on which you have hung your *collection* of guitars. If there are no guitars, the collection is empty. Each guitar on the wall has a lot of information associated with it. For example, the data includes the

- manufacturer
- model name
- year first introduced
- year manufactured
- body type
- neck length
- neck material

What is your strategy for placing new guitars on the wall? Starting on the left, you could take every new guitar you get and put it to the right of the next older guitar. Or you could insert the guitar among the others so that the collection is in alphabetical order by the model name. Or you might not care much and just put it anywhere on the wall, but only if a copy of the guitar is not already there. That is, the order is random, but you want no duplicates.

These different schemes affect how easy it is to retrieve a specific guitar from the collection. In the first and third cases, you might need to look at every guitar to find the one you need. In the second case, you only need to start at the left and move to the right until you either find the guitar or see one where the model name is alphabetically later than the one you seek.

You could be even smarter and look at the guitar in the middle of the collection. If that is the guitar you seek, you are finished. If your model is before it, you only need to search the first half of the guitars. If it is after, you search the second half. Repeat this kind of search in your chosen half.

As with guitars, collections in Python have different storage and retrieval schemes. They are more or less efficient for a given use. In this chapter, we look at the three core collection types in Python: lists, dictionaries, and sets. Along the way, we encounter *tuples*, which we saw when we did multiple assignments in section 2.4.

Topics covered in this chapter

3.1 The big three

In section 2.3, I informally introduced lists. These are objects of type **list** and can contain items of any type. We access them via an **int** index. They are mutable and in a defined order. In Figure 3.1, the list on the left contains six items, two of which are duplicates.

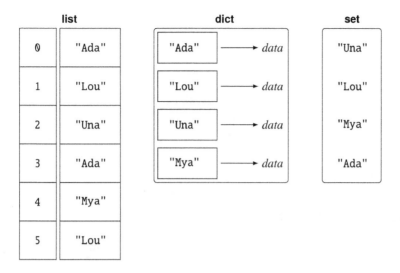

Figure 3.1: List, dictionary, and set collections

A *dictionary* is an object of type **dict**. A dictionary has a set of unique keys that are associated with data values. The collection is mutable, but the keys are not, and so are often numbers or strings. In the middle of Figure 3.1, the four unique names are the keys pointing to data values.

Suppose you belong to a local library and have a library card number to check out books. That number is unique to you, but can be used by the library staff to find your name and phone number. We can use a **dict** to implement this.

A *set* is an object of type **set** and contains unique immutable items. The overall collection is mutable because you can add and remove items. The order in which you access these objects is not guaranteed. In Figure 3.1, the set is on the right. The four unique names are in the set, and I placed them in random order.

In this chapter, we work with information about guitars. We use these models from these manufacturers:

- Fender Stratocaster, first introduced in 1954
- Fender Telecaster, first introduced in 1950
- Gibson Les Paul, first introduced in 1952
- Gibson Flying V, first introduced in 1958
- Ibanez RG, first introduced in 1987

3.2 Lists

Lists are a powerful and versatile type of object in Python.

3.2.1 Creating lists

The primary method to create a list of items is to put square brackets around them. Separate the items with commas.

```
brands = ["Fender", "Gibson", "Ibanez"]
brands
```

```
['Fender', 'Gibson', 'Ibanez']
```

A list can be empty, which means it contains no items. The *list* function with no arguments creates an empty list.

```
[]
```

```
[]
```

```
list()
```

```
[]
```

Use *len* to get the number of items in a list, also known as the *length*.

```
len(brands)
```

```
3
```

```
len([])
```

```
0
```

Python allows you to break lines within parentheses, brackets, or braces.

```
brands = [
    "Fender",
    "Gibson",
    "Ibanez"
    ]
```

The *Style Guide for Python Code* discusses formatting conventions like these to make your code readable. [PEP008]

3.2.2 Accessing list items

You get an item in a list via an **int** index. For a non-empty list, the first item is at index 0, and the last is at index one less than the list length.

```
brands
```

```
['Fender', 'Gibson', 'Ibanez']
```

```
brands[0]
```

```
'Fender'
```

```
brands[len(brands) - 1]
```

```
'Ibanez'
```

Python allows negative indices as well. The last item is at index -1, and the first is at the negative of the length of the list.

```
brands[-1]
```

```
'Ibanez'
```

```
brands[-len(brands)]
```

```
'Fender'
```

For our example list, a valid_index is in the range

$$-\texttt{len(brands)} \le \texttt{valid\_index} < \texttt{len(brands)}.$$

Python raises an error exception if you use an invalid index.

```
brands[10]
```

```
IndexError: list index out of range
```

3.2.3 Exceptions

The error exception we just saw is final. Computation stopped, and Python returned to waiting for the next user expression to evaluate. What if we want to keep processing?

```
try:
    print(brands[10])
except:
    print("Whoops, something went wrong.")
```

```
print("We are still working after try!")
```

```
Whoops, something went wrong.
We are still working after try!
```

A try / except code block allows you to attempt something and then recover if something goes wrong. try *catches* an *exception* that some code *raised* within the attempted action.

Add an else clause to perform some action if Python did not raise an exception.

```
try:
    print(brands[2])
except:
    print("Whoops, something went wrong.")
else:
    print("We got the list item!")

print("We are still working after try!")
```

```
Ibanez
We got the list item!
We are still working after try!
```

Python provides finally so can you perform an action whether or not an exception is raised. The finally action will always run, even if your exception handling is in a function, and you have a return statement in except or else clauses.

```
def try_try(n):
    try:
        print(brands[n])
    except:
        print("Whoops, something went wrong.")
        return 1
    else:
        print("We got the list item!")
        return 0
    finally:
        print("We can finalize processing here.")

try_try(2)
```

```
Ibanez
We got the list item!
We can finalize processing here.
0
```

```
try_try(20)
```

```
Whoops, something went wrong.
We can finalize processing here.
1
```

Here except: catches any kind of exception raised within the evaluated code. You should be as specific as possible in naming the exception you want to catch. In addition to the several dozen built-in Python exceptions, you can define your own. Let's write a simple function with several except clauses.

```python
def error_tester(index, divisor):
    try:
        print(brands[index // divisor])
    except IndexError as error:
        print(f"Index problem:\n'{error}'")
    except ZeroDivisionError:
        print("Did you mean to divide by 0?")
    except:
        print("Something unexpected happened.")
```

```
error_tester(6, 1)
```

```
Index problem:
'list index out of range'
```

```
error_tester(2, 0)
```

```
Did you mean to divide by 0?
```

```
error_tester(4, "hello")
```

```
Something unexpected happened.
```

Exercise 3.1

Evaluate brands[2/1]. What kind of error is raised?

The Python Standard Library lists the full set of built-in exceptions. [P.Y.L] Note how I use as with error to store the information that Python provides with the exception. The value of error can help you determine why the exception was raised.

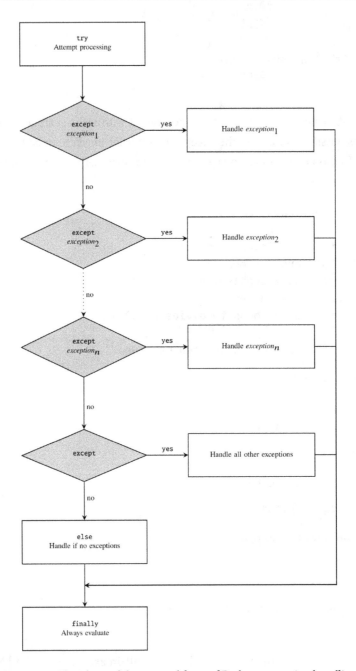

Figure 3.2: Flowchart of the general form of Python exception handling

Figure 3.2 shows the general flow of Python exception handling. As we see in the next section, you can use exceptions as a style of coding in addition to catching bad behavior.

3.2.4 Removing list items

Let's recall our list of electric guitar brand names and make a copy of it so we can use the original later:

```
brands = [
    "Fender",
    "Gibson",
    "Ibanez"
    ]
brands_2 = brands

print(f"{brands = }")
print(f"{brands_2 = }")

brands = ['Fender', 'Gibson', 'Ibanez']
brands_2 = ['Fender', 'Gibson', 'Ibanez']
```

I also assigned brands to brands_2, and we'll see why in a moment.

Let's remove the list entry "Ibanez". We can see by inspection that this is at index 2. Python uses del to remove a list item.

```
del brands[2]
print(f"{brands = }")
print(f"{brands_2 = }")

brands = ['Fender', 'Gibson']
brands_2 = ['Fender', 'Gibson']
```

brands changes and so does brands_2! When I did the assignment brands_2 = brands, I was not making two lists. I was giving two names to the same list. Therefore, when I changed the underlying list, we saw it whether we referred to it by brands or brands_2.

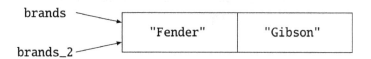

Figure 3.3: Two variables referring to the same list object

As an alternative to del, use the *pop* list method.

```
brands.pop(0)
```

```
'Fender'
```

```
brands
```

```
['Gibson']
```

```
brands.pop()
```

```
'Gibson'
```

pop returns the removed item while del does not. If you use *pop* with no arguments, it returns the last item if the list is not empty.

Since I'm going to show several techniques for removing items, let's make two lists with the same guitar brands.

```
brands = [
    "Fender",
    "Gibson",
    "Ibanez"
    ]

original_brands = brands.copy()
```

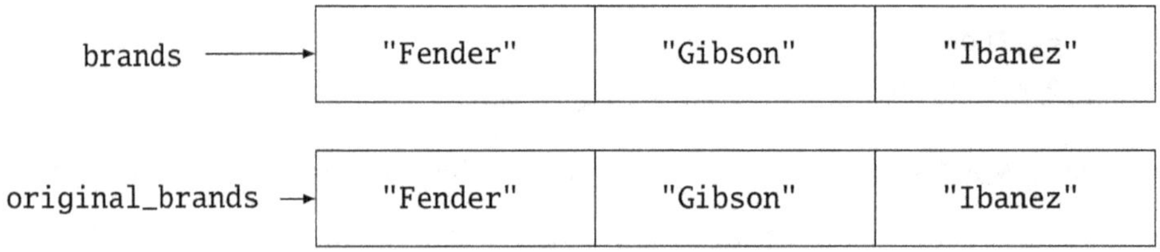

| brands ——————▶ | "Fender" | "Gibson" | "Ibanez" |

| original_brands → | "Fender" | "Gibson" | "Ibanez" |

Figure 3.4: Variables referring to two copies of a list

If we don't know if `"Ibanez"` is in the list or where it is, we can use the *index* method to locate the first occurrence.

```
brands.index("Ibanez")

2
```

If the item is not in the list, *index* raises an exception:

```
brands.index("Gretsch")

ValueError: 'Gretsch' is not in list
```

One way of removing `"Ibanez"` from the list is to catch the exception:

```
try:
    del brands[brands.index("Ibanez")]
    print("'Ibanez' was present and was removed")
except ValueError:
    print("'Ibanez' was not present and so not removed")


brands

'Ibanez' was present and was removed
['Fender', 'Gibson']

brands = original_brands.copy()

try:
    del brands[brands.index("Gretsch")]
    print("'Gretsch' was present and was removed")
except ValueError:
    print("'Gretsch' was not present and so not removed")


brands

'Gretsch' was not present and so not removed
['Fender', 'Gibson', 'Ibanez']
```

Exercise 3.2

Write and use a function that works for the last two examples.

This does what we want and shows how *index* works. A more direct way to remove an item is with *remove*:

```
brands = original_brands.copy()

try:
    brands.remove("Ibanez")
    print("'Ibanez' was present and was removed")
except:
    print("'Ibanez' was not present and so not removed")

brands

'Ibanez' was present and was removed
['Fender', 'Gibson']
```

You can avoid the exception entirely by first testing list membership with in.

```
brands = original_brands.copy()

if "Ibanez" in brands:
    brands.remove("Ibanez")

brands

['Fender', 'Gibson']
```

Exercise 3.3

Which is the more efficient technique, raising an exception or using in?

Exercise 3.4

The *clear* function removes all items from a list. Explain the difference in behavior between

```
a = [1, 2, 3]
b = a
a = []
print(f"{a = }")
print(f"{b = }")

a = []
b = [1, 2, 3]
```

and

```
a = [1, 2, 3]
b = a
a.clear()
print(f"{a = }")
print(f"{b = }")

a = []
b = []
```

How does this show the mutability of lists?

3.2.5 Changing list items

We can change lists, and that's what "mutable means." The same syntax we use to refer to a list item works on the left-hand side of an assignment.

```
brands = original_brands.copy()

brands[2] = "Gretsch"
brands

['Fender', 'Gibson', 'Gretsch']
```

Exercise 3.5

What exception is raised if you use an invalid index in an assignment like this?

If an item in a list is a list itself, we reference it using multiple indices and square brackets.

```
array = [[1, 2], [3, 4]]
array
```

```
[[1, 2], [3, 4]]
```

```
array[0]
```

```
[1, 2]
```

```
array[0][1]
```

```
2
```

```
array[0][1] = 2.001
```

```
array
```

```
[[1, 2.001], [3, 4]]
```

Exercise 3.6

Optional: If you know linear algebra, define a 2 by 2 matrix *m*, its determinant, and its trace by

$$m = \begin{bmatrix} a & b \\ c & d \end{bmatrix}$$

$$\text{determinant}(m) = ad - bc$$

$$\text{trace}(m) = a + d$$

Define the functions *determinant* and *trace*, each of which has a single parameter m. m must be a list of two items, each of which is a list of two numbers.

Within each function, use list index notation to compute the respective values. Hint: each should mention m[0][0].

3.2.6 Slicing: Referencing sub-lists

Slice notation is a convenient way to specify a group of zero or more items in a list. When you first start slicing, it can appear both powerful and mystifying. Over time, you'll appreciate its elegance, and your comfort will increase.

We start by looking at how we use slicing to pull out a sub-list of items from the original list. The items might not be next to each other in the original, though they are picked out from the usual or reverse item sequence. For a list called `the_list`, a slice looks like

```
the_list[start=0 : end=len(the_list)] : step=1]
```

There are three parameters separated by colons, and they each have default values.

Exercise 3.7

For each of the following examples, shade the item boxes in a diagram like that in Figure 3.5 to show the list items in the slice.

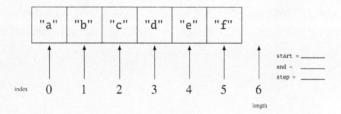

Figure 3.5: List, index, and slice example

Fill in the values of `start`, `end`, and `step`.

I now show several examples of how to slice lists.

```
the_list = ["a", "b", "c", "d", "e", "f"]
the_list
```

```
['a', 'b', 'c', 'd', 'e', 'f']
```

- `the_list[0:]` and `the_list[:]` return a copy of `the_list`.

  ```
  the_list[0:]
  ```

  ```
  ['a', 'b', 'c', 'd', 'e', 'f']
  ```

```
the_list[:]
```

```
['a', 'b', 'c', 'd', 'e', 'f']
```

- the_list[the_index:] returns a new list containing the items in the_list from position the_index to the end. If the_index is positive and ≥ len(the_list), we get an empty list.

```
the_list[2:]
```

```
['c', 'd', 'e', 'f']
```

```
the_list[10:]
```

```
[]
```

- the_list[:the_index] returns a new list from the items in the original list up to but not including the item at the_index.

```
the_list[:2]
```

```
['a', 'b']
```

- the_list[the_first_index:the_second_index] returns a new list from the items in the original list starting with the item at position the_first_index up to but not including the item at the_second_index.

```
the_list[1:3]
```

```
['b', 'c']
```

- the_list[::2] returns a new list from the items in the original list starting with the first, and then including every second item.

```
the_list[::2]
```

```
['a', 'c', 'e']
```

- the_list[1::3] returns a new list from the items in the original list starting with the second, and then including every third item.

```
the_list[1::3]
```

```
['b', 'e']
```

- the_list[::-1] returns a copy of the_list in **reverse order**.

```
the_list[::-1]
```

```
['f', 'e', 'd', 'c', 'b', 'a']
```

I think the technique of reversing the list is marvelous.

Exercise 3.8

Write a Python function to generate all possible questions for this book with negative, zero, and positive integer values of start, end, and step. Just kidding, but do experiment.

Exercise 3.9

Devise formal rules for how slicing works with allowable values of start, end, and step.

3.2.7 Slicing: Changing sub-lists

Just as we saw that we could change and remove list items, we can use slice notation with assignment to alter a list's structure.

The idea is that we create a slice expression to specify part of the list, and then use assignment to substitute another list for the slice.

```
the_list = ["a", "b", "c", "d", "e", "f"]

the_list[:1] = [1, 2, 3]
the_list

[1, 2, 3, 'b', 'c', 'd', 'e', 'f']

the_list[6:] = [7, 8, 9]
the_list

[1, 2, 3, 'b', 'c', 'd', 7, 8, 9]

the_list[4:6] = ["alpha", "beta", "gamma"]
the_list

[1, 2, 3, 'b', 'alpha', 'beta', 'gamma', 7, 8, 9]
```

Exercise 3.10

While these examples work fine, Python requires the slice and the list to the right of the assignment to have the same length when you use a step other than one. Experiment with positive and negative steps to see what works and what doesn't.

3.2.8 Extending lists

How can we add items to an existing list? There are several things we might like to do:

1. Append an item to the end of our list.

2. Prepend an item at the beginning of our list.

3. Insert an item at a specific index for our list.

4. Extend our list at the end with items from another list.

5. Add items from another list to the beginning of our list.

6. Add items from another list at a specific index for our list.

The first four are easy, but the last two require more work on your part. We again work with guitar brands.

```
brands = [
    "Fender",
    "Gibson",
    "Ibanez"
    ]

original_brands = brands.copy()
```

To add an item at the end of our list, use *append*.

```
brands.append("Gretsch")
brands

['Fender', 'Gibson', 'Ibanez', 'Gretsch']
```

To add an item at the beginning of our list, use *insert* with index 0.

```
brands = original_brands.copy()
brands.insert(0, "Gretsch")
brands

['Gretsch', 'Fender', 'Gibson', 'Ibanez']
```

More generally, use the index where you want to place the new item with *insert*.

```
brands = original_brands.copy()
brands.insert(1, "Gretsch")
brands

['Fender', 'Gretsch', 'Gibson', 'Ibanez']
```

Exercise 3.11

Suppose you want to put an item at the beginning of an empty list. Should you use *append* or *insert*? Try both approaches.

Exercise 3.12

How does *insert* behave when you use an invalid index for your original list?

The *extend* method is like *append*, but it takes all items of a given list and puts them at the end of our list.

```
brands = original_brands.copy()
brands.extend(["Epiphone", "Paul Reed Smith"])
brands
```

```
['Fender', 'Gibson', 'Ibanez', 'Epiphone', 'Paul Reed Smith']
```

It does the same as this loop:

```
brands = original_brands.copy()

for brand in ["Epiphone", "Paul Reed Smith"]:
    brands.append(brand)

brands
```

```
['Fender', 'Gibson', 'Ibanez', 'Epiphone', 'Paul Reed Smith']
```

This technique gives us a hint at how to insert the items at the beginning. Let's try this:

```
brands = original_brands.copy()

for brand in ["Epiphone", "Paul Reed Smith"]:
    brands.insert(0, brand)

brands
```

```
['Paul Reed Smith', 'Epiphone', 'Fender', 'Gibson', 'Ibanez']
```

Whoops, they are in the wrong order! There are several ways to fix the problem, but this works:

```
brands = original_brands.copy()
index = 0

for brand in ["Epiphone", "Paul Reed Smith"]:
    brands.insert(index, brand)
    index += 1

brands

['Epiphone', 'Paul Reed Smith', 'Fender', 'Gibson', 'Ibanez']
```

Exercise 3.13

Write code that works with our guitar brand lists to perform the sixth activity from the start of this section: "Add items from another list at a specific index for our list."

3.2.9 Unpacking lists

We've seen how we can access individual items and sub-lists of a list. Via assignment, we can unpack a list into its parts. Given this list,

```
models = [
    "Stratocaster",
    "Telecaster",
    "Les Paul",
    "Flying V",
    "RG"
    ]
```

we extract the first and remaining items by

```
first_model, *remaining_models = models

first_model

'Stratocaster'

remaining_models

['Telecaster', 'Les Paul', 'Flying V', 'RG']
```

A variable with an asterisk in front of it collects list items into a new list. You are allowed only one per unpacking expression. A variable without an asterisk can hold only one item from the original list.

```
*starting_models, last_model = models

starting_models

['Stratocaster', 'Telecaster', 'Les Paul', 'Flying V']

last_model

'RG'
```

You can have more than one un-starred variable.

```
first_model, *middle_models, last_model = models

first_model

'Stratocaster'

middle_models

['Telecaster', 'Les Paul', 'Flying V']

last_model

'RG'
```

You must have at least as many list items as you have un-starred variables. Most of the exceptions you see are related to incorrect item and variable counts.

Exercise 3.14

What exceptions does Python raise in each of the following?

```
first_model, second_model = []
first_model, second_model = models
*first_models, middle_model, *last_model = models
```

What other combinations cause errors?

Exercise 3.15

Optional: If you know the Lisp or Scheme programming languages, how would you define *car* or *cdr*? Do one version of each that uses list unpacking and another that uses slicing.

I think list unpacking is an underused feature in Python.

3.2.10 Reversing and sorting lists

Let's create a list of years when our guitar models were introduced.

```
years = [1954, 1950, 1952, 1958, 1987]
years
```

```
[1954, 1950, 1952, 1958, 1987]
```

The *reverse* method flips the order of items in the **list**. It changes the list itself; it doesn't make a copy.

```
years.reverse()
years
```

```
[1987, 1958, 1952, 1950, 1954]
```

By default, *sort* puts the list elements in ascending order.

```
years.sort()
years
```

```
[1950, 1952, 1954, 1958, 1987]
```

With an optional keyword second parameter, you can sort them in descending order.

```
years.sort(reverse=True)
years
```

```
[1987, 1958, 1954, 1952, 1950]
```

Though I've shown methods and operations that physically change the **list** and its contained items, some similar functions leave the original intact.

```
years = [1954, 1950, 1952, 1958, 1987]
years

[1954, 1950, 1952, 1958, 1987]

sorted(years)

[1950, 1952, 1954, 1958, 1987]

years

[1954, 1950, 1952, 1958, 1987]
```

3.3 The joy of O(1)

How long should it take for *len* to compute the length of the list? One way is to iterate through the list:

1. Initialize a counter to 0.

2. If the list is not empty, go to the first item and add one to the counter. Otherwise, return the value of the counter, which is 0.

3. Try going to the next item. If it is not present, we have reached the end and return the counter. Otherwise, go to the item and increment the counter.

4. Go back to the previous step and repeat until we are finished.

There is a computing time overhead in moving from item to item and incrementing the counter. Let's call this *C*. If our list has *n* items, then

$$\text{time to compute the length by this method} \le C \times n.$$

If your computer is faster than mine, your *C* might be smaller. *C* also depends on the implementation of lists and Python itself.

This algorithm is *O(n)*: its time is proportional to the length of the list. We are using "Big O" notation.

You probably don't know how Python implements lists. In addition to somehow referencing the items, it also happens to store the list length. When you append an item, Python increases the length counter by 1. When you extend a list by a second list, it adds the length of the second to the length of the first. When you clear a list, Python sets the list counter to 0.

It takes the same amount of time to get the length, whether the list is tiny or immense. There is still some system- and implementation-dependent overhead, but now

$$\text{the time to compute the length} \leq C \times 1 = C.$$

This method is $O(1)$. It's wonderful!

Exercise 3.16

What does it mean for a list algorithm to be $O(n^2)$? We say that the algorithm is *quadratic* or has *quadratic complexity.*

Show that the selection sort from section 2.12.3 is $O(n^2)$.

Instead of analyzing how much time an algorithm uses, we could also ask how much memory it needs and use Big O for that. We analyze the best case, worst case, and average case for execution time and memory usage.

An algorithm such as $O(2^n)$ is exponential and should be avoided unless n is small and you have no alternative. The "2" might be another number greater than 1.

I cover complexity and Big O notation in section 2.8 of *Dancing with Qubits.* [DWQ]

Exercise 3.17

I give you an $O(3^n)$ algorithm and tell you that $C = 17.4$. What is the upper bound on the time it takes for the algorithm's implementation to run for $n = 1$, $n = 10$, and $n = 100$?

3.4 Tuples

Tuples are like lists except:

- You use parentheses instead of square brackets, and the parentheses are often optional.
- You can't change the structure of a tuple: they are *immutable.*

We saw tuples when we did multiple assignments. We did not need parentheses then, but we could have included them if we wished.

```
a, b = 3, -2
print(f"{a = }    {b = }")

a, b = (9, 0.3)
print(f"{a = }    {b = }")

(a, b) = ("a", "b")
print(f"{a = }    {b = }")

a = 3     b = -2
a = 9     b = 0.3
a = 'a'    b = 'b'
```

Exercise 3.18

What exception does Python raise if you try to assign a value to an item in a tuple via an index?

Use *tuple* to turn a list into a tuple and *list* to turn a tuple into a list.

```
t = 2, -4, 9
t
list(t)
tuple([-16, 25, -36])

(-16, 25, -36)
```

Exercise 3.19

What are three ways to make an empty tuple?

The syntax gets confusing if you try to create a tuple with a single element.

```
(3.14)

3.14
```

These parentheses grouped a simple expression. Instead, add a trailing comma to make the tuple.

(3.14,)

(3.14,)

Tuples show up in many ways in Python coding, and you may not always realize you are using them. Coders often return multiple values from functions via tuples. In the next example, I define *quo_rem* to compute the quotient and remainder of two integers. The second integer must not be zero.

```
def quo_rem(dividend, divisor):
    if not isinstance(dividend, int):
        raise TypeError("dividend must be an integer")
    if not isinstance(divisor, int):
        raise TypeError("divisor must be an integer")

    if divisor == 0:
        raise ZeroDivisionError("divisor is 0")

    return dividend // divisor, dividend % divisor

quo_rem(11, 4)

(2, 3)
```

Though I could have coded only the last line containing the `return`, I added three conditionals to ensure that the arguments were valid. In the first two on lines 2 and 4, I used *isinstance* to make sure that the arguments were integers. That is, I checked that they were *instances* of type **int**. It should not surprise you that a value having the wrong type should raise a `TypeError` exception.

Had I tried to divide by zero, Python would have raised a `ZeroDivisionError` itself. I wanted to be explicit in my function, so I added the test on line 7.

Exercise 3.20

Should I have used `elif` for the second and third conditionals in *quo_rem*?

Exercise 3.21

Verify that *len, max,* and *min* all work for tuples.

Sometimes you want to iterate over a list and get both the index and its corresponding list item. This works:

```
tens = [10, 20, 30, 40]

index = 0
for item in tens:
    print(f"[{index}] {item}")
    index += 1
```

```
[0] 10
[1] 20
[2] 30
[3] 40
```

As does this:

```
for index in range(len(tens)):
    print(f"[{index}] {tens[index]}")
```

```
[0] 10
[1] 20
[2] 30
[3] 40
```

It can be inefficient for you to access the items by index repeatedly. Use *enumerate* to get the index and the item as a tuple.

```
for index, item in enumerate(tens):
    print(f"[{index}] {item}")
```

```
[0] 10
[1] 20
[2] 30
[3] 40
```

3.5 Comprehensions

We have seen two ways of making a copy of a list. The first is to use the aptly named *copy* method.

```
brands = ["Fender", "Gibson", "Ibanez"]
brands.copy()
```

```
['Fender', 'Gibson', 'Ibanez']
```

The second technique is copying via slicing.

```
brands[:]
```

```
['Fender', 'Gibson', 'Ibanez']
```

We now look at a third technique that generalizes well beyond copying.

```
[brand for brand in brands]
```

```
['Fender', 'Gibson', 'Ibanez']
```

This is a *list comprehension*, and this is an elementary example of it. It is a simpler way of building the list than with a full loop and *append*.

```
brands_copy = []

for brand in brands:
    brands_copy.append(brand)

brands_copy
```

```
['Fender', 'Gibson', 'Ibanez']
```

Let's add a filter in the form of a conditional and create a new list.

```
[brand for brand in brands if brand[1] in "aeiou"]
```

```
['Fender', 'Gibson']
```

The only guitar brands placed in the new list are those whose second character is an English vowel. Compare the comprehension with

```
brands_filtered = []

for brand in brands:
    if brand[1] in "aeiou":
        brands_filtered.append(brand)

brands_filtered

['Fender', 'Gibson']
```

I wasn't careful enough in the comprehension code. This is better:

```
[brand for brand in brands
    if len(brand) > 1 and brand[1] in "aeiou"]

['Fender', 'Gibson']
```

Python let me break the long expression over two lines because we were within square brackets.

We can apply a function or method to brand at the start of the comprehension.

```
[brand.upper() for brand in brands
    if len(brand) > 1 and brand[1] in "aeiou"]

['FENDER', 'GIBSON']
```

We uppercased the brand names that have a vowel as their second character. *upper* is a method of **str**.

Exercise 3.22

Define a function *vowel_count* that counts and returns the number of vowels in a string. Use a list comprehension to create a list of the number of vowels in each of our brand names.

3.6 What does "Pythonic" mean?

You've now seen enough of Python to see that it is a powerful and elegant programming language. You are coding in a *Pythonic* way when you take advantage of that elegance and further create something that beautifully and succinctly accomplishes your task. Examples of being Pythonic include:

- You use `for` in a loop such as `for item in my_list:`, instead of defining and using an index to access list members.
- You interchange the values of two variables using simultaneous assignment.
- You make a new list in reverse order from an existing list using slicing: `my_list[::-1]`.
- You code a comprehension to build a new list instead of using *append*:

  ```
  [x**2 for x in range(13) if x % 3 == 0]

  [0, 9, 36, 81, 144]
  ```

You'll see other examples as we move through this introduction to Python.

Tim Peters summed up Pythonic design and use of the language in *The Zen of Python*:

Beautiful is better than ugly.
Explicit is better than implicit.
Simple is better than complex.
Complex is better than complicated.
Flat is better than nested.
Sparse is better than dense.
Readability counts.
Special cases aren't special enough to break the rules.
Although practicality beats purity.
Errors should never pass silently.
Unless explicitly silenced.
In the face of ambiguity, refuse the temptation to guess.
There should be one—and preferably only one—obvious way to do it.
Although that way may not be obvious at first unless you're Dutch.[*]
Now is better than never.
*Although never is often better than **right** now.*

If the implementation is hard to explain, it's a bad idea.
If the implementation is easy to explain, it may be a good idea.
Namespaces are one honking great idea – let's do more of those!

[*] Guido van Rossum, the creator of Python, is Dutch. [PEP020] [ZEN]

3.7 Nested comprehensions

You can nest comprehensions to create lists within lists. Our task in this section is to explore several ways of making this list of four lists, each of which has four numbers:

$$
\begin{array}{ll}
[& [1, 0, 0, 0], \\
  & [0, 1, 0, 0], \\
  & [0, 0, 1, 0], \\
  & [0, 0, 0, 1] \quad]
\end{array}
\qquad
\begin{bmatrix}
1 & 0 & 0 & 0 \\
0 & 1 & 0 & 0 \\
0 & 0 & 1 & 0 \\
0 & 0 & 0 & 1
\end{bmatrix}
$$

On the left is the list of lists, and on the right is a matrix. In particular, it is a 4 by 4 identity matrix, as you see in linear algebra. The matrix is diagonal: the only non-zero entries are on the main diagonal. Coders frequently use lists of lists to implement matrices in Python.

Linear algebra is at the heart of quantum computing and many other disciplines. For now, we look at how to build the list.

In this case, it is not onerous to type in the list, but we use matrices of size 2^n by 2^n in quantum computing. When n is 15, we get a 32768 by 32768 matrix, for example.

Let's begin by creating a 4 by 4 list of lists filled with zeros.

```
m = [[0 for i in range(4)] for j in range(4)]
m

[[0, 0, 0, 0], [0, 0, 0, 0], [0, 0, 0, 0], [0, 0, 0, 0]]
```

Note how there are two for loops, one acting on the outer list, and the other on the four contained lists. That's what makes this a *nested* comprehension. Now we set the diagonal elements.

```
for i in range(4): m[i][i] = 1
m

[[1, 0, 0, 0], [0, 1, 0, 0], [0, 0, 1, 0], [0, 0, 0, 1]]
```

Since the loop and the action are so small, I put them both on the same line. I can define a function and do everything in the nested list comprehension.

```
def one_or_zero(x, y):
    if x == y:
        return 1
    return 0
```

```
[[one_or_zero(i, j) for i in range(4)] for j in range(4)]
```

```
[[1, 0, 0, 0], [0, 1, 0, 0], [0, 0, 1, 0], [0, 0, 0, 1]]
```

I can be more Pythonic and improve the function:

```
def one_or_zero(x, y): return 1 if x == y else 0
```

```
[[one_or_zero(i, j) for i in range(4)] for j in range(4)]
```

```
[[1, 0, 0, 0], [0, 1, 0, 0], [0, 0, 1, 0], [0, 0, 0, 1]]
```

We haven't seen this before! Python allows you to put a conditional in the middle of an expression. Finally, I skip the function entirely.

```
[[1 if i == j else 0 for i in range(4)] for j in range(4)]
```

```
[[1, 0, 0, 0], [0, 1, 0, 0], [0, 0, 1, 0], [0, 0, 0, 1]]
```

Using a conditional in this way is an additional technique that allows you to write concise and readable code.

Exercise 3.23

What is happening in the following comprehension?

```
[f"i + j = {i+j} for {i = } and {j =}"
    for i in range(2) for j in range(3)]
```

```
['i + j = 0 for i = 0 and j =0',
 'i + j = 1 for i = 0 and j =1',
 'i + j = 2 for i = 0 and j =2',
 'i + j = 1 for i = 1 and j =0',
 'i + j = 2 for i = 1 and j =1',
 'i + j = 3 for i = 1 and j =2']
```

Is this nested or are the for loops running in parallel?

3.8 Parallel traverse

Given the data about guitars from the beginning of this chapter:

```
brands = ["Fender", "Fender", "Gibson",
    "Gibson", "Ibanez"]
models = ["Stratocaster", "Telecaster",
    "Les Paul", "Flying V", "RG"]
years = [1954, 1950, 1952, 1958, 1987]
```

how can we print out a sentence like

```
        The Fender Stratocaster was first introduced in 1954.
```

for each model?

We can use an index via *range* and do it with four nested `for` loops, but this is extremely slow and inefficient.

Exercise 3.24

Write code to accomplish this task using an index via *range* and four nested `for` loops. Why is it slow and inefficient?

Instead, we use *zip* to move through the four lists at the same time:

```
for brand, model, year in zip(brands, models, years):
    print(f"The {brand} {model} was first introduced in {year}.")
```

```
The Fender Stratocaster was first introduced in 1954.
The Fender Telecaster was first introduced in 1950.
The Gibson Les Paul was first introduced in 1952.
The Gibson Flying V was first introduced in 1958.
The Ibanez RG was first introduced in 1987.
```

I "zipped" the lists together so I could traverse them in parallel. If the lists have different lengths, the zipping stops when we reach the end of the shortest list.

You can use *zip* with *range*.

```
for number, brand, model, year in \
    zip(range(len(brands)), brands, models, years):
        print(f"{number + 1}. The {brand} {model} "
            + f"was first introduced in {year}.")
```

1. The Fender Stratocaster was first introduced in 1954.
2. The Fender Telecaster was first introduced in 1950.
3. The Gibson Les Paul was first introduced in 1952.
4. The Gibson Flying V was first introduced in 1958.
5. The Ibanez RG was first introduced in 1987.

Exercise 3.25

In this example, I coded two forms of line continuation and concatenated two strings. Identify where I did these. Why are both strings f-strings?

3.9 Dictionaries

I went into detail about lists to illustrate the main collection operations you need to know:

- Creating a collection
- Testing for membership in a collection
- Accessing items in a collection
- Adding items to a collection, if it is mutable
- Removing items from a collection, if it is mutable
- Iterating over items in the collection
- Extracting a sub-collection
- Building a new collection by calling functions on items filtered from an existing collection

For dictionaries and sets, I move faster over these core operations and focus on the distinguishing features of the collection types.

3.9.1 What is a dictionary?

Let's begin by putting our information about guitars into the table in Figure 3.6.

Brand	Model	Year Introduced
Fender	Stratocaster	1954
	Telecaster	1950
Gibson	Les Paul	1952
	Flying V	1958
Ibanez	RG	1987

Figure 3.6: Table of guitar brands, models, and years introduced

A Python dictionary is a collection of *key-value* pairs. For example, I could have the keys

```
"Fender Stratocaster", "Fender Telecaster",
    "Gibson Les Paul", "Gibson Flying V",
                "Ibanez RG"
```

and associate them with the values

```
1954, 1950, 1952, 1958, 1987,
```

respectively. The keys here are Python strings, and the values are integers. A key that is not present in a dictionary is *invalid*.

Keys are unique in a given dictionary and must be of an immutable type. In particular, the keys cannot be lists or sets, but they can be tuples if the tuples do not contain mutable objects.

Exercise 3.26

Which of the following are valid keys for a Python dictionary?

```
2023
"GPA"
(2, 3)
[2, 3]
[2, (3, 4)]
{2, 3}
[{2, 3}]
(2, [3])
(2, (3,))
```

The keys do not need to all be of the same type, for example, **str** or **int**. If you are mixing key types, think carefully about the design of your data and if this is something you should be doing.

The values in a dictionary can be of any type; it's common to see dictionary values be dictionaries themselves. The type of a dictionary is **dict**.

3.9.2 Creating a dictionary

Create an empty dictionary with braces or *dict*.

```
{}

{}

dict()

{}
```

When using braces, separate the key and the value with a colon, and separate the pairs with commas.

```
{"Fender Stratocaster": 1954, "Fender Telecaster": 1950}

{'Fender Stratocaster': 1954, 'Fender Telecaster': 1950}
```

Remember this: **dictionaries use colons**.

3.9.3 Testing for membership in a dictionary

Let's create the full dictionary of our guitars and the years they were introduced.

```
guitars = {
    "Fender Stratocaster": 1954,
    "Fender Telecaster": 1950,
    "Gibson Les Paul": 1952,
    "Gibson Flying V": 1958,
    "Ibanez RG": 1987
    }
```

It's Pythonic to use the dictionary name alone in a test to see if it is empty:

```
if guitars:
    print("Our guitar dictionary is not empty.")
```

```
Our guitar dictionary is not empty.
```

You retrieve the number of key-value pairs with *len*.

```
len(guitars)
```

```
5
```

Use in and not in to see if a key is in the dictionary.

```
"Fender Stratocaster" in guitars
```

```
True
```

```
"Martin BC-16E" not in guitars
```

```
True
```

3.9.4 Accessing items in a dictionary

For a list my_list, I access the item at index my_index by my_list[my_index]. Similarly, for a dictionary my_dictionary, I access the value at my_key by my_dictionary[my_key].

```
guitars["Gibson Les Paul"]
```

```
1952
```

If that key is not in the dictionary, Python raises a **KeyError** exception.

```
guitars["Martin BC-16E"]
```

```
KeyError: 'Martin BC-16E'
```

There is another way to attempt to retrieve the value, but first we must look at a type we have not seen before.

3.9.5 None

Have you ever asked someone what they were thinking, only to have them reply, "Nothing"? Well, None of type **NoneType** is the explicit "nothing" of Python. Here are two example coding situations in which you might use it:

- You are writing a function to search a new collection type. When the user looks for an item that isn't there, you don't want to raise an exception. The search can return **None**. Just make sure users can't look for None!

- You are writing code to format text, and you use True or False to state whether the text should be italicized. However, you want a way to preserve whatever the surrounding paragraph is doing for italics. Use None to mean, "Don't turn italics on or off, leave it the way it is."

Exercise 3.27

Try None in a conditional. Does it behave as True or False?

Empty strings, lists, tuples, dictionaries, and sets all behave like False.

We use is or is not for testing for equality or inequality with None. Don't test with "==" or "!=".

```
3 is None

False

"None" is not None

True

None is None

True
```

The only thing equal to None is None itself.

3.9.6 Accessing items in a dictionary, again

The *get* method for dictionaries returns the value for a key or returns None if the key is not valid.

```
guitars.get("Gibson Les Paul")
```

```
1952
```

```
guitars.get("Martin BC-16E") is None
```

```
True
```

You can state what should be returned for an invalid key via a second argument to *get*. None is the default value for this second argument. In our example dictionary, the values are the years in which the guitar models were introduced.

```
guitars.get("Martin BC-16E", "Sometime in the 20th century")
```

```
'Sometime in the 20th century'
```

3.9.7 Iterating over items in a dictionary

If you iterate across a dictionary using a for loop, you get the keys. In our example, either guitars or guitars.keys() works.

```
for model in guitars: print(model)
```

```
Fender Stratocaster
Fender Telecaster
Gibson Les Paul
Gibson Flying V
Ibanez RG
```

You can also get the values:

```
for year in guitars.values(): print(year)
```

```
1954
1950
1952
1958
1987
```

To get the key-value pairs as tuples, use *items*.

```
for model, year in guitars.items():
    print(f"The {model} was introduced in {year}")
```

```
The Fender Stratocaster was introduced in 1954
The Fender Telecaster was introduced in 1950
The Gibson Les Paul was introduced in 1952
The Gibson Flying V was introduced in 1958
The Ibanez RG was introduced in 1987
```

The dictionary and its keys are ordered in the sequence in which the key-value pairs are added. This is different from the previous behavior where the order was random and unspecified. If you are writing code that may be run on older versions of Python, you should not depend on the order being consistent.

I've found that *sorted* and *reversed* are useful functions for working with dictionaries. Here I use them both:

```
for model in reversed(sorted(guitars)): print(model)
```

```
Ibanez RG
Gibson Les Paul
Gibson Flying V
Fender Telecaster
Fender Stratocaster
```

3.9.8 Adding items to a dictionary

Use the dictionary square-bracket syntax on the left-hand side of an assignment to add key-value pairs.

```
numbers = {1: "one", 2: "two"}
```

```
numbers[3] = "three"
```

```
numbers
```

```
{1: 'one', 2: 'two', 3: 'three'}
```

If you assign again using the same key, the previous value is replaced.

```
numbers[3] = "THREE"
```

```
numbers
```

```
{1: 'one', 2: 'two', 3: 'THREE'}
```

If other_numbers is another dictionary, we create a new dictionary, merging it and numbers by using double asterisks "**".

```
other_numbers = {4: "four", 5: "five"}
```

```
{**numbers, **other_numbers}
```

```
{1: 'one', 2: 'two', 3: 'THREE', 4: 'four', 5: 'five'}
```

The "**" here means "expand everything out into a collection of key-value pairs."

Exercise 3.28

Starting with an empty dictionary, iterate across numbers and other_numbers to create a dictionary containing all key-value pairs on both.

In general, does it matter in which order you merge two dictionaries? Why or why not?

3.9.9 Removing items from a dictionary

We delete a key-value pair from a dictionary with del.

```
numbers
```

```
{1: 'one', 2: 'two', 3: 'THREE'}
```

```
del numbers[2]
numbers
```

```
{1: 'one', 3: 'THREE'}
```

Exercise 3.29

What exception does Python raise if you try to remove a key-value pair from a dictionary using an invalid key? Is it better to test for a valid key with **in** first, or catch the exception if there is a problem?

Empty the dictionary with *clear*.

```
numbers.clear()
numbers
```

```
{}
```

3.9.10 Dictionary comprehensions

In a list comprehension, we use a for loop within a square-bracketed expression to construct a list. We do the same with dictionaries but use braces instead. Remember that we need to include both the new key and the new value.

```
{model : year for model, year
    in guitars.items() if 1950 < year and year < 1958}
```

```
{'Fender Stratocaster': 1954, 'Gibson Les Paul': 1952}
```

Exercise 3.30

What is happening in this code?

```
{year : model.upper() for model, year
    in guitars.items() if 1955 < year}
```

```
{1958: 'GIBSON FLYING V', 1987: 'IBANEZ RG'}
```

What would happen if two different guitar models were introduced in the same year?

Exercise 3.31

From `guitars`, programmatically create the lists `models` and `years`. Use *zip* to create a copy of `guitars` from the lists.

3.10 Sets

A set is a collection of items like a list, but the primary purpose is to have a structure that contains no duplicates and where testing for membership is fast. There is no order to the members of a set, so we cannot talk about "the first member in the set." **The items in a set must be immutable.**

You likely encountered sets in your mathematical studies, and the Python implementation of them supports operations like union and intersection. I used sets recently in a coding project when I processed data for over two hundred companies. A Python set made it easy for me to get the collection of countries in which they were headquartered with no duplicates.

3.10.1 Creating a set

Use braces "{ }" to create a set. Suppose I have two Fender guitars, two Ovations, and a Gibson, while my friend has two Gibsons, a Gretsch, and a Fender.

```
my_brands = {"Fender", "Fender", "Ovation", "Ovation", "Gibson"}
their_brands = {"Gibson", "Gibson", "Gretsch", "Fender"}
```

If these were lists, the first would have five items, and the second would have four. They are sets, though, and they contain no duplicates.

```
my_brands
```

```
{'Fender', 'Gibson', 'Ovation'}
```

```
len(my_brands)
```

```
3
```

```
their_brands
```

```
{'Fender', 'Gibson', 'Gretsch'}
```

```
len(their_brands)
```

```
3
```

You can *copy* a set with a function call, but **slicing does not work with sets**. This makes sense since there is no sequencing order for the members of a set. You may want to copy a set if you intend to add or remove members but need the original for some other purpose.

```
my_original_brands = my_brands.copy()
their_original_brands = their_brands.copy()
```

You create an empty set with a call to *set*.

```
set()
```

```
set()
```

If you use empty braces, you get a dictionary.

```
isinstance({ }, dict)
```

```
True
```

You can also verify this with *type*.

```
type({ })
```

```
dict
```

Exercise 3.32

What do you get when you use *type* on integers, floating-point numbers, lists, tuples, and sets?

set also converts a list or tuple into a set.

```
set(["b", "o", "b"])
```

```
{'b', 'o'}
```

```
set(("b","o","o","k","k","e","e","p","i","n","g"))
```

```
{'b', 'e', 'g', 'i', 'k', 'n', 'o', 'p'}
```

The Python type of a set is **set**.

3.10.2 Testing for membership in a set

Use in to check if an item is a member of a set.

```
"Gibson" in my_brands
```

```
True
```

```
"Ibanez" in my_brands
```

```
False
```

not in does the opposite:

```
"Gretsch" not in my_brands
```

```
True
```

3.10.3 Adding items to a set

My sister gives me a new Ibanez AZ guitar. I *add* it to my set.

```
my_brands.add("Ibanez")
my_brands
```

```
{'Fender', 'Gibson', 'Ibanez', 'Ovation'}
```

My friend decides to focus on banjo playing, so sells me their guitars for €750. I *update* my set to reflect this.

```
my_brands.update(their_brands)
my_brands
```

```
{'Fender', 'Gibson', 'Gretsch', 'Ibanez', 'Ovation'}
```

3.10.4 Removing items from a set

To complete the transaction with my friend, I must *clear* their set.

```
their_brands.clear()
their_brands
```

```
set()
```

My sister gets annoyed with me and takes the Ibanez back. I *discard* it from my set.

```
my_brands.discard("Ibanez")
my_brands
```

```
{'Fender', 'Gibson', 'Gretsch', 'Ovation'}
```

Exercise 3.33

What does *discard* do if the item is not in the set? You can also use *remove*. Describe its behavior when you try to remove an item that is not a set member.

Suppose `my_set` is a set. To understand the difference between `my_set.clear()` and `my_set = set()`, consider this analogy. Suppose you have a box of apples. *clear* removes all apples from the box, but you keep the same box. The assignment takes away your box and gives you a new empty one.

3.10.5 Accessing set members

For a list, you can say, "give me the item at this index." For a dictionary, you can request, "give me the value for this key." What similar question makes sense for a set? How can you access the members?

First, you can loop across the set members.

```
for brand in my_brands: print(brand)
```

```
Gretsch
Gibson
Fender
Ovation
```

Second, you can *pop*, remove, and return a random member.

```
my_brands
my_brands.pop()
```

```
'Gretsch'
```

```
my_brands
```

```
{'Fender', 'Gibson', 'Ovation'}
```

Exercise 3.34

What kind of exception is raised if you try to *pop* from an empty set?

3.10.6 Subsets and supersets

Let's recover our original guitar collections.

```
my_brands = my_original_brands.copy()
my_brands
```

```
{'Fender', 'Gibson', 'Ovation'}
```

```
their_brands = their_original_brands.copy()
their_brands
```

```
{'Fender', 'Gibson', 'Gretsch'}
```

Use "==" and "!=" to test whether two sets have the same members.

```
print(my_brands == my_brands)
print(my_brands == their_brands)
print(my_brands != their_brands)
```

```
True
False
True
```

If S_1 and S_2 are sets, S_1 is a *subset* of S_2 if all members of S_1 are members of S_2. It is a *proper subset* if it is a subset but not equal to all of S_2. The mathematical notations for these two are $S_1 \subseteq S_2$ and $S_1 \subset S_2$. The Python operators to test for these are "≤" and "<".

```
print({"Gibson"} <= my_brands)
print(my_brands <= my_brands)
print(my_brands < my_brands)
```

```
True
True
False
```

S_2 is a *superset* of S_1 if all members of S_1 are members of S_2. It is a *proper superset* if it is a superset but not equal to all of S_1. The mathematical notations for these two are $S_2 \supseteq S_1$ and $S_2 \supset S_1$. The Python operators to test for these are "≥" and ">".

```
print(my_brands > {"Gibson", "Ovation"})
print(my_brands >= {"Gibson", "Ovation"})
print(my_brands > my_brands)
```

```
True
True
False
```

3.10.7 Set union, intersection, and difference

Just as we manipulate subsets and supersets, we also need Python operations for union, intersection, and difference.

If S_1 and S_2 are sets, the *union* $S_1 \cup$ of S_2 is the set of all items in *either* S_1 or S_2.

```
my_brands.union({"Martin", "Taylor"})
```

```
{'Fender', 'Gibson', 'Martin', 'Ovation', 'Taylor'}
```

You can use "|" instead of *union*. The advantage of this form is that you then have operation assignment.

```
guitars = my_brands.copy()
guitars | {"Martin", "Taylor"}
```

```
{'Fender', 'Gibson', 'Martin', 'Ovation', 'Taylor'}
```

```
guitars |= {"Martin", "Taylor"}
guitars
```

```
{'Fender', 'Gibson', 'Martin', 'Ovation', 'Taylor'}
```

The *intersection* $S_1 \cap$ of S_2 is the set of all items that are in *both* S_1 and S_2. You may use *intersection* or "&".

```
my_brands.intersection(their_brands)
```

```
{'Fender', 'Gibson'}
```

```
my_brands & their_brands
```

```
{'Fender', 'Gibson'}
```

The *difference* $S_1 - S_2$ is the set of all items in S_1 that are not in S_2. You may use *difference* or "-".

```
my_brands.difference(their_brands)
```

```
{'Ovation'}
```

```
my_brands - their_brands
```

```
{'Ovation'}
```

Exercise 3.35

The *symmetric difference* of S_1 and S_2 is the set of all items in S_1 and S_2 that are not in both. You may use *symmetric_difference* or "^".

Give examples as above for the symmetric difference. If *symmetric_difference* was not already provided by Python, how would you define it using *intersection* and *difference*?

3.10.8 Set comprehensions

You can create a set using a comprehension in the same way as you would a list; just use braces instead of square brackets. Python removes any duplicates.

```
[i * j for i in range(1,5) for j in range(2,5)]

[2, 3, 4, 4, 6, 8, 6, 9, 12, 8, 12, 16]

{i * j for i in range(1,5) for j in range(2,5)}

{2, 3, 4, 6, 8, 9, 12, 16}
```

3.11 Summary

In this chapter, we looked at four Python collection types: lists, tuples, dictionaries, and sets. They are each optimized for particular applications. I hope this gives you insight for looking at a coding problem and thinking something like, "this is a job for a dictionary whose keys are strings and whose values are lists." That is, these collection types are useful for a broad range of coding solutions, and your job is to learn to use the right one for the right application.

In the next chapter, we look again at strings and the many ways of creating and manipulating them. You should begin to think of what we do with strings as text processing, a topic we return to in *Chapter 12, Searching and Changing Text*.

4

Stringing You Along

The world is all gates, all opportunities, strings of
tension waiting to be struck.

—Ralph Waldo Emerson

This chapter builds on our discussion of strings in section 2.2. It may help if you think about the similarities between lists of items and strings of characters, but you must remember that **strings are not mutable**.

Topics covered in this chapter

4.1 Single, double, and triple quotes

You may use single quotes or double quotes to delimit a literal string. Python usually prints a string with single quotes.

```
"Franz"

'Franz'

'Ferdinand'

'Ferdinand'
```

If your string contains one kind of quote, use the other as the delimiter or escape the quote mark.

```
"That is Franz's cat toy."

"That is Franz's cat toy."

'The ball is Ferdinand\'s.'

"The ball is Ferdinand's."
```

Use the character '\n' character to embed a newline in your string.

```
"You can see the\nnewline character here."

'You can see the\nnewline character here.'

print("But 'print' breaks\nthe line at the character.")

But 'print' breaks
the line at the character.
```

For very long strings, you can concatenate them together or use triple quotes. You may use either three single quotes or three double quotes, though the *Style Guide for Python Code* recommends the latter. [PEP008]

```
"""
I've broken
this string
over several lines.
"""

"\nI've broken\nthis string\nover several lines.\n"
```

Note how I did not have to use newline characters explicitly. I also inadvertently included extra newlines at the beginning and end.

```
fix = """This fixes my
    problem with extra newlines."""

fix

'This fixes my\n    problem with extra newlines.'

print(fix)

This fixes my
    problem with extra newlines.
```

Python kept the extra spaces I inserted at the beginning of the second line.

When we delimit with triple quotes, we don't escape single or double quotes within the string. We use triple quotes when we document our code, as we will see in section 7.8.

```
"""A single quote (') and a double quote (")."""

'A single quote (\') and a double quote (").'
```

If you want to include a backslash in a string, you can escape it with another backslash.

```
"C:\\src\\code\\chapter-01"

'C:\\src\\code\\chapter-01'
```

Alternatively, use a *raw string* by putting an r before the first quote.

```
r"C:\src\code\chapter-01"

'C:\\src\\code\\chapter-01'

print(r"There is no newline \n in this string.")

There is no newline \n in this string.
```

4.2 Testing for substrings

Use not and not in to see if the content of one string is within another.

```
"Quantum" in "Quantum computing"

True
```

```
"QUANTUM" not in "Quantum computing"
```

```
True
```

To test without worrying about uppercase and lowercase, we use the *casefold* method on each string first.

```
"QUANTUM".casefold() in "Quantum computing".casefold()
```

```
True
```

Python provides *upper* and *lower* to change the case of strings.

```
"Charles Darwin".upper()
```

```
'CHARLES DARWIN'
```

```
"CAUTION: HELMETS MUST BE WORN".lower()
```

```
'caution: helmets must be worn'
```

Use *capitalize* to put the first character in uppercase and the remaining ones in lowercase. Note that this may not give you what you want if the text contains characters that should remain in uppercase.

```
"this NEEDS to look like A Sentence!".capitalize()
```

```
'This needs to look like a sentence!'
```

Use *casefold* if you want to normalize and compare two strings in a case-independent way. *casefold* does everything that *lower* does but works harder for characters outside the usual English collection. For example, *casefold* changes the German character "ß" to "ss".

4.3 Accessing characters

Valid indices for strings range from the negative string length to one less than the string length.

```
guitars = "Fender Gibson Taylor"
guitars[0]
```

```
'F'
```

```
guitars[-1]
```

```
'r'
```

```
guitars[-5]
```

```
'a'
```

Use *find* to locate a substring within an index range in a string. If `my_string` is a string, then *find* can have from one to three arguments.

```
my_string.find(substring, start_index=0, end_index=len(my_string))
```

This notation means that one argument is required, which is the substring for which you are looking. If you give no other arguments, *find* starts looking at index 0. If you do not give the third argument, it defaults to the length of the string.

```
guitars.find("o")
```

```
11
```

```
guitars.find("o", 12)
```

```
18
```

```
guitars.find("o", 2, 9)
```

```
-1
```

If Python cannot locate the substring, *find* returns -1.

You can use *index* also, but it raises an exception if the substring is not present.

4.4 Creating strings

We've now seen many examples where we have entered literal strings between single, double, or triple quotes. Let's look at additional methods to construct strings. You always create new strings and you cannot modify the originals.

```
letters = "abcd"
letters + "efg"
```

```
'abcdefg'
```

```
letters
```

```
'abcd'
```

```
letters += "efg"
letters
```

```
'abcdefg'
```

The "+" operator concatenates two strings. "Concatenate" is a fancy word meaning "stick them together, one after the other." You can assign the new value to the variable or use operator assignment, as I did here.

String concatenation is very powerful, allowing you to mix text with other strings and other data types.

```
brands = [
    "Fender",
    "Gibson",
    "Ibanez"
    ]
```

```
"I own " + str(len(brands)) + " brands of guitars"
```

```
'I own 3 brands of guitars'
```

4.4.1 f-strings

It's usually simpler and clearer to use an f-string.

```
f"I own {len(brands)} brands of guitars"
```

```
'I own 3 brands of guitars'
```

Sometimes you need to format data, especially numbers. You can include the formatting details within the braces. [PYF]

```
f"This is 1/7 to 3 decimal places: {(1/7):.3f}"
```

```
'This is 1/7 to 3 decimal places: 0.143'
```

```
f"This is 1/7 to 6 decimal places: {(1/7):.6f}"
```

```
'This is 1/7 to 6 decimal places: 0.142857'
```

For floating-point scientific notation, use an e or E instead of f within the braces.

```
f"{(1/7000):.5E}"
```

```
'1.42857E-04'
```

If you are presenting numbers in a table, you might want to specify the field width and pad on the left with spaces or zeros.

```
for j in [1, 141, 1441, 14441]:
    print(f"| {j:5} | {j*j:010} |")
```

```
|     1 | 0000000001 |
|   141 | 0000019881 |
|  1441 | 0002076481 |
| 14441 | 0208542481 |
```

j:5 means "format the value of j right-justified in a field 5 characters wide, and pad with blanks." j*j:010 means "format the value of j*j right-justified in a field 10 characters wide, and pad with zeros."

4.4.2 From a list of strings to a string

If you have a list of strings, you can concatenate them together with text between them using *join*.

```
words = ['If', 'you', 'have', 'a', 'list']

print("".join(words))
```

```
Ifyouhavealist
```

```
print(' '.join(words))
```

```
If you have a list
```

```
print('\n'.join(words))
```

```
If
you
have
a
list
```

I've always liked *join* and how it is a method on the string rather than the list.

Another way to make a string is to replace a substring with something else.

```
"My favorite guitar is a Fender.".replace("Fender", "Gibson")
```

```
'My favorite guitar is a Gibson.'
```

By default, this method replaces all occurrences. You can restrict the number of changes to a given maximum by using a third argument.

```
"My favorite guitar is a Fender.".replace("e", "E")
```

```
'My favoritE guitar is a FEndEr.'
```

```
"My favorite guitar is a Fender.".replace("e", "E", 1)
```

```
'My favoritE guitar is a Fender.'
```

Another use of *replace* is removing a substring. Just substitute an empty string "" for the substring.

```
"My favorite guitar is a Fender.".replace(" guitar", "")
```

```
'My favorite is a Fender.'
```

Exercise 4.1

Redo the last example with *find* to locate " guitar", and then slice and concatenate.

4.5 Strings and iterations

Another way to access the characters in a string is to iterate across it with a `for` loop.

```
characters = []

for c in "Fender":
    characters.append(c)

characters

['F', 'e', 'n', 'd', 'e', 'r']

guitars

'Fender Gibson Taylor'
```

```
for c in guitars:
    if c.isupper():
        print(c)
```

```
F
G
T
```

While we cannot do a string comprehension in Python, we can create a list of strings and *join* them together.

```
"".join([c for c
    in "OX63 5WC - SL46 3AP - BN96 0VU"
    if c.isdigit()])
```

```
'635463960'
```

Strings can participate in parallel traverse via *zip* with lists and other collections that support iteration.

```
for a, b in zip("123", [1, 2, 3]):
    print(f"{a = }    {b = }")
```

```
a = '1'    b = 1
a = '2'    b = 2
a = '3'    b = 3
```

4.6 Strings and slicing

The slicing rules we saw for lists in section 3.2.6 also work for strings, except that you cannot put a string slice on the left-hand side of an assignment.

```
word = "sesquipedalianism"
len(word)
```

```
17
```

```
print(word[3:9])
print(word[:7])
print(word[7:])
```

```
quiped
sesquip
edalianism
```

```
print(word[::3])
print(word[::-1])
```

```
sqpaas
msinailadepiuqses
```

```
word[0] = 'S'
```

```
TypeError: 'str' object does not support item assignment
```

Exercise 4.2

Look up the English definition of "sesquipedalianism."

4.7 String tests

What does this function do?

```
def validate(name):
    if name:
        if name[0] != '_' and not name[0].isalpha():
            return False
        for character in name[1:]:
            if character != '_' and not character.isalnum():
                return False
        return True
    return False
```

Empty names are not good.

```
validate("")
```

```
False
```

Names that have a single underscore are okay.

```
validate("_")
```

```
True
```

Names that are digits and letters can be good. These are "alphanumeric."

```
validate("income2021")
```

```
True
```

Names that contain underscores and alphanumeric characters can be good.

```
validate("income_2021")
```

```
True
```

Names that contain other characters are bad.

```
validate("income-2021")
```

```
False
```

Names that start with a digit are bad.

```
validate("2021_income")
```

```
False
```

Our *validate* function tests if name is a valid Python identifier name. Identifiers are used, for example, for variable, function, and method names.

Python provides methods to test if a string contains only characters of a certain kind. In my work, these strings are often only single characters.

- *isalnum* tests if the string contains only alphabetic characters or digits.
- *isalpha* tests if the string contains only alphabetic characters.
- *isdigit* tests if the string contains only the digits 0 through 9.
- *islower* tests if all alphabetic characters in the string are lowercase.
- *isspace* tests if all characters in the string are white space. These include spaces ' ' and newlines \n.
- *isupper* tests if all alphabetic characters in the string are uppercase.

Exercise 4.3

How would you use these methods if you were writing a calculator in Python? For example, how would you process `"-1+ 2 ** 14/7 +84.56"`?

Another usual method is *startswith*, which tests whether one string begins with another.

```
"##hashtaghashtag".startswith("##")
```

```
True
```

```
"##hashtaghashtag".startswith("##H")
```

```
False
```

Exercise 4.4

Write a function using slicing that implements *startswith*. Of course, don't use *startswith* itself!

Exercise 4.5

What does *endswith* do? Write a function to mimic its functionality.

4.8 Splitting and stripping

Suppose you have the text

```
text = "  SPECIAL  #  SALE  #  TODAY  "
```

and you want to break it into a list containing the three words. First, use *split* to break the string at the hash marks.

```
text.split('#')
```

```
['  SPECIAL  ', '  SALE  ', '  TODAY  ']
```

Second, remove the whitespace from the left and right of each word with *strip*. In one line:

```
[word.strip() for word in text.split('#')]
```

```
['SPECIAL', 'SALE', 'TODAY']
```

Exercise 4.6

Try to rewrite *strip* using *replace*. Does it work for embedded spaces as in " How are you? "?

lstrip and *rstrip* remove the white space from only the left and right sides, respectively.

```
[word.lstrip() for word in text.split('#')]

['SPECIAL  ', 'SALE  ', 'TODAY  ']

[word.rstrip() for word in text.split('#')]

['  SPECIAL', '  SALE', '  TODAY']
```

If you do not provide an argument to *split*, it breaks at any whitespace. If you give it an optional **int** argument count, it splits the string a maximum of count times.

```
"936g9477-A9y7-23j9-zz8r-387".split('-', 2)

['936g9477', 'A9y7', '23j9-zz8r-387']
```

rsplit starts at the end of the string and works backward.

```
"936g9477-A9y7-23j9-zz8r-387".rsplit('-', 2)

['936g9477-A9y7-23j9', 'zz8r', '387']
```

Exercise 4.7

Rewrite *strip* using *isspace* and one or more loops.

4.9 Summary

In this chapter, we saw many ways of creating and manipulating strings. These are important for working with text, be it for websites, document processing, messaging, report generation, or general app coding. We return to them in *Chapter 12, Searching and Changing Text*, when we look more deeply into text processing. In particular, we discuss and use regular expressions in *section 12.2*.

In the next chapter, we return to numbers and go beyond basic arithmetic to explore the wide range of Python's computational features.

5

Computing and Calculating

It is not in numbers, but in unity, that our great
strength lies ...

—Thomas Paine, *Common Sense*

Programming languages vary in their support for different kinds of numbers and
mathematical structures. It's not unusual to see floating-point numbers and integers that are
limited in size by the hardware. For many computations in the sciences and engineering, we
need larger numbers than those that fit in the fixed sizes.

In this chapter, I introduce many forms of numbers and other objects with which
you can calculate in Python. We look at integers, fractions, floating-point numbers,
random numbers, complex numbers, qubits (quantum bits), and symbolic objects like
polynomials.

Topics covered in this chapter

5.1 Using Python modules

So far, all the code we have seen or used has been built into Python, or we wrote it ourselves. Python uses *modules* to collect together function, class, and constant definitions so that you can reuse them by selectively bringing them into your environment. We can group modules in a hierarchical, tree-like structure to create *packages*. Together, these provide the functionality of libraries that we saw in section 1.4.

5.1.1 How do you get a module?

More than 200 modules come with the Python Standard Library. [PYL] Other than loading them into your environment via import, you don't need to install anything extra to use them. The **math**, **fractions**, **random**, and **cmath** modules we use in this chapter are part of the standard library.

That's not the full extent of pre-written code you can use. More than *200,000* modules are available from the Python Package Index. [PYPI]

In appendix section A.3, I discuss the **pip** application and how to install the modules and packages you need. As I write this, I have over 140 Python packages on my computer. Packages can depend on other packages, so when you install one, it may trigger the installation of several others.

The **sympy** package that I introduce in this chapter is from the Python Package Index.

5.1.2 How do you import and use a module?

The **math** module contains five constant definitions and about 50 functions. An example of a constant in the module is Euler's number e, which is approximately 2.718281828459045. *factorial* is one of the functions, and it computes $n! = 1 \times 2 \times \cdots \times n$ for $n \geq 0$.

The import statement brings in these constant and function definitions and allows you to use them if you precede their names with "math.". math is a namespace, and you are qualifying a constant or function name with it.

```
import math

math.e
```

```
2.718281828459045
```

```
math.factorial(10)
```

```
3628800
```

Once you do the import, you can use any module constant or function in this way. If I define my version of *factorial*:

```
def factorial(n):
    print("Using my factorial")

    if n < 2:
        return 1

    return n * factorial(n - 1)
```

I now have both functions available to me:

```
factorial(5)
```

```
Using my factorial
Using my factorial
Using my factorial
Using my factorial
Using my factorial
120
```

```
math.factorial(5)
```

```
120
```

In this way, I have avoided having one name cover the other. We also say that we do not have a *namespace collision*. This is good practice: you use what you need, and you know its source. I put in the call to *print* in my function so we can tell when Python is running this specific code.

A variation of this is to use `as` to create a new namespace for what you import. This is convenient if the module name is long or it is traditional to use a shortened name. You can also use a word that has more meaning to you for your application. Don't abuse this feature. Where there is a convention for naming an imported module or package, follow that. For example, it is traditional to use `import numpy as np`. Don't bother doing something like this:

```
import math as maths

maths.cos(maths.pi)

-1.0
```

What if you do not want to qualify an import? In this case, use `from` together with `import`.

```
from math import factorial as fac

fac(3)

6
```

We can use *fac* without "`math.`" and the *factorial* we coded is still available.

```
factorial(3)

Using my factorial
Using my factorial
Using my factorial
6
```

If we don't use `as`, our *factorial* is covered by the one we imported.

```
from math import factorial

factorial(5)

120
```

Since Python doesn't print anything, this is the version from the **math** module.

You can import several definitions at once.

```
from math import pi, sin
```

```
sin(pi/4)
```

```
0.7071067811865476
```

Finally, you can import everything at once from a module, and you do not have to qualify the names.

```
from math import *
```

```
log2(1024)
```

```
10.0
```

Don't do this if you can avoid it unless the module contains only a few definitions. Otherwise, you are bringing many named definitions into your environment that may cover others.

The preferred order of importing definitions from modules or packages is:

- Use a plain `import` and then qualify all names. Pick a new name for the namespace if you really must.
- Use `from` and `import` for a small number of names that won't clash with other definitions. Rename them to avoid naming collisions.
- Use `from` and `import` `*` if there are only a few definitions, and you know there won't be collisions you don't want.

5.2 Integers

Integers are numbers like – 4, 0, and –123. They do not include fractions such as 2/3 nor numbers with decimal points such as 2.71828. Unlike many programming languages, where an integer's size is limited by the so-called hardware "word size," a Python integer can be as large as your system memory can support. The type of a Python integer is `int`. [PIW]

If you don't need to use negative integers, the largest hardware integer that a 64-bit processor can represent is usually $2^{64} - 1$. (Here, 64 bits is the word size of that processor.)

```
2**64 - 1
```

```
18446744073709551615
```

The symbol ** is the exponentiation or "power" operator. It raises the number on its left to the power on its right.

Exercise 5.1

Check the maximum signed integer on your system by issuing

```
import sys
```

```
sys.maxsize
```

This is $2^n - 1$ for some n. What is n?

For many calculations in mathematics, the sciences, and AI, we need much larger numbers. We don't necessarily want to go to approximations that include decimals, with the resulting loss of accuracy.

```
2.0**150
```

```
1.42724769270596e+45
```

That answer is 1.42724769270596 times 10^{45}. We lose information in the calculation, but we keep it all when we use **int**.

```
2**150
```

```
1427247692705959881058285969449495136382746624
```

A number that includes a decimal point is a floating-point number, and we cover these in section 5.3. Python integers and floating-point numbers represent real numbers, so it makes sense to consider values like `sin(3)`. However, in this section, we only look at functions that must have integers as arguments.

I cover integers in section 3.3 of *Dancing with Qubits*. [D.W.Q]

5.2.1 Permutations and combinations

Suppose my daughter grows heirloom tomatoes, and she gives me four plants she grew from seed. They are of the varieties V_1, V_2, V_3, and V_4.

I have space in my garden for only four plants, and I number the locations where I can put them L_1, L_2, L_3, and L_4. In how many different ways can I put the four plants in the ground?

We begin with L_1. Since this is the first location, I have my choice of any of the four varieties. Once chosen, there are three varieties left to put in L_2. I have two options for L_3, and finally, just one for L_4.

The number of *permutations* in which I can plant the tomatoes is

$$4 \times 3 \times 2 \times 1 = 4! = 24 .$$

```
import math

math.factorial(4)
```

```
24
```

What if I only have two locations available, L_1 and L_2? Then I have $4 \times 3 = 4! / (4 - 2)!$ permutations of plant placements. We use *perm* to compute the answer.

```
math.perm(4, 2)
```

```
12
```

Figure 5.1 shows the 12 permutations.

L_2	V_2	V_3	V_4	V_1	V_3	V_4	V_1	V_2	V_4	V_1	V_2	V_3
L_1	V_1	V_1	V_1	V_2	V_2	V_2	V_3	V_3	V_3	V_4	V_4	V_4

Figure 5.1: Permutations of 4 items taken 2 at a time

Exercise 5.2

Verify that this works for four plants and one and three locations.

In general, if I have *n* objects with no duplicates, and I look at ordered sub-collections of *k* objects, I have *n*! / *k*! permutations.

For a *combination*, the order doesn't count. If I have *n* objects with no duplicates, and I want to look at unordered sub-collections of *k* objects, I have *n*! / (*n* − *k*)! *k*! combinations.

For four plants and four locations, I have just one combination. I put all four plants in the ground, and I don't care about the sequence.

```
math.comb(4, 4)
```

```
1
```

Exercise 5.3

How many combinations would there be if I had five unique varieties and three planting locations?

Exercise 5.4

Multiply out the polynomial $(x + 1)^5$ and look at the coefficients. What is the relationship between those numbers and values returned by *comb* when the first argument is 5?

5.2.2 Factoring

An integer *n* > 0 is *prime* if its only integer factors are 1 and itself. So 11 is prime, but

$$48 = 2 \times 2 \times 2 \times 2 \times 3 = 2^4\,3$$

is not. For 48, its prime factors are 2 and 3, and 2 occurs with *multiplicity* 4.

In general, any non-zero integer can be factored uniquely into ± 1 times zero or more primes. ± 1 are called *units* because they have an absolute value of 1. The factorization of 0 is just 0.

Exercise 5.5

Factor each of the following: -111; 10,000,000; 1,024; and −210.

Python does not provide any built-in function for factoring integers. Integer factorization can be very hard, and this difficulty is the basis for several encryption protocols, including RSA. Peter Shor's 1997 paper on gaining a significant improvement for factoring using quantum computers has been a topic of interest, concern, and confusion for many years. Should anyone eventually use Shor's Algorithm for factoring large integers, we do know it will require a quantum system with hundreds of thousands, if not millions, of qubits. [PTA]

Nevertheless, there are some tools we can use that may help us factor specific integers or tell if they are prime. The methods I show are basic and often not suitable for large integers.

Our first and probably most important tool is the remainder operation "%". If n and m are integers, then n is divisible by m if $n \% m = 0$.

```python
for m in (2, 3, 5, 7):
    if 14 % m == 0:
        print(f"14 is divisible by {m}")
    else:
        print(f"14 is not divisible by {m}")
```

```
14 is divisible by 2
14 is not divisible by 3
14 is not divisible by 5
14 is divisible by 7
```

It's necessary to have a list of primes if we want to perform these zero-remainder checks. This function uses the ancient Sieve of Eratosthenes algorithm to create a list of all prime numbers less than or equal to a positive integer.

```python
 1  def sieve_of_eratosthenes(n):
 2      # Return a list of the primes less than or equal to n.
 3      # First check that n is an integer greater than 1.
 4
 5      if not isinstance(n, int) or not n > 1:
 6          print(
 7              "The argument to sieve_of_eratosthenes "
 8              "must be an integer greater than 1.")
 9          return []
10
11      # Make a list holding the integers from 0 to n.
12
13      potential_primes = list(range(n + 1))
14
```

```
15      # If index is not prime, set potential_primes[index] to 0.
16      # We start with 0 and 1
17
18      potential_primes[0] = 0
19      potential_primes[1] = 0
20
21      p = 2  # 2 is prime, so start with that.
22
23      while p <= n:
24          # If at an index whose value in potential_primes
25          # is not 0, it is prime.
26
27          if potential_primes[p]:
28              i = 2 * p
29
30              # Mark p+p, p+p+p, etc. as not prime.
31              while i <= n:
32                  if i != p:
33                      potential_primes[i] = 0
34                  i += p
35          p += 1
36
37      # The only non-zero integers left in potential_primes
38      # are primes. Return a list of those.
39
40      return [prime for prime in potential_primes if prime]

sieve_of_eratosthenes(35)

[2, 3, 5, 7, 11, 13, 17, 19, 23, 29, 31]
```

Since I have comments in the code, I only want to mention several Python features I use, all at the beginning.

- I explicitly check that the argument n has an acceptable value on line 5.
- *isinstance* tests whether n is an **int**.
- I check that n is greater than 1.
- I concatenated literal strings by putting them next to each other on lines 7 and 8.
- I remembered to return an empty list after printing the error message on line 9.

Exercise 5.6

Rewrite this to raise one or more exceptions. Where might you use a **TypeError** or a **ValueError**?

Exercise 5.7

How and where do I test if an **int** is not 0 in the main loop?

Before we leave primes and factoring, there is one more function that mathematicians use in studying the properties of numbers. *isqrt* computes the "integer square root" of a non-negative **int**. It returns the largest integer less than or equal to the real square root.

```
import math

math.isqrt(63)

7

math.isqrt(64)

8

math.isqrt(65)

8
```

Exercise 5.8

Write a function *is_perfect_square* that accepts a non-negative integer and returns **True** or **False** based on whether it is another integer multiplied by itself. For example, `is_perfect_square(36)` should be **True**, but `is_perfect_square(41)` should be **False**. Your function must use *isqrt* from **math**.

Exercise 5.9

Analyze and add comments to this simple factorization function:

```
1  def simple_factor(n):
2      # factor the integer n by trial division
3
4      if not isinstance(n, int) or not n > 1:
5          print(
6              "The argument to simple_factor "
7              "must be an integer greater than 1.")
8          return []
9
10     primes_to_check = sieve_of_eratosthenes(math.isqrt(n))
11     prime_factors = []
12
13     for prime in primes_to_check:
14         while n % prime == 0:
15             prime_factors.append(prime)
16             n = n // prime
17
18     return prime_factors
```

```
f10 = math.factorial(10)

print(f"The prime factors of 10! = {f10} are\n"
      f"{simple_factor(f10)}")

The prime factors of 10! = 3628800 are
[2, 2, 2, 2, 2, 2, 2, 2, 3, 3, 3, 3, 5, 5, 7]
```

Why is the argument to *sieve_of_eratosthenes* on line 10 equal to `math.isqrt(n)` and not n?

I cover integer factorization in section 10.2 of *Dancing with Qubits*. [D.W.Q]

5.2.3 GCDs and LCMs

Consider 12 and 15. The largest integer I can divide into each of them is 3. This is their *greatest common divisor*, or GCD. The smallest integer that is a multiple of each of them into is 60. This is their *least common multiple*, or LCM.

```
math.gcd(12, 15)
```

```
3
```

```
12 * 15 // math.gcd(12, 15)
```

```
60
```

Exercise 5.10

Research and understand Euclid's algorithm for computing the GCD. Write a Python function that implements this. Be sure to include simultaneous assignment in the main loop.

Exercise 5.11

Explain why the LCM of two positive integers is the product of the integers divided by their GCD.

5.2.4 Binary and hexadecimal

In section 1.1, I introduced the binary representation of positive integers. In particular,

$$\text{decimal: } 72 = 7 \times 10^1 + 2 \times 10^0$$
$$\text{binary: } 1001000 = 1 \times 2^6 + 0 \times 2^5 + 0 \times 2^4 + 1 \times 2^3 + 0 \times 2^2 + 0 \times 2^1 + 0 \times 2^0$$

72 in base 10 is the same as 1001000 in base 2. The *bin* function provides the binary form of an integer as a **str**.

```
bin(72)
```

```
'0b1001000'
```

Since this is a string, it cannot be used directly in computation.

```
bin(72) + 1
```

```
TypeError: can only concatenate str (not "int") to str
```

Python thought we were trying to append an integer to a string with "+"! Use the `0b` prefix to type a binary number directly.

```
0b1001000
```

```
72
```

Python and its libraries use the *int* function in many places to try to convert objects to integers. With one argument, *int* assumes you are giving it an expression in base 10.

```
int("-32856")
```

```
-32856
```

You can give it the base in the second argument. If b is expressed in base 2:

```
b = bin(100)
b
```

```
'0b1100100'
```

then we tell *int* about the base in the second argument:

```
int(b, base=2)
```

```
100
```

In addition to binary base 2 and decimal base 10, coders frequently use *hexadecimal* base 16. Binary uses 0 and 1, decimal uses 0 through 9, and hexadecimal uses 0 through F (or f). This table shows the hexadecimal, decimal, and binary representations of the numbers 0 through 15.

```
for n in range(8):
    m = n + 8
    print(f"{n:01x}    {n:2}    {n:04b}          "
          f"{m:01x}    {m:2}    {m:04b}")
```

0	0	0000	8	8	1000
1	1	0001	9	9	1001
2	2	0010	a	10	1010
3	3	0011	b	11	1011
4	4	0100	c	12	1100
5	5	0101	d	13	1101
6	6	0110	e	14	1110
7	7	0111	f	15	1111

The *hex* function returns a string representing an integer in hexadecimal.

```
hex(100)
```
```
'0x64'
```
```
hex(252)
```
```
'0xfc'
```

Exercise 5.12

From the previous code, describe how to use f-strings to print numbers in binary and hexadecimal.

Coders usually shorten "hexadecimal" to "hex." Hex is often used in RGB color definitions within HTML and CSS for the web. For example, `#fff68f` is a light yellow. The first two "hexits" are the red component, 3 and 4 are the green component, and 5 and 6 are the blue component. `#000000` is black, `#ffffff` is white, and gray shades have all three parts equal. You may use lower or uppercase letters in hexadecimal.

5.2.5 Unicode

This section is about characters in strings, but I delayed introducing the topic until we had seen hexadecimal.

A byte is eight bits or two hexits. The largest integer you can fit into a byte is 255 decimal or FF hex, so if you use single bytes for characters in a string, you are limited to 256 characters. The original character encoding was ASCII, and it used 7 bits for 128 characters.

Since the English language has 52 uppercase and lowercase letters, 10 digits, and less than 20 punctuation marks, 256 may seem more than enough. However, there are many languages, mathematical symbols, and emoji. The Unicode standard represents all these characters. As of May 2021, there are 144,697 of them in draft version 14.0. [UNI]

A mapping from characters to sequences of bytes is called an *encoding*. The most common encoding, and the one used by the majority of websites, is UTF-8. It uses one to four bytes. Its specification includes the ASCII characters in the first byte. The sequences of bytes represent integers, and we call them *code points*.

By default, Python uses Unicode and UTF-8 in its strings and identifiers.

```
right_arrow = "→"
```

The hexadecimal code point for the right arrow is 2192, and you can use this within a string by escaping it with \u.

```
"A right arrow looks like '\u2192'"

"A right arrow looks like '→'"
```

Python expects four hexits after the \u.

The Python documentation describes how to use additional encodings and byte representations of strings. [P.Y.U]

5.3 Floating-point numbers

Floating-point numbers in Python are usually written with a decimal point or in scientific notation.

```
-3.9005

-3.9005

1.1456e-4

0.00011456

f = -9E100
f

-9e+100

type(f)

float
```

For scientific notation, you may use either "e" or "E", followed by an integer. The integer is the base 10 exponent. -9E100 means -9×10^{100}.

Exercise 5.13

What does it mean when the integer after "e" or "E" is negative, as in `2.8e-3`?

Python cannot express in floating-point the rational numbers that have infinite decimal expressions because of hardware truncation. In general, Python truncates numbers with too many digits.

$$\frac{1}{7} = 0.142857142857142857142857\cdots = 0.\overline{142857}$$

```
1/7
```

```
0.14285714285714285
```

```
0.1010101010101010101010101010101010101
```

```
0.10101010101010101
```

The **math** module contains many functions for floating-point numbers in addition to those we saw in section 5.2 for integers. Besides those I cover below, **math** provides the hyperbolic (*sinh*, *cosh*, and *tanh*) and special functions (*erf*, *gamma*, and *lgamma*), plus several others.

I cover real numbers, including floating-point numbers, in section 3.5 of *Dancing with Qubits*. [DWQ]

5.3.1 inf and nan

Now let's create a very large **float**.

```
1e100000000000000000000000000000
```

```
inf
```

```
float("inf")
```

```
inf
```

This is *infinity* and is larger than any number Python can express. There is also a negative version.

```
-1e1000000000000000000000000000
```

```
-inf
```

```
float("-inf")
```

```
-inf
```

Working with infinities gives you some interesting and unusual algebra and comparisons.

```
float("inf") + 3
```

```
inf
```

```
2e8383747474847474747 == 9e38383992272273898383
```

```
True
```

```
float("inf") + float("inf")
```

```
inf
```

```
float("inf") - float("inf")
```

```
nan
```

The result from the last calculation is nan, which stands for "Not A Number." Python can represent zero, positive, and negative floating-point numbers, the two infinities, and nan. Here is another example that returns nan.

```
float("inf") / float("inf")
```

```
nan
```

Exercise 5.14

Can you use nan in any useful calculations? What are the rules for algebra with
float("inf") and float("-inf")?

The **math** module provides *isfinite. isinf,* and *isnan* so you can test for the different kinds
of floating-point numbers. In what follows, I won't always discuss the "edge cases" of working
with the infinities and nan, but you should think about where they might appear and how you
should handle them.

Exercise 5.15

Write a function *isnegativeinf* that returns True if its argument is -inf, and False
otherwise.

Exercise 5.16

Define a function *sgn(x)* that returns -1 if x < 0, 0 if x == 0, and 1 if 0 < x.
Make sure your function works with inf and -inf. How should it handle nan?

5.3.2 Rounding

We saw the *round* function in section 2.6.2. It takes one or two arguments. The first is the
number to be rounded. The optional second argument is how many decimal places to use.

```
round(8356.92665, 2)
```

```
8356.93
```

```
round(8356.92665, -2)
```

```
8400.0
```

To throw away everything after the decimal point, use *trunc.* The two graphs in Figure 5.2 compare *round* and *trunc.* See how they handle numbers like 1.5 that are midway between two integers? The dashed diagonal line is $y = x$.

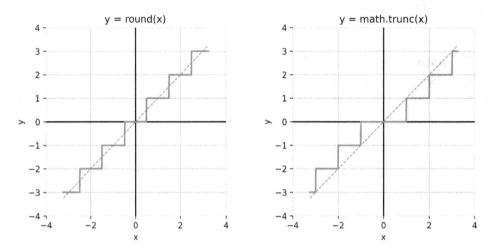

Figure 5.2: Plots of the round and trunc functions

floor and *ceil* round floating-point numbers down or up. They return integers.

```
[math.floor(-1.2,), math.ceil(-1.2), math.floor(1.2), math.ceil(1.2)]
[-2, -1, 1, 2]
```

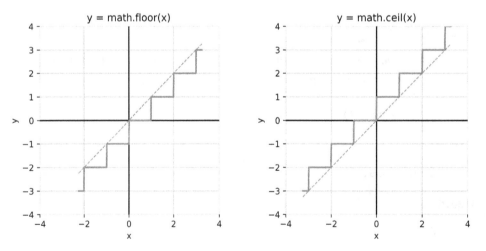

Figure 5.3: Plots of the floor and ceil functions

Exercise 5.17

How is the graph of *trunc* similar to parts of the graphs of *floor* and *ceil*?

5.3.3 Trigonometry

Trigonometry is the study of triangles, angles, and length. The most important constant in trigonometry is π, an irrational real number. It cannot be expressed exactly as the fraction of two integers.

```
math.pi
```

3.141592653589793

The **math** module also supplies τ = *tau*, which is equal to 2π.

```
math.tau
```

6.283185307179586

Many people learn about degrees when they first encounter the geometry of circles. We use radians more often in trigonometry for areas of math, science, and computer graphics. 360° is equal to 2π radians. Python provides the *degrees* and *radians* functions to switch back and forth.

```
math.degrees(2 * math.pi)
```

360.0

```
math.radians(270) / math.pi
```

1.5

It's common for coders and scientists to use θ (theta) and φ (phi) for the names of angles. Unless otherwise noted, I use radians in this book.

The **math** module provides the core trigonometric functions *sin*, *cos*, and *tan*. These are the sine, cosine, and tangent, respectively. **math** also includes their inverse functions *asin*, *acos*, and *atan*.

Figure 5.4 shows the graphs of the sine *sin* and inverse sine *asin* functions.

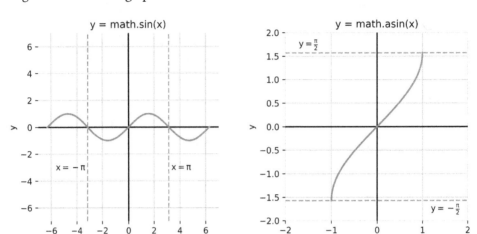

Figure 5.4: Plots of the sine and inverse sine functions

There are three essential things to note:

1. *sin* can take any real number represented by a **float** as its argument, but only produces values between −1.0 and 1.0. inclusive. Said another way, the *domain* of *sin* is the set of real numbers, and its *range* is the set of real numbers y such that $-1 \le y \le 1$.

2. The sine function is periodic: given a real number x, $sin(x) = sin(x + 2k\pi)$ for any integer k. That's why *sin* repeats itself as you move from one side of the graph to the other. We say that the sine is periodic with period 2π.

3. The inverse function *asin* has the domain $-1 \le x \le 1$ for any real number x, but we restrict the range to $-\pi/2 \le y \le \pi/2$.

```
math.asin(-1.0)
```

```
-1.5707963267948966
```

```
math.asin(-1.5)
```

```
ValueError: math domain error
```

Exercise 5.18

The *cos* function also has a period of 2π. What is its domain and range? What are the domain and range for *acos*?

Mathematically, the tangent function is the quotient of the sine and cosine. In Python, rounding errors may cause these values to be slightly different.

```
math.tan(1.0)
```

```
1.5574077246549023
```

```
math.sin(1.0) / math.cos(1.0)
```

```
1.557407724654902
```

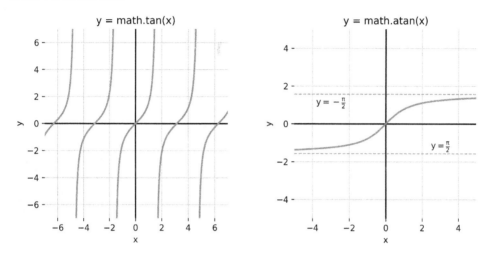

Figure 5.5: Plots of the tangent and inverse tangent functions

You should use *tan* directly. Remember that the tangent is not defined where the cosine is 0, such as at $\pi/2$. You may not get a **ValueError** because of a round-off error or **ZeroDivisionError**, but you should check for this case.

Computed values that you expect to be 0.0 might only be close to it.

```
math.cos(math.pi/2)
```

```
6.123233995736766e-17
```

This is a good time to use the *isclose* function:

```
math.isclose(math.cos(math.pi/2), 0.0, abs_tol=1e-15)
```

```
True
```

This result means that `math.cos(math.pi/2)` and `0.0` are the same to approximately 15 decimal places. The exact value you use for the absolute tolerance `abs_tol` depends on your application. There is no general default value that makes sense in all cases. [PEP 485]

The **math** module does not provide direct support for the secant, cosecant, and cotangent functions. You can implement them yourself, but you must consider where they are not defined.

Exercise 5.19

The graph in Figure 5.6 is of which trigonometric function?

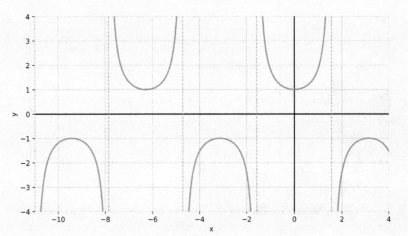

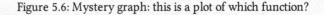

Figure 5.6: Mystery graph: this is a plot of which function?

Where is the function undefined, and why?

 I cover trigonometry in section 4.3 of *Dancing with Qubits*. [DWQ]

5.3.4 Exponentials and logarithms

An exponential function has the form

$$c\,a^x$$

for $a > 0$, $a \neq 1$, and $c \neq 0$. We frequently use e for a, where e is *Euler's number*. The Python function that computes e^x is *exp*.

```
math.e
```

```
2.718281828459045
```

```
math.exp(0.0)
```

```
1.0
```

```
math.exp(1.0)
```

```
2.718281828459045
```

```
math.exp(4.5)
```

```
90.01713130052181
```

Suppose you have a bank account that compounds interest continuously. This function computes the amount of money you have after `years` if you start with `principal` and the interest at `annual_rate`.

```python
def compound_continuously(principal, annual_rate, years):
    return principal * math.exp(annual_rate * years)
```

```
compound_continuously(100, 0.02, 1)
```

```
102.02013400267558
```

```
compound_continuously(100, 0.02, 35)
```

```
201.37527074704767
```

In these examples, we used 2% interest and started with 100 units of currency. You doubled your money after 35 years. That's not very impressive!

Exercise 5.20

Modify *compound_continuously* to round to two decimal places.

Exercise 5.21

If the annual interest rate is 5%, how long will it take to double your money? What if it is 10%?

For more general exponential functions, we use *pow*.

```
y = math.pow(8.0, 3.0)
y
```

```
512.0
```

```
math.pow(y, 0.333333333333333)
```

```
7.999999999999982
```

The second value approximates the cube root.

We use square roots often, so **math** includes a dedicated function for it.

```
math.sqrt(105.0)
```

```
10.246950765959598
```

We are working with real numbers and the argument to *sqrt* must be non-negative.

```
math.sqrt(-1.0)
```

```
ValueError: math domain error
```

Figure 5.7 compares exponential functions where $c = 1$, and a takes on each of the values 2, e, and 0.5. It uses different scales for the horizontal and vertical axes. When $0 < a < 1$, we have *exponential decay*. When $1 < a$, we have *exponential growth*.

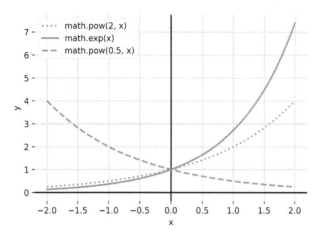

Figure 5.7: Plots of three exponential functions

The logarithm is the inverse of the exponential. If

$$y = ca^x \qquad \text{then} \qquad \log_a(\frac{y}{c}) = x.$$

This means

$$x = a^{\log_a(x)} \qquad \text{and} \qquad x = \log_a(a^x).$$

We say, "$\log_a$ is the base a logarithm." It is not defined for an argument less than or equal to 0. When $c = 1$ and $a = 2$, $\log_2(16)$ is the power to raise 2 to get 16.

```
math.log2(16)
```

```
4.0
```

The **math** module provides the *log*, *log2*, and *log10* functions. Mathematicians call *log* the *natural logarithm* and sometimes write it as "ln." *log2* is the *binary logarithm*.

```
math.log10(10**8)
```

8.0

```
math.log2(2*2*2*2*2)
```

5.0

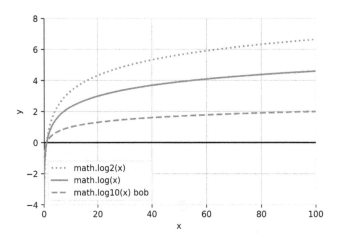

Figure 5.8: Plots of three logarithm functions

Exercise 5.22

Which grows faster, x^2 or 2^x? What about $\sqrt{x}$ or $\log_2(x)$?

Exercise 5.23

What does the function

```
def mystery_function(n):
    return math.trunc(math.log2(n))
```

do for n, a positive **int**? Write a function that returns the same result without using any floating-point numbers or functions from any imported Python module.

Suppose we have an algorithm **A** that runs in $O(K)$ time, as discussed in section 3.3. For example, K might be n, n^2, or 2^n. Compared to **A**, a new algorithm **B** is an *exponential improvement* if it runs in $O(\log(K))$ time. It is a quadratic improvement if it runs in $O(\sqrt{K})$ time.

Quantum computing is increasing in importance because there are algorithms that are quadratically and exponentially better than anything classical that we have now. We do not yet have quantum computers that are powerful enough to implement those algorithms to gain an advantage. If a task runs in one million $= 10^6$ seconds, a quadratic improvement means we could do it in about one thousand $= 10^3$ seconds. An exponential improvement would let us complete the job in about six seconds.

Exercise 5.24

Research the formula for radioactive decay and write a Python function to compute it. For extra credit, use your function for Carbon-14 and Iodine-131.

I cover exponentials and logarithms in section 4.2.4 of *Dancing with Qubits*. [DWQ]

5.4 Rational numbers

A rational number is a fraction. You learned about fractions early in your mathematical education. We use *Fraction* from the **fractions** module to create them.

```
from fractions import Fraction

Fraction(6, 5)

Fraction(6, 5)

f = Fraction(20, 30)
f

Fraction(2, 3)
```

Python stores a fraction in the lowest terms, which means the numerator and denominator have no common prime factors. Another way to say this is that the numerator and denominator have a greatest common divisor equal to 1.

```
f.numerator
```

```
2
```

```
f.denominator
```

```
3
```

```
import math
```

```
math.gcd(f.numerator, f.denominator)
```

```
1
```

Look carefully at the last expression and note that I wrote `f.numerator` instead of `f.numerator()`. Instead of calling *numerator* as a method, I asked for the "numerator property" of the fraction. A *property* is an intrinsic part or characteristic of an object. We look at how to define properties within classes in section 7.5.

Python normalizes a negative fraction so its numerator is negative.

```
Fraction(1, -5)
```

```
Fraction(-1, 5)
```

```
- Fraction(3, 7)
```

```
Fraction(-3, 7)
```

Fraction supports the usual algebraic operators.

```
a = Fraction(2, 3)
b = Fraction(7, 4)
```

```
2 * a
```

```
Fraction(4, 3)
```

```
a + b
```

```
Fraction(29, 12)
```

```
a - b
```

```
Fraction(-13, 12)
```

```
a * b
```

```
Fraction(7, 6)
```

```
1 / b
```

```
Fraction(4, 7)
```

```
a / b
```

```
Fraction(8, 21)
```

```
a ** 4
```

```
Fraction(16, 81)
```

```
a ** -2
```

```
Fraction(9, 4)
```

Use *float* to convert a fraction to a floating-point number.

```
f = float(a)
f
```

```
0.6666666666666666
```

In this case, we got a **float** approximation to the fraction. The correct answer has an infinite number of sixes after the decimal point. You can also see this when we convert back to a fraction.

```
Fraction(f)
```

```
Fraction(6004799503160661, 9007199254740992)
```

You may see inexact conversions because of the binary internal representation of floating-point.

```
Fraction(0.125)
```

```
Fraction(1, 8)
```

```
Fraction(0.1)

Fraction(3602879701896397, 36028797018963968)
```

 I cover fractions, also known as rational numbers, in section 3.4 of *Dancing with Qubits*. [D.W.Q]

5.5 Complex numbers

The *complex numbers* extend the real numbers by adding the square root of −1. Mathematicians typically call this square root *i*, but people in fields like electrical engineering use *j*. Python uses *j*. I strongly disagree with this choice, but it is part of the language now. As you work with Python and read technical documents, you may need to translate mentally back and forth between *i* and *j*.

The **cmath** module provides many functions and methods for working with complex numbers. These include complex versions of inf, nan, e, pi, tau, and the trigonometric, hyperbolic, exponential, and logarithmic functions.

 I cover complex numbers in sections 3.9 and 4.5 of *Dancing with Qubits*. [D.W.Q]

5.5.1 Creating complex numbers

For *a* and *b* real numbers, a complex number *z* looks like $a + b\,i$. The *real part* Re(*z*) of *z* is *a*, and the *imaginary part* Im(*z*) is *b*. We use the *complex* function or j-syntax to create a complex number.

```
import cmath

cmath.sqrt(-1.0)

1j
```

```
z = complex(1.5, -2)
z
```

```
(1.5-2j)
```

The *real* and *imag* methods extract the real and imaginary parts.

```
z.real
```

```
1.5
```

```
z.imag
```

```
-2.0
```

```
3.4 - 8.03j
```

```
(3.4-8.03j)
```

When you type in a complex number with j, you must precede the j with a number. Otherwise, Python thinks you are trying to use a variable named j.

Note the difference in the results returned by functions from **math** and **cmath.**

```
print(f"{math.sqrt(4)}    {cmath.sqrt(4)}")
```

```
2.0    (2+0j)
```

```
print(f"{math.sin(0)}    {cmath.sin(0)}")
```

```
0.0    0j
```

The type of a complex number is **complex.**

```
type(1j)
```

```
complex
```

```
isinstance(1j, complex)
```

```
True
```

5.5.2 Algebra of complex numbers

The algebraic operations on complex numbers are

$$
\begin{aligned}
0 &= 0 + 0i \\
1 &= 1 + 0i \\
-(a + bi) &= a - bi \\
(a + bi) + (c + di) &= (a + c) + (b + d)i \\
(a + bi) - (c + di) &= (a - c) + (b - d)i \\
(a + bi) \times (c + di) &= (ac - bd) + (ad + bc)i
\end{aligned}
$$

Conjugation negates the imaginary part of a complex number. The Python method that implements this is *conjugate*.

```
(3 + 4j).conjugate()
```

```
(3-4j)
```

In mathematical texts, you may see conjugation indicated by a line over the complex number or expression.

The absolute value $|z|$ of a complex number z is the square root of the sum of the squares of the real and imaginary parts. This sum is the complex number times its conjugate.

$$
\begin{aligned}
|a + bi| &= \sqrt{a^2 + b^2} \\
(a + bi) \times \text{conjugate}(a + bi) &= (a + bi) \times \overline{(a + bi)} \\
&= (a + bi) \times (a - bi) \\
&= a^2 + b^2
\end{aligned}
$$

```
abs(0.5 - 0.25j)
```

```
0.5590169943749475
```

Exercise 5.25

Show that $1/(a + b\,i)$ is

$$
\frac{a}{a^2 + b^2} - \frac{b}{a^2 + b^2}i
$$

Re-express this using the conjugate.

Exercise 5.26

Verify the preceding algebraic rules with

```
z = 3 + 4j
w = 2 - 1j
```

What is the formula for $(a + b\,i)^2$? Verify it for z and w.

5.5.3 Geometry of complex numbers

We also have the geometry of complex numbers, which we graph in a plane. The real part is on the horizontal axis, and the imaginary part is on the vertical.

Conjugation reflects the point representing the complex number across the horizontal axis. In the graph, I connect two conjugate values on the right-hand side with a dashed line.

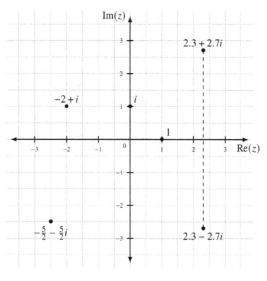

The representation $a + b\,i$ is called the *rectangular form* of a complex number. We also have the *polar form*, which uses two numbers too. Instead of being coordinates as in the rectangular form, they are an angle and a length. The formal names for these are *phase* and *modulus*, which we often write as φ and *r*.

Imagine a line segment **L** connecting the origin $(0, 0)$ with the complex number. *r* is the length of **L**. φ is measured in radians and is the angle from the right-hand side of the horizontal axis to **L**.

```
r, phi = cmath.polar(3 + 4j)
```

Let $z = -5\sqrt{2}/4 + -5\sqrt{2}/4\ i$.

```
z = complex(-5.0 * math.sqrt(2.0) / 4.0,
            -5.0 * math.sqrt(2.0) / 4.0)
```

```
r, phi = p = cmath.polar(z)

print(f"{z = }\n{r = }\n{phi = }")

z = (-1.7677669529663689-1.7677669529663689j)
r = 2.5
phi = -2.356194490192345

-3 * math.pi / 4

-2.356194490192345
```

The *polar* method converts from the rectangular form, and the *rect* method converts from the polar form.

```
cmath.rect(r, phi)

(-1.7677669529663687-1.7677669529663689j)
```

Some authors give the phase φ of a non-zero complex number in the range $0 \le \varphi < 2\pi$, and others in $-\pi < \varphi \le \pi$. Python does the first, and my graph shows the second. The phase is sometimes called the *argument* of a complex number.

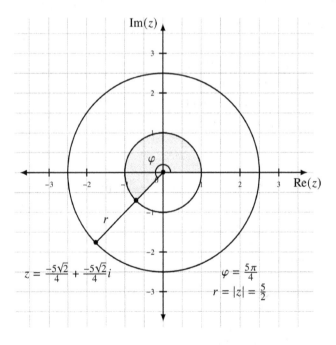

Figure 5.9: Complex numbers with rectangular and polar coordinates

Exercise 5.27

How would you convert between the two choices for the range of the phase?

Exercise 5.28

cmath includes a *phase* method but does not provide a *modulus* function or method. Write a function that computes the modulus. Is this the same as *abs*?

Complex numbers have many applications across mathematics and the sciences. They are fundamental to quantum mechanics, and hence quantum computing. We can write a qubit's state as a pair of complex numbers (a, b) with $|a|^2 + |b|^2 = 1$. These are the "two pieces of information" held by the qubit. This representation is mathematically equivalent to the qubit's state being a point on the Bloch sphere introduced in section 1.11.

5.6 Symbolic computation

In mathematics, we see more than numbers and numeric vectors and matrices. Polynomials are used in many disciplines, though they do not have a native Python implementation. In a polynomial, we use an "indeterminate" symbolically without requiring a numeric value. For example,

$$p(x) = (x - 3)(x - 1)(x + 2) = x^3 - 2x^2 - 5x + 6$$

The computer science discipline that deals with these and more advanced mathematical objects is called "computer algebra" or "symbolic mathematical computation." [AXM]

The **sympy** package implements symbolic computation tools, and its documentation showcases its broad functionality. It is not part of the Python Standard Library, so you must install it via a command from the operating system command line:

```
pip install sympy
```

Several examples give you an idea of what **sympy** can do.

We first define an indeterminate symbol x and a polynomial p.

```
import sympy as sym

x = sym.Symbol('x')
p = x**3 - 2*x**2 - 5*x + 6
p
```

```
x**3 - 2*x**2 - 5*x + 6
```

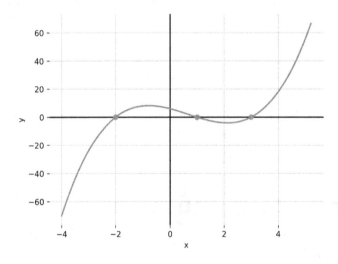

Figure 5.10: Plot of the cubic polynomial $x^3 - 2x^2 - 5x + 6$

Though not all polynomials can be factored over the integers, p can.

```
sym.factor(p)
```

```
(x - 3)*(x - 1)*(x + 2)
```

By inspection of the factorization, we see that $p(3) = 0$. We confirm it by substituting 3 for x.

```
p.subs(x, 3)
```

```
0
```

We can be slightly more sophisticated and evaluate the limit as $x \to 3$.

```
sym.limit(p, x, 3)
```

```
0
```

The derivative is straightforward,

```
sym.diff(p, x)

3*x**2 - 4*x - 5
```

as is the integral.

```
sym.integrate(p, x)

x**4/4 - 2*x**3/3 - 5*x**2/2 + 6*x
```

5.7 Random numbers

Here is a task for you: give me five random integers between 1 and 10, inclusive. It's okay if there are duplicates.

Python provides *randint* in the module **random** to give you a random integer within a range.

```
import random

[random.randint(1, 10) for i in range(6)]

[5, 8, 5, 10, 10, 1]
```

Are these really random? Let's do it again:

```
[random.randint(1, 10) for i in range(6)]

[9, 6, 4, 5, 6, 8]
```

The lists are different, but let me insert one extra call before the comprehension and again do it twice.

```
random.seed(10)
[random.randint(1, 10) for i in range(6)]

[10, 1, 7, 8, 10, 1]

random.seed(10)
[random.randint(1, 10) for i in range(6)]

[10, 1, 7, 8, 10, 1]
```

These sequences don't look so random anymore.

The function *random* computes a pseudo-random **float** greater than or equal to 0 and less than 1.

```
[random.random() for i in range(3)]
```

```
[0.20609823213950174, 0.81332125135732, 0.8235888725334455]
```

5.7.1 Pseudo-randomness

In classical computing, we start with some *seed* number and then perform a series of arithmetic and bit operations to get "random" numbers. If we begin with the same seed, we get the same sequence of numbers. By default, Python uses the computer system time as the seed, though you can set it via *seed*. For this reason, we call these *pseudo-random* numbers.

If I know your seed and I know your random number generation algorithm, I may be able to replicate the numbers you produce. For this reason, you must be very careful about using pseudo-random numbers for cryptography. The Python documentation states this in its discussion of **random**:

> **Warning:** The pseudo-random generators of this module should not be used for security purposes.
> For security or cryptographic uses, see the **secrets** module.

With this in mind, we continue with **random**.

5.7.2 Distributions

The function *random* provides a *uniform distribution* of numbers in the interval, meaning that any value is as likely to appear as any other. More generally, uniform(r, s) provides a random **float** n in the range $s \leq n \leq s$.

```
[random.uniform(0, 2) for i in range(3)]
```

```
[1.3069450678023515, 0.3204591130376393, 1.0413387192798491]
```

Those are three random numbers between 0 and 2 inclusive. You can't tell that this is a uniform distribution with only three values, but you get a better idea with a million.

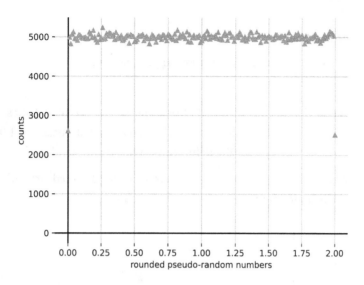

Figure 5.11: One million pseudo-random numbers in a uniform distribution

I created the graph in Figure 5.11 with **matplotlib**, which we discuss in *Chapter 13, Creating Plots and Charts*. This is what I did:

- In a loop, I called `random.uniform(0, 2)` one million times and rounded each pseudo-random number to two decimal places.
- I created a dictionary where the keys were the rounded pseudo-random numbers, and the values were the counts of how many times the numbers occurred.
- I plotted the pseudo-random numbers on the horizontal axis and the counts on the vertical.

Forgetting about the artifacts at the endpoints, you can see that I generated each value approximately the same number of times. This is the "uniform" part of the distribution and is often what people want when they ask for a "random number."

Exercise 5.29

Describe how the graph would look if I used many more or many less than one million pseudo-random numbers.

Exercise 5.30

Research where uniform distributions appear naturally.

There are several other kinds of distributions that people see in statistics. The graph in Figure 5.12 is a *Bell curve* for a *Gaussian* or *normal distribution* with mean μ = 1.0 and standard deviation σ = 0.2. See section 14.1 for the definitions of mean and standard deviation.

Note how many of the pseudo-random numbers are clustered around the middle. I generated each pseudo-random number with `random.gauss(1.0, 0.2)`.

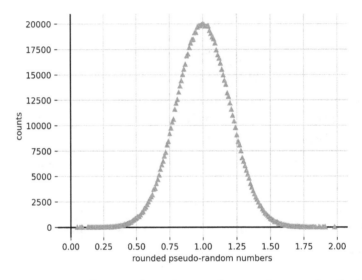

Figure 5.12: One million pseudo-random numbers in a Gaussian distribution

Exercise 5.31

Research where Bell curves and normal distributions appear naturally.

5.7.3 Sampling

People often use random numbers in statistics and data analysis for sampling. Let's generate a deck of 52 playing cards. I hold them in the list `card_deck`.

```
card_deck = []
```

```
for suit in ("C", "D", "H", "S"):
    # create cards with rank 2 through 10
    for rank in range(2, 11):
        card_deck.append(f"{rank}{suit}")

    # create cards for jacks, queens, kings, and aces
    for rank in ("J", "Q", "K", "A"):
        card_deck.append(f"{rank}{suit}")

def print_cards(cards):
    # print the cards with up to 13 cards per row

    for card, n in zip(cards, range(1, len(cards) + 1)):
        print(f"{card:3}", end="  ")
        if n % 13 == 0:
            print()

print_cards(card_deck)
```

```
2C   3C   4C   5C   6C   7C   8C   9C   10C  JC   QC   KC   AC
2D   3D   4D   5D   6D   7D   8D   9D   10D  JD   QD   KD   AD
2H   3H   4H   5H   6H   7H   8H   9H   10H  JH   QH   KH   AH
2S   3S   4S   5S   6S   7S   8S   9S   10S  JS   QS   KS   AS
```

The suits are "C", "D", "H", and "S". These stand for "Clubs," "Diamonds," "Hearts," and "Spades," respectively. The card "2C" is the Two of Clubs, and QD is the Queen of Diamonds.

Now we use *shuffle* from **random** to mix up the cards in the list in place.

```
random.shuffle(card_deck)
print_cards(card_deck)
```

```
5S   8H   4H   10H  AD   7H   6S   5C   9S   9H   7C   10D  AS
3H   3S   2D   6H   QC   4S   QH   6D   JC   8C   KS   9C   2C
2S   3C   AC   JH   3D   QS   KH   7S   JS   8D   KC   5H   5D
7D   8S   KD   JD   AH   10C  2H   4C   QD   10S  4D   6C   9D
```

When I deal a hand, I am taking a *sample* of cards, a random selection from the collection. In this case, I am dealing five cards, perhaps for a game of Five Card Stud poker.

```
random.sample(card_deck, 5)
```

```
['5D', 'JD', 'AD', 'JS', '7H']
```

The *sample* function does not remove the cards from the deck. You must remove them yourself.

Exercise 5.32

Write a function that deals a hand of *n* cards and then removes them from the deck. Be sure to check that there are enough cards to deal.

If I want one random card, I use *choice*. This function also does not modify the list.

```
random.choice(card_deck)
```

```
'9H'
```

choice and *sample* work for lists, but do not work directly for sets or dictionaries. You must do a conversion to a list first.

```
random.choice(list({1, 2, 3, 4, 5}))
```

```
4
```

```
numbers = {0: "zero", 1: "one", 2: "two"}
key = random.choice(list(numbers.keys()))
key
```

```
2
```

```
numbers[key]
```

```
'two'
```

5.8 Quantum randomness

In this chapter, we've seen that there are many ways of writing different kinds of numbers. For integers, we have decimal, binary, and hexadecimal. For floating-point, there are digits with a decimal point and scientific notation. For complex numbers, we use Python's j-syntax.

Qubits also have their special representation. Mathematically, we write the quantum state of a qubit as

$$a\,|0\rangle + b\,|1\rangle$$

where a and b are complex numbers, and the sum of the squares of their absolute values is 1:

$$|a|^2 + |b|^2 = 1 .$$

An expression enclosed in the "vertical bar–greater than" symbol pair is called a *ket*. For example, you might see $|\varphi\rangle$ in a text or an article. We pronounce this "ket-phi." $|0\rangle$ is ket-0 and $|1\rangle$ is ket-1.

For computation, qubits don't exist as independent, free-floating objects. Instead, they underlie quantum circuits, where we apply 1- and 2-qubit gates and operations to the qubits to change their states based on algorithms. Think of a gate as a reversible function on one or two qubits. Whatever result it produces, we can run the function backward to get the original arguments.

The most widely used quantum computing development software is Qiskit, an open source, community-led project on GitHub. [**QIS**] I use the **qiskit** Python package exclusively for the quantum computations in this book. [**LQCQ**][**QCP**]

```
import qiskit
```

Exercise 5.33

List at least three different ways to pronounce "Qiskit."

A circuit operation that is not reversible is *measurement*. If a qubit is in the state $a\,|0\rangle + b\,|1\rangle$, then when we measure it, it will result in either $|0\rangle$ or $|1\rangle$. We can interpret these as the bits 0 and 1. When do we get $|0\rangle$, and when do we get $|1\rangle$? If b is 0, we always get $|0\rangle$, and if a is 0, we get $|1\rangle$. Otherwise, we only know the chance of getting one or the other!

The probability of getting $|0\rangle$ is $|a|^2$. Similarly, the likelihood we will get $|1\rangle$ is $|b|^2$. Since these are the only two possible outcomes, these probabilities must sum to one.

In general, measurement is non-deterministic. Except in the special cases of a or b being 0, when we measure a qubit multiple times, we can get different answers.

How is this useful? Are we getting random results? Quantum algorithms use multiple qubits and many gates to produce results that have a high probability of being the "correct" answer we want. We start that discussion in earnest in section 9.6.

If I have a qubit in the state

$$(\sqrt{3}/2)\,|0\rangle + (1/2)\,|1\rangle$$

then the probability I will get $|0\rangle$ is $(\sqrt{3}/2)^2 = 3/4 = 0.75$. The chance I will get $|1\rangle$ is $(1/2)^2 = 1/4 = 0.25$. If I measure a qubit in that state many times, I should see $|0\rangle$ three times more than I see $|1\rangle$.

While most of our discussion of gates will come later, the Hadamard **H** gate is one of the most widely used in quantum computing. It operates by

$$\mathbf{H}\,|0\rangle \longrightarrow (\sqrt{2}/2)\,|0\rangle + (\sqrt{2}/2)\,|1\rangle$$
$$\mathbf{H}\,|1\rangle \longrightarrow (\sqrt{2}/2)\,|0\rangle - (\sqrt{2}/2)\,|1\rangle$$

When I apply the Hadamard **H** gate to either $|0\rangle$ or $|1\rangle$, I get a result with an equal probability of yielding $|0\rangle$ or $|1\rangle$ when measured.

Exercise 5.34

Prove the last statement.

This is a random number generator, or at least the start of one! When you use a quantum computer, the result is not driven by a random number seed but depends solely on nature's quantum mechanical properties and processes. If you are using a quantum computer *simulator*, which is classical software and hardware, you are only producing pseudo-random numbers by a quantum-like process.

To do real quantum computing,
you require real quantum computing hardware.

Let's end this chapter by using quantum computing to generate random 2-bit numbers. There are four of these numbers: 00, 01, 10, and 11. I begin by creating a 2-qubit circuit and then add code to apply the **H** gate to each. I am not running this circuit yet. I am defining the circuit that I will execute later.

```
circuit = qiskit.QuantumCircuit(2)

circuit.h(0)
circuit.h(1)

<qiskit.circuit.instructionset.InstructionSet at 0x247703f3370>
```

Next I add the measurement operators and see what the circuit looks like.

```
circuit.measure_all()
```

```
circuit.draw(output="mpl")
```

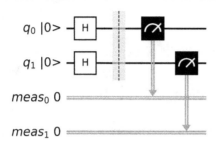

Under the hood, **qiskit** can use **matplotlib** to draw circuits. The qubits are q_0 and q_1. The Hadamard gates are the squares with "**H**" in them, and the white squares with the black dials are measurement operators.

Now I'm going to represent the quantum computer by the variable backend. In section 9.7.5, I will show you how to do this for both hardware and simulators, but I'm skipping the details for now. Trust me.

Finally, I run our circuit. The argument shots=8000 means the computer will execute the circuit 8,000 times.

```
job_sim = qiskit.execute(circuit, backend, shots=8000)
```

```
job_sim.result().get_counts(circuit)
```

```
{'11': 2000, '10': 2002, '00': 2000, '01': 1998}
```

The result of *get_counts* is a dictionary. The keys are the string representations of the 2-bit numbers, and the values are the counts of how many times the computer computed each number. We expect to see each value show up approximately 2,000 times, though since this is a small sample, the counts may be quite different from this average.

Exercise 5.35

Do you expect the result to be a uniform distribution? Why or why not? What would cause it not to be?

I start the discussion of qubits in Chapter 7 of *Dancing with Qubits*. [D.W.Q]

5.9 Summary

In this chapter, we saw many kinds of numbers and ways to represent them. I introduced symbolic computational objects like polynomials via **sympy**. We ended our discussion with computing and using random and pseudo-random numbers. I began using **qiskit** to demonstrate a simple way of generating 2-bit integers on a quantum computer. Quantum computing is probabilistic and this is an essential distinction from the classical case for how we design and implement algorithms.

Though we have used and defined Python functions many times, we return to them in depth in the next chapter. We'll discuss the many function definition features that Python provides, along with the correct ways for you to document what your functions do.

6
Defining and Using Functions

We are what we repeatedly do. Excellence, therefore, is
not an act, but a habit.

—Will Durant, *The Story of Philosophy*

We have already seen many examples of simple functions. Functions let you write and package code so that it may be used more than once. Suppose you want to print out the odd integers greater than 0 and less than 6. This works:

```
for i in range(6):
    if i % 2 == 1:
        print(f"{i} ", end="")
```

    1 3 5

Instead of 6, what if you wanted to use 9 or 16 or 1,036? You do not want to repeat the above every time you change the upper bound. Instead, you code a *function* that is called repeatedly with different values.

```
def odd_numbers_less_than(n):
    # this function prints out all odd numbers greater
    # than 0 and less than n
    for i in range(n):
        if i % 2 == 1:
            print(f"{i} ", end="")
```

```
odd_numbers_less_than(9)
```

```
1 3 5 7
```

```
odd_numbers_less_than(16)
```

```
1 3 5 7 9 11 13 15
```

In this chapter, we explore the many features Python provides for writing functions. Functions are valuable in their own right, and we will use the techniques again when we look at methods within classes in section 7.2.

Topics covered in this chapter

6.1 The basic form

A minimal function definition is

```
def f():
    pass
```

This definition begins with the def keyword. Then comes the name of the function, *f,* an open-close parentheses pair, and a colon, ":". The next line is indented and begins the *function body.* Here it is pass.

We use pass where we must include an action but do not have anything useful to do. Think of it as saying, "look how I am drawing attention to not doing anything." For example, I can use pass in a conditional statement to indicate that I am aware of a processing option and choose to do nothing. You may use three periods, "...", instead of pass.

Starting from this minimal form, we begin to add and use additional Python features. The function body can be any collection of expressions we want to package together and call repeatedly. Let's define a function that prints the current time.

I begin by importing two Python modules that work with time: **datetime** and **time**.

```
import datetime
import time
```

Now I define a function that gets the current time and creates a formatted string. I use *strftime* for the hour, minutes, and seconds but it also provides other options to include the year, the day, the month, and the day of the week.

```
def print_the_time():
    print(datetime.datetime.now().strftime("%H:%M:%S"))

print_the_time()

12:47:43
```

Because computers are so fast, if I call *print_the_time* immediately, I will likely get the same result. I pause two seconds with *sleep,* and then again print the time.

```
time.sleep(2.0)

print_the_time()

12:47:45
```

I do not give *print_the_time* any additional data to let it do its work. Functions usually have *parameters* to vary their behavior and computation.

6.2 Parameters and arguments

Let's redefine *print_the_time* with the parameter `delay_in_seconds`.

```
def print_the_time(delay_in_seconds):
    # print the current time after waiting delay_in_seconds
    time.sleep(delay_in_seconds)
    print(datetime.datetime.now().strftime("%H:%M:%S"))

print_the_time(0.0)

12:47:46

print_the_time(1.0)

12:47:47
```

In the first call to *print_the_time*, the *argument* `0.0` is substituted for the *parameter* `delay_in_seconds`. In the second call, the argument is `1.0`, which results in Python sleeping for one second and then computing and printing the time.

Exercise 6.1

Add a type check to ensure that `delay_in_seconds` is a non-negative **float**.

Python allows you to define functions with multiple arguments.

```
def format_text(text, make_uppercase):
    if make_uppercase:
        print(text.upper())
    else:
        print(text)

format_text("this text is not uppercase", False)

this text is not uppercase

format_text("this text is uppercase", True)

THIS TEXT IS UPPERCASE
```

Exercise 6.2

Change *format_text* so that the second argument determines whether the string should be uppercased, lowercased, or left as-is.

Rather than having too many parameters, you should consider grouping together values that your function handles in similar ways. Instead of using five parameters that we collect into a list:

```
def sort_and_print_5(z1, z2, z3, z4, z5):
    for z in sorted([z1, z2, z3, z4, z5]):
        print(f"{z} ", end="")

sort_and_print_5(4, 2, -9, 0, 3)

-9 0 2 3 4
```

pass in a list as a single argument:

```
def sort_and_print(numbers):
    for z in sorted(numbers):
        print(f"{z} ", end="")

sort_and_print([4, 2, -9, 0, 3])

-9 0 2 3 4
```

On the other hand, Python provides *-notation to group arguments into a tuple and assign it to a single named parameter.

```
def sort_and_print(*numbers):
    for z in sorted(numbers):
        print(f"{z} ", end="")

sort_and_print(4, 2, -9, 0, 3)

-9 0 2 3 4
```

Exercise 6.3

Carefully compare these three styles of parameter definitions and the use of arguments in function calls.

You can mix a starred parameter with non-starred ones.

```python
def sort_and_print(ascending, *numbers):
    if ascending:
        for z in sorted(numbers, reverse=False):
            print(f"{z} ", end="")
    else:
        for z in sorted(numbers, reverse=True):
            print(f"{z} ", end="")
```

```python
sort_and_print(False, 4, 2, -9, 0, 3)
```

```
4 3 2 0 -9
```

```python
sort_and_print(True, 4, 2, -9, 0, 3)
```

```
-9 0 2 3 4
```

Exercise 6.4

Rewrite *sort_and_print* with the `ascending` parameter at the end. Does it work? Can you use more than one starred parameter?

If you use a "*" on a list argument to a function, Python expands it to match multiple parameters.

```python
def expander3(a, b, c):
    print(f"{a = }  {b = }  {c = }")
```

```python
numbers = [10, 20, 30]
expander3(*numbers)
```

```
a = 10  b = 20  c = 30
```

The length of the list must match the number of parameters.

```python
numbers = [10, 20]
expander3(*numbers)
```

```
TypeError: expander3() missing 1 required positional
          argument: 'c'
```

We return to parameters and arguments in section 6.5, "Keyword arguments," and section 6.6, "Default argument values."

6.3 Naming conventions

In section 2.4.1, I gave the rules for Python identifiers. These include variables, function names, and parameters. In brief, you may use lowercase and uppercase letters, digits, and underscores, but you cannot begin an identifier with a digit.

These rules prescribe how you can name an identifier, but not what you should use as a name. *Naming conventions* help you make that choice.

If you look through Python packages, you will see several conventions in use, especially if you go back to parts of the library that have been around for many years. If you are extending a module, you should probably follow its conventions. Similarly, your organization may have its own rules.

The source of naming convention "truth" is *PEP 8 – Style Guide for Python Code*, from the Python Software Foundation, though usage may vary as I described above. [PEP008] In this section, I summarize the conventions for functions, parameters, and variables. We return to naming for classes and their methods in section 7.6.

Function, parameter, and variable names should use digits, lowercase letters, and inner underscores. Do not start a name with an underscore except when you don't need to use the variable.

```
text = "can   you see   multiple   spaces?"
text
```
```
'can   you see   multiple   spaces?'
```
```
for _ in range(3):
    text = text.replace("  ", " ")
```
```
text
```
```
'can you see multiple spaces?'
```

I ran the same code three times to replace any occurrence of two consecutive spaces with a single space.

Exercise 6.5

Use in and call *replace* the minimum times necessary to remove all double spaces.

If you unpack a tuple or list, use "_" for unneeded values. You may use it more than once.

```
quo, _ = divmod(37, 5)
quo
```

```
7
```

```
_, b, _, d = 10, 20, 30, 40
b + d
```

```
60
```

Use a trailing underscore if you want to avoid using a Python keyword as a name.

```
import_ = "to bring items into a country for sale"
```

When you view code on a screen or printed on paper, it can be hard to distinguish some characters, especially with some fonts. Is that a capital "O" or the digit "0"? What about lowercase letters "l" and "i" versus the digit "1"? Be careful using these characters in names if there could be any confusion.

It's common for coders to use the variables i, j, and k for indices and counters in loops. We often use n and m for integers and z and w for complex numbers. These reflect authors' use of the names in technical books and articles.

I use generic function names like *f*, *g*, and *h* in this book when I do not need to be descriptive. Similarly, I use x and y for generic variable and parameter names.

Make your code as self-documenting as possible by using descriptive names when it makes sense. In section.7.8, we look at conventions for explicitly documenting your work.

6.4 Return values

A Python function can return nothing, one thing, or many things. That covers all possibilities!

In truth, a Python function always returns one object. Let me explain. Our simple do-nothing function from the beginning of this chapter appears to produce nothing.

```
def f():
    pass
```

But if we check the type of the result, we see that it is **NoneType**.

```
type(f())
```

```
NoneType
```

```
f() is None
```

```
True
```

The returned object is None, which we discussed in section 3.9.5. We get the same result if we use a bare return in *f*.

```
def f():
    return
```

```
f() is None
```

```
True
```

Use return to have the function pass a value back to whatever called the function.

```
def g(x, y):
    return x + y
```

```
g(10, 23)
```

```
33
```

If you forget the return, you will get None.

```
def bad_g(x, y):
    x + y
```

```
bad_g(10, 23) is None
```

```
True
```

It's a common mistake for coders to forget a return statement, especially with conditionals.

```
def bad_abs(x):
    if x < 0:
        return -x
    x
```

```
bad_abs(-2)**3
```

```
8
```

```
bad_abs(5) is None
```

```
True
```

```
bad_abs(5)**3
```

```
TypeError: unsupported operand type(s) for ** or pow():
            'NoneType' and 'int'
```

If you see an error like this, there's a good chance a function is not returning what you think it should. If you ever return a value for some execution path through your function, you should return a value for every other path. That is, do not implicitly default to returning None if you elsewhere return an **int**, for example. Be explicit and return None.

To pass back "multiple" values, construct a list or a tuple, and unpack it. We are once again producing a single object.

```
def h(x, y):
    return x + y, x *y
```

```
type(h(10, 15))
```

```
tuple
```

```
sum, product = h(10, 15)
```

```
print(f"{sum = }   {product = }")
```

```
sum = 25   product = 150
```

Code is *unreachable* if your function can never execute it.

```
def f(x):
    return x * x
    return x + x
```

The second return will never run. This looks like a silly mistake that you would avoid, but there could be a lot of code between those two return statements. A good Python editing environment like Visual Studio Code helps you identify unreachable code.

Exercise 6.6

Write a function *adder* that takes two arguments and tries to use "**+**" with them. If the operation is successful, return the result. If it fails, return None.

When I write a function, I use `return` as soon as I have a value I want to pass back. Sometimes this can be a question of style, but I think it is better to write less code and proceed through the function logic with fewer possibilities to consider.

Compare these two versions of a function that tests whether a string contains only vowels.

```
def isvowel(text):
    # return True if text is empty or composed only of English vowels

    result = True
    for c in text:
        if c not in "aeiouAEIOU":
            result = False
            break
    return result

isvowel("Eeeuuu"), isvowel("Woohoo")

(True, False)

def isvowel(text):
    # return True if text is empty or composed only of English vowels

    for c in text:
        if c not in "aeiouAEIOU":
            return False
    return True

isvowel("Eeeuuu"), isvowel("Woohoo")

(True, False)
```

In both implementations, if we get to the end of the outer loop, there is one possible answer, and we return it.

6.5 Keyword arguments

Let's define a function that encapsulates *pow* from **math**:

```
import math

def raiser(base, exponent):
    return math.pow(base, exponent)
```

I can call *raiser* using *positional arguments*, where Python assigns the first argument to base and the second to exponent.

```
raiser(2, 3)
```

```
8.0
```

I can also explicitly use the names of the parameters in *keyword arguments*.

```
raiser(base=2, exponent=3)
```

```
8.0
```

By itself, that's not very exciting, but it is clearer what the arguments are providing to the function. You can use keyword arguments in any order.

```
raiser(exponent=3, base=2)
```

```
8.0
```

Positional arguments must come before keyword arguments. This works:

```
raiser(2, exponent=3)
```

```
8.0
```

but this does not:

```
raiser(base=2, 3)
```

```
SyntaxError: positional argument follows keyword argument
```

If you want to force users of your code to pass keyword arguments, put a "*" in the parameter list.

```
def raiser(*, base, exponent):
    return math.pow(base, exponent)
```

```
raiser(2, 3)
```

```
TypeError: raiser() takes 0 positional arguments
            but 2 were given
```

Though you cannot use a positional argument after a starred parameter, you can use a keyword argument.

```
def pick_an_argument(*args, k):
    return args[k]

pick_an_argument("one", "two", "three", k=1)

'two'

pick_an_argument("one", "two", "three", 1)

TypeError: pick_an_argument() missing 1 required
            keyword-only argument: 'index'
```

We use "*" before a parameter name to collect positional arguments into a tuple, and we use "**" to collect keyword arguments into a dictionary.

```
def show_keyword_args(**kwargs):
    for key, value in kwargs.items():
        print(f"{key = }   {value = }")

show_keyword_args(x=1, y=2, z=3)

key = 'x'   value = 1
key = 'y'   value = 2
key = 'z'   value = 3
```

I would not use this method for required arguments as in

```
def raiser(**kwargs):
    if "base" in kwargs and "exponent" in kwargs:
        return math.pow(kwargs["base"], kwargs["exponent"])
    return "I do not know what you are trying to do"

raiser(exponent=3, base=2)

8.0

raiser(sugar=3, flour=2)

'I do not know what you are trying to do'
```

but I would consider it for optional arguments. You may use any valid variable name you wish instead of kwargs.

In this **non-working** example, I am drawing a line between two points. Python collects the keyword arguments that change the default values for the line's color, width, style, and other attributes into `formatting_info`.

```python
def draw_line(x1, y1, x2, y2, **formatting_info):
    # Draw a line from the point (x1, y1) to (x2, y2).
    # Use the information in the dictionary formatting_info
    # to change the style of the line

    # do something ...
    return

draw_line(0, 0, 2, 1)
draw_line(0, 0, 2, 1, color="red")
draw_line(0, 0, 2, 1, width=4, style="dotted")
```

Let's look again at *print*, which we first began examining in section 2.9. In its simplest form, *print* takes an object, converts it to a string if necessary, and then displays it and a newline. It does not display quotes around a string unless you add them explicitly.

```python
print("a string")
```
```
a string
```
```python
print("'a string with quotes'")
```
```
'a string with quotes'
```
```python
print(57.4 / 8)
```
```
7.175
```

If you give *print* several arguments, it displays them separated by spaces.

```python
print('q', 'u', 'a', 'n', 't', 'u', 'm')
```
```
q u a n t u m
```

You change the separator to something else with the `sep` keyword argument.

```python
print('q', 'u', 'a', 'n', 't', 'u', 'm', sep='_')
```
```
q_u_a_n_t_u_m
```

By default, *print* ends the displayed line with a newline. Use the keyword argument end to change that.

```
for i in range(10, 0, -1):
    print(i, end=' ')
print("launch!")
```

```
10 9 8 7 6 5 4 3 2 1 launch!
```

Exercise 6.7

Write a function that prints the numbers 1 through 20 in four columns, with five numbers in each column. Your function should accept an argument that determines whether Python displays the numbers in ascending order from left to right or top to bottom. The columns must line up neatly.

6.6 Default argument values

You use a keyword argument when you *call* a function. You give default argument values when you *define* a function.

Here is a function that prints a string. By default, it displays the string as-is. However, you can pass a second argument that tells the function to put the characters in uppercase.

```
def display_string(the_string, put_in_uppercase=False):
    if put_in_uppercase:
        print(the_string.upper())
    else:
        print(the_string)
```

```
display_string("This is the default behavior")
```

```
This is the default behavior
```

```
display_string("This overrides the default behavior", True)
```

```
THIS OVERRIDES THE DEFAULT BEHAVIOR
```

You can have the second argument be the same as the default, and you can give it as a keyword argument.

```
display_string("This is not uppercased", put_in_uppercase=False)
```

```
This is not uppercased
```

Parameters with default values must follow parameters without them.

```
def f(a=2, b):
    print(a + b)
```

```
SyntaxError: non-default argument follows default argument
```

6.7 Formatting conventions

The Python parser that reads the code you type is forgiving but does follow a strict set of core rules. Beyond that, formatting conventions make it easier for others to read and understand what you have written. Think of these as strong suggestions that you should try to follow.

Review your code on a set schedule for correct behavior and adherence to naming and formatting conventions. Read and use *PEP 8 – Style Guide for Python Code*, from the Python Software Foundation. [PEP008] Your organization may require or recommend some variations. If you work in a group of programmers, help each other to follow the rules.

Here are my "Top Ten" formatting conventions from PEP 8 and my own experience:

1. Indent 4 spaces. Don't use tabs. **Never use tabs.** (It is standard convention to indent in multiples of four. It is my personal convention to never uses tabs. If you are editing existing code that uses tabs, you should continue using them.)

2. Use single quotes around strings of length 1, double quotes otherwise.

3. Put all `import` statements at the top of the file containing your code. Sort the lines by the module or package names. Although you *can* import two modules on the same line as in `import math, time`, **don't**.

4. **Do** put two, and only two, blank lines before and after function definitions. (To conserve space in this book, I have usually not followed this convention.)

5. **Don't** use extra spaces immediately after opening parentheses, brackets, or braces. Similarly, **don't** use extra spaces immediately before closing parentheses, brackets, or braces.

6. **Don't** put a space between the function name and opening parenthesis in a function call or definition.

7. **Do** put a space after a comma in a list, tuple, or set expression. **Do** put a space after a comma in a parameter or argument list. **Don't** put spaces before commas.

8. **Do** put a space on each side of an equal sign in an assignment or operator assignment like "+=". **Do** put a space to each side of comparison operators like "<" and ">=". **Do** put spaces around not, and, and or.

9. **Don't** include spaces around the equal sign in a keyword or default argument.

10. In arithmetic expressions, **don't** put a space between a negative sign and a number, variable, or expression. **Don't** put spaces around "**" for exponentiation. **Do** put spaces around "+", "-", and "*" for addition, subtraction, and multiplication. However, you may leave out the spaces when multiplying a variable by an integer or when grouping terms closely makes your expression easier to read. For example, 2*x - x*y. Use your judgment, but be consistent.

6.8 Nested functions

You can define a function inside another function. We say that the inner function is *nested* inside the outer.

```
1  def f(x, y):
2      w = 10
3      def g(z):
4          return z**3 + w
5      return  g(x + y)

f(2, 3)

135
```

The nested function *g* on line 3 can see and use w outside its definition but within *f*.

You cannot call a nested function by name outside its enclosing function.

```
g(3)

NameError: name 'g' is not defined
```

To avoid confusion, do not use a parameter name for a nested function that you also use for the containing function. You cannot return from a nested function all the way out of the function in which you defined it. You may code this behavior using an exception via try and except.

You may define more than one nested function, and a nested function may contain additional function definitions within it.

Exercise 6.8

Consider the function skeleton

```
def f(x, y):
    def g():
        def h():
            # if x + y == 10, return all the way out
            # of f with True, otherwise return False to g
            ...

        h()
    g()
    return None
```

Add the necessary code to the definitions of *f* and its nested functions to implement the behavior I describe in the comments for *h*.

6.9 Variable scope

When I assign a variable, what other code can access and use it? What else can change it? The answers to these questions define the variable's *scope*.

6.9.1 Global versus local scopes

Global scope is outside every function definition. Let's look at how the value of my_variable changes inside and outside functions.

```
my_variable = "SET AT GLOBAL SCOPE"
print(f"Global: {my_variable}")
```

```
Global: SET AT GLOBAL SCOPE
```

```
def f():
    print(f"Inside f: {my_variable}")
```

```
f()
```

```
Inside f: SET AT GLOBAL SCOPE
```

```
print(f"Global: {my_variable}")
```

```
Global: SET AT GLOBAL SCOPE
```

I use my_variable but do not assign to it. Once I assign to my_variable, it changes from global to a *local* scope.

```
my_variable = "SET AT GLOBAL SCOPE"
print(f"Global: {my_variable}")
```

```
Global: SET AT GLOBAL SCOPE
```

```
def g():
    my_variable = "SET IN g"
    print(f"Inside g: {my_variable}")
```

```
g()
```

```
Inside g: SET IN g
```

```
print(f"Global: {my_variable}")
```

```
Global: SET AT GLOBAL SCOPE
```

In the case of *f*, we have one my_variable. For *g*, we have two: the one defined at global scope and the one defined at local scope inside the function. The local version "covered" the outer version.

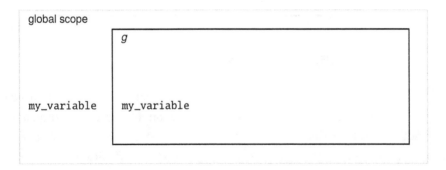

Figure 6.1: Global and local variable scopes

6.9.2 global

We force *g* to use the outer my_variable with the global keyword.

```
my_variable = "SET AT GLOBAL SCOPE"
print(f"Global: {my_variable}")
```

```
Global: SET AT GLOBAL SCOPE
```

```
def g():
    global my_variable
    my_variable = "SET IN g WITH 'global'"
    print(f"Inside g: {my_variable}")

g()

Inside g: SET IN g WITH 'global'

print(f"Global: {my_variable}")

Global: SET IN g WITH 'global'
```

I reset the value of the outer (and only) my_variable inside *g*.

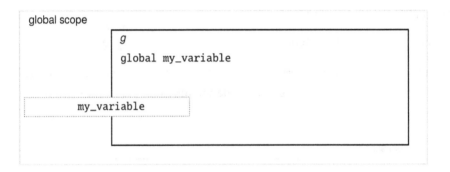

Figure 6.2: Using a variable in global scope

Try not to use global. If you define a global variable and a function that reassigns the variable, it limits where and how you can call the function. Multiple users of your function may change the global variable unexpectedly, and errors can be tough to trace and debug. If you can't avoid the concept altogether, consider using class variables. I introduce these in section 7.11.

A variable that you only assign once at global scope is a *constant*. For example, I can define constants to represent standard RGB colors:

```
RGB_BLACK = (0x00, 0x00, 0x00)
RGB_WHITE = (0xFF, 0xFF, 0xFF)
RGB_RED   = (0xFF, 0x00, 0x00)
RGB_GREEN = (0x00, 0xFF, 0x00)
RGB_BLUE  = (0x00, 0x00, 0xFF)
```

By convention, we fully uppercase the names of global constants. To make my code more readable, I align the equal signs in blocks of assignments.

6.9.3 nonlocal

There is one case that `global` cannot fix. How do I handle a variable defined in a function that I want to reassign in a nested function?

```
my_variable = "SET AT GLOBAL SCOPE"
print(f"Global: {my_variable}")
```

```
Global: SET AT GLOBAL SCOPE
```

```
def h():
    my_variable = "SET IN h"
    print(f"Inside h: {my_variable}")
    def inner_h():
        my_variable = "SET IN inner_h"
        print(f"Inside inner_h: {my_variable}")
    inner_h()
    print(f"Inside h: {my_variable}")
```

```
h()
```

```
Inside h: SET IN h
Inside inner_h: SET IN inner_h
Inside h: SET IN h
```

```
print(f"Global: {my_variable}")
```

```
Global: SET AT GLOBAL SCOPE
```

The nesting is acting as we expect, and we have three variables called `my_variable` in different scopes. How can I get `my_variable` in *inner_h* to be the same as `my_variable` in *h*? Let's see what happens when we use `global`:

```
my_variable = "SET AT GLOBAL SCOPE"
print(f"First global: {my_variable}")
```

```
First global: SET AT GLOBAL SCOPE
```

```
def h():
    my_variable = "SET IN h"
    print(f"In h: {my_variable}")
```

```
    def inner_h():
        global my_variable
        my_variable = "SET IN inner_h"
        print(f"In inner_h: {my_variable}")
    inner_h()
    print(f"In h: {my_variable}")

h()
```

```
In h: SET IN h
In inner_h: SET IN inner_h
In h: SET IN h
```

```
print(f"Global: {my_variable}")
```

```
Global: SET IN inner_h
```

This behavior is interesting! The global my_variable and the one inside *inner_h* are the same, and it is different from the my_variable defined at the top level of *h*.

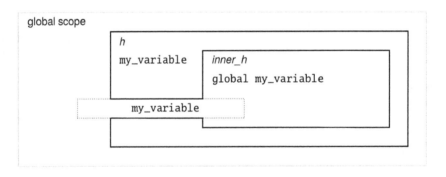

Figure 6.3: Using "global" in a nested function

We use nonlocal instead of global to get the behavior we are after.

```
my_variable = "SET AT GLOBAL SCOPE"
print(f"Global: {my_variable = }")
```

```
Global: my_variable = 'SET AT GLOBAL SCOPE'
```

```
def h():
    my_variable = "SET IN h"
    print(f"In h: {my_variable = }")
```

```
    def inner_h():
        nonlocal my_variable
        my_variable = "SET IN inner_h"
        print(f"In inner_h: {my_variable = }")
    inner_h()
    print(f"In h: {my_variable = }")

h()

In h: my_variable = 'SET IN h'
In inner_h: my_variable = 'SET IN inner_h'
In h: my_variable = 'SET IN inner_h'

print(f"Global: {my_variable = }")

Global: my_variable = 'SET AT GLOBAL SCOPE'
```

Figure 6.4: Using "nonlocal" in a nested function

Exercise 6.9

Define a function *inner_inner_h* inside *inner_h*. Assign a value to my_variable in *inner_inner_h* so that it is the same my_variable in *inner_h*. Can you make the my_variable in *inner_inner_h* be the same as the one in *h* but different from the my_variable in *inner_h*?

As you can see, this can get very confusing, so keep your code simple. Consider using function parameters and returned tuples to access and pass back changes to variables. Python provides nonlocal for completeness, but I have never used it in the code I have written. The choice of the language features I use reflects my software design philosophy and style.

6.10 Functions are objects

I can assign a Python function to a variable, store it in a data structure, and later recall and apply it to arguments.

```python
def f():
    print("I am inside the original definition of 'f'")

f()

I am inside the original definition of 'f'

g = f
g()

I am inside the original definition of 'f'

def f():
    print("I am inside the new definition of 'f'")

f()

I am inside the new definition of 'f'

g()

I am inside the original definition of 'f'
```

Let's create a calculator. The function that implements this takes a single argument, a string with three characters. The first and last characters must be digits, and the middle must be one of "+", "-", "*", and "/".

We define four helper functions:

```python
def adder(x, y): return x + y
def subtracter(x, y): return x - y
def multiplier(x, y): return x * y
def divider(x, y): return x / y
```

and put these in a dictionary:

```python
operations = {
    "+": adder,
    "-": subtracter,
    "*": multiplier,
    "/": divider,     # this trailing comma is ok
}
```

I define the *calculator* function:

```
def calculator(digit_op_digit):
    # The left operand is a single digit. Convert it to an int.
    left = int(digit_op_digit[0])

    # Get the operation helper function
    operation = operations[digit_op_digit[1]]

    # The right operand is a single digit.
    right = int(digit_op_digit[2])

    # Apply the retrieved function to the arguments.
    return operation(left, right)
```

Now let's try it out on a few examples.

```
calculator("2+3")
```

5

```
calculator("7-9")
```

-2

```
calculator("4*6")
```

24

```
calculator("8/4")
```

2.0

In *calculator*, I retrieve the arithmetic functions from the dictionary and apply them using function syntax.

Exercise 6.10

calculator is scary in that it does no error checking on its argument whatsoever. Rewrite the function using the *isdigit* method and anything else you need to ensure that Python raises no exceptions.

Exercise 6.11

Improve *calculator* so that it takes a string argument that is a sequence of one or more digits, followed by one of the four arithmetic operations, followed by a string of one or more digits. Do full error checking in this and the following exercises.

Exercise 6.12

Improve *calculator* to allow at most one decimal point to appear within the first and second numbers. For example, in addition to the above cases, it should also accept strings like ".345*97", "89.345-990", and "0.0003/2.98".

Exercise 6.13

Improve *calculator* to allow zero or more negative signs to appear before each number. What is --3?

Exercise 6.14

Improve *calculator* to support the exponentiation operation "**" and integer division "//".

6.11 Anonymous functions

In the last section, I defined four helper arithmetic functions and stored them in a dictionary. After that, I never needed their names again. Python provides `lambda` to let you create a function anonymously.

Instead of

```
def adder(x, y): return x + y
def subtracter(x, y): return x - y
def multiplier(x, y): return x * y
def divider(x, y): return x / y

operations = {
    "+": adder,
```

```
        "-": subtracter,
        "*": multiplier,
        "/": divider
    }
```

do this:

```
operations = {
    "+": lambda x, y: x + y,
    "-": lambda x, y: x - y,
    "*": lambda x, y: x * y,
    "/": lambda x, y: x / y
}

operations["-"](12.4, 0.4)

12.0
```

A lambda function should contain one expression. Python evaluates the expression and returns the result. **Don't** use return, and **don't** put parentheses around the parameters.

If you plan to give a function a name, **don't** use lambda. Use def.

```
# Don't do this even though it works
f = lambda x: x**2

f(100)

10000
```

Don't use lambda simply to call another function with the same arguments.

```
from math import sin

# Don't do this even though it works

[(lambda x: sin(x))(y) for y in range(4)]

[0.0, 0.8414709848078965, 0.9092974268256817, 0.1411200080598672]

# This works as-is

[sin(y) for y in range(4)]

[0.0, 0.8414709848078965, 0.9092974268256817, 0.1411200080598672]
```

Do use `lambda` to wrap a Python operator when you need a function: `lambda x, y: x + y`. **Do** use `lambda` for small functions used briefly, passed to other functions, or stored for later use.

Developers used lambda functions in Python in a style of coding called *functional programming*. Comprehensions have replaced many applications of functional programming. The *reduce* function from the **functools** module is quite useful.

Exercise 6.15

What is the following function computing?

```
from functools import reduce

def caf(n):
    return reduce(lambda x, y: x * y, range(1, n + 1))

[(n, caf(n)) for n in range(1, 6)]

[(1, 1), (2, 2), (3, 6), (4, 24), (5, 120)]
```

Have we seen something similar before?

6.12 Recursion

A recursive function is one that calls itself. At first thought, this could be a problem because it might seem that the execution would never stop: the function would call itself, which would call itself, which would call itself, and so on. Indeed, this simple example function does just that:

```
def f(x):
    f(x)

f(2)
```

Eventually, Python will give up and raise an error:

```
File "<stdin>", line 1, in <module>
File "<stdin>", line 1, in f
File "<stdin>", line 1, in f
File "<stdin>", line 1, in f
[Previous line repeated 996 more times]
RecursionError: maximum recursion depth exceeded
```

Python is saying: "Look, I tried doing the same thing a thousand times, but enough is enough, and you probably didn't intend to do it." The recursion depth can vary from system to system. In what follows, I assume it is 1,000.

I give this example because though you likely did not mean to do something like this, it is easy to make a mistake and do it anyway. We are missing a properly working *terminating condition.*

6.12.1 The factorial function

There are two classic recursive functions that every coding book includes. The first is *factorial,* and we saw that in section 5.1.2. The mathematical notation for factorial is a non-negative integer or variable followed by an exclamation point.

$$n! = n \times (n - 1) \cdots \times 2 \times 1$$

We can use recursion because $n! = n \times (n - 1)!$

```
1  def recursive_factorial(n):
2      if not isinstance(n, int) or n < 0:
3          raise ValueError("Argument must be a non-negative integer.")
4      if n < 2:
5          return 1
6      return n * recursive_factorial(n - 1)
```

```
recursive_factorial(12)
```

The terminating condition for the recursion is if n < 2: return 1. *recursive_factorial* stops calling itself and begins returning values.

Exercise 6.16

Show that this implementation of factorial is $O(n)$.

It's costly for Python to set up and run all these function calls. Iteration is much more efficient, so try to use it instead of recursion.

```
def iterative_factorial(n):
    if not isinstance(n, int) or n < 0:
        raise ValueError("Argument must be a non-negative integer.")
```

```
    product = 1
    for i in range(1, n + 1):
        product *= i
    return product

iterative_factorial(12)

479001600

iterative_factorial(1000)
```

402387260077093773543702433923003985719374864210714632543799910429938512
398629020592044208486969404800479988610197196058631666872994808558901323
829669944590997424504087073759918823627727188732519779505950995276120874
975462497043601418278094646496291056393887437886487337119181045825783647
849977012476632889835955735432513185323958463075557409114262417474349347
553428646576611667797396668820291207379143853719588249808126867838374559
731746136085379534524221586593201928090878297308431392844403281231558611
036976801357304216168747609675871348312025478589320767169132448426236131
412508780208000261683151027341827977704784635868170164365024153691398281
264810213092761244896359928705114964975419909342221566832572080821333186
116811553615836546984046708975602900950537616475847728421889679646244945
160765353408198901385442487984959953319101723355556602139450399736280750
137837615307127761926849034352625200015888535147331611702103968175921510
907788019393178114194545257223865541461062892187960223838971476088506276
862967146674697562911234082439208160153780889893964518263243671616762179
168909779911903754031274622899880051954441428201218736174599264295658174
662830295557029902432415318161721046583203678690611726015878352075151628
422554026517048330422614397428693306169089796848259012545832716822645806
652676995865268227280707578139185817888965220816434834482599326604336766
017699961283186078838615027946595513115655203609398818061213855860030143
569452722420634463179746059468257310379008402443243846565724501440282188
525247093519062092902313649327349756551395872055965422874977401141334696
271542284586237738753823048386568897646192738381490014076731044664025989
949022222176590433990188601856652648506179970235619389701786004081188972
991831102117122984590164192106888438712185564612496079872290851929681937
238864261483965738229112312502418664935314397013742853192664987533721894
069428143411852015801412334482801505139969429015348307764456909907315243
327828826986460278986432113908350621709500259738986355427719674282224875
758676575234422020757363056949882508796892816275384886339690995982628095
612145099948717012445164612603790293091208890869420285106401821543994571
568059418727489980094254742173582401063677404597417851608292301353580816

```
840096996372524230560855903700624271243416909004153690105933983835777939
410970027753472000000000000000000000000000000000000000000000000000000000
000000000000000000000000000000000000000000000000000000000000000000000000
000000000000000000000000000000000000000000000000000000000000000000000000
0000000000000000000000000000000000000000000000000
```

```
recursive_factorial(1000)
```

```
File "<stdin>", line 1, in <module>
File "<stdin>", line 5, in recursive_factorial
File "<stdin>", line 5, in recursive_factorial
File "<stdin>", line 5, in recursive_factorial
[Previous line repeated 995 more times]
File "<stdin>", line 2, in recursive_factorial
RecursionError: maximum recursion depth exceeded while
               calling a Python object
```

6.12.2 Greatest common divisor

In section 5.2.3, I introduced the greatest common divisor (GCD) of two integers, x and y.

$gcd(x, y) =$

- $|x|$ if $y = 0$
- $|y|$ if $x = 0$
- The largest positive integer g that divides both x and y
- $gcd(|x|, |y|)$ (so we may assume x and y are non-negative)
- $gcd(y, x)$ (so we may assume $x \geq y$)
- y if y divides x
- 1 if there does not exist a prime number that divides both x and y

How do we compute g?

We'll begin with a recursive algorithm that does subtraction. This algorithm works because if p divides x and y, then p divides $x - y$.

Exercise 6.17

Why will the recursion terminate?

```python
def gcd(x, y, step=1):
    print(f"-> {step:>2}: {x = }   {y = }")
    x, y = abs(x), abs(y)

    if x < y:
        x, y = y, x

    if y == 0:
        return x

    # at this point, we know that x and y and positive integers > 1
    # with x >= y > 0

    return gcd(x - y, y, step + 1)
```

We can now compute the GCD of any two integers, so let's try it for −24 and 400.

```python
print(f"\ngcd = {gcd(-24, 400)}")
```

```
->  1: x = -24  y = 400
->  2: x = 376  y = 24
->  3: x = 352  y = 24
->  4: x = 328  y = 24
->  5: x = 304  y = 24
->  6: x = 280  y = 24
->  7: x = 256  y = 24
->  8: x = 232  y = 24
->  9: x = 208  y = 24
-> 10: x = 184  y = 24
-> 11: x = 160  y = 24
-> 12: x = 136  y = 24
-> 13: x = 112  y = 24
-> 14: x = 88  y = 24
-> 15: x = 64  y = 24
-> 16: x = 40  y = 24
-> 17: x = 16  y = 24
-> 18: x = 8  y = 16
```

```
-> 19: x = 8  y = 8
-> 20: x = 0  y = 8
```

```
gcd = 8
```

We can do better with a version using division. Recall that divmod(x, y) returns the tuple (x // y, x % y). If we let q, r = divmod(x, y), then x == q*y + r and x - q*y = r.

Exercise 6.18

Show that this equality implies that if prime *p* divides *x* and *y*, then *p* divides *r*.

```
def gcd(x, y, step=1):
    print(f"-> {step:>2}: {x = }  {y = }")
    x, y = abs(x), abs(y)

    if x < y:
        x, y = y, x

    if y == 0:
        return x

    # at this point, we know that x and y and positive integers > 1
    # with x >= y > 0

    return gcd(y, x % y, step + 1)

print(f"\ngcd = {gcd(-24, 400)}")
```

```
->  1: x = -24  y = 400
->  2: x = 24  y = 16
->  3: x = 16  y = 8
->  4: x = 8  y = 0
```

```
gcd = 8
```

See how we replaced the difference "–" with the remainder "%"? The two algorithms are nearly identical, but the second reduces the size of the arguments much faster. The next example shows how quickly it works for large integers.

```
print(f"\ngcd = {gcd(2**20 * 3**30, 2**30 * 3**20)}")
```

```
->   1: x = 215892499727278669824   y = 3743906242624487424
->   2: x = 3743906242624487424   y = 2489843897682886656
->   3: x = 2489843897682886656   y = 1254062344941600768
->   4: x = 1254062344941600768   y = 1235781552741285888
->   5: x = 1235781552741285888   y = 18280792200314880
->   6: x = 18280792200314880   y = 10968475320188928
->   7: x = 10968475320188928   y = 7312316880125952
->   8: x = 7312316880125952   y = 3656158440062976
->   9: x = 3656158440062976   y = 0

gcd = 3656158440062976
```

Exercise 6.19

Estimate how many steps our first version of *gcd* would need for the same arguments.

In the last algorithm, the terminating condition is when y becomes 0. If this is not true, we call *gcd* with smaller versions of x and y. **There is no reason to use recursion to do this!** With a loop and reassignment to x and y, we can efficiently calculate the greatest common divisor.

```
def gcd(x, y, step=1):
    print(f"-> {step:>2}: {x = }  {y = }")
    x, y = abs(x), abs(y)

    if x < y:
        x, y = y, x

    if y == 0:
        return x

    while y != 0:
        x, y = y, x % y

    return x
```

```
print(f"\ngcd = {gcd(-24, 400)}")

->  1: x = -24  y = 400

gcd = 8
```

The non-recursive implementation is known as the *Euclidean algorithm* and dates from approximately 300 BC.

6.12.3 The Fibonacci sequence

The other "must discuss" recursive function computes the numbers F_n in the Fibonacci sequence for argument $n \geq 0$:

$$F_0 = 0$$
$$F_1 = 1$$
$$F_n = F_{n-1} + F_{n-2} \text{ for } n \geq 2$$

This function is not just recursive; it is *doubly recursive*.

```
def recursive_fibonacci(n):
    if not isinstance(n, int) or n < 0:
        raise ValueError("Argument must be a non-negative integer.")

    if n < 2:
        return n

    return recursive_fibonacci(n - 1) + recursive_fibonacci(n - 2)

for j in range(16):
    print(f"{recursive_fibonacci(j)} ", end="")

0 1 1 2 3 5 8 13 21 34 55 89 144 233 377 610
```

Exercise 6.20

For $n = 0$ to 9, compute how many calls C_n to *recursive_fibonacci* we must do to calculate F_n.

There are two obvious issues with this translation of the mathematical definition to the Python implementation:

1. There are many recursive calls, and, in particular,

2. in the process of computing F_{n-1}, we compute F_{n-2} which we then compute again separately.

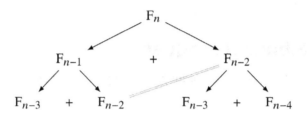

Figure 6.5: Fibonacci doubly recursive function calls

Exercise 6.21

Find a number c_2 such that

$$C_n \le c_2 2^n$$

for F_0 through F_9. What is the smallest c_2 you can find?

Let $a = (1 + \sqrt{5}) / 2$. Find a number c_a such that $C_n \le c_a a^n$ for F_0 through F_9. What is the smallest c_a you can find?

Is the recursive Fibonacci function $O(2^n)$? Is it $O(a^n)$?

One way to be more efficient is to cache the values we compute in a dictionary. "Caching" is saving a result for later reuse.

```python
def cached_fibonacci(n, cache=None):
    if not isinstance(n, int) or n < 0:
        raise ValueError("Argument must be a non-negative integer.")

    if n < 2:
        return n

    if cache is None:
```

```
        cache = {0:0, 1:1}

    # Compute Fibonacci(n - 1) first, thereby calculating all
    # Fibonacci values for values less than n - 1

    if n - 1 in cache:
        fibonacci_n_minus_1 = cache[n - 1]
    else:
        cache[n - 1] = fibonacci_n_minus_1 = cached_fibonacci(n - 1,
cache)

    # I know we have computed Fibonacci(n - 2)

    result = cache[n] = fibonacci_n_minus_1 + cache[n - 2]
    return result

for j in range(16):
    print(f"{cached_fibonacci(j)} ", end="")

0 1 1 2 3 5 8 13 21 34 55 89 144 233 377 610
```

Note how I use None and the default argument value for cache.

Exercise 6.22

For $n = 0$ to 9, compute how many calls C_n to *cached_fibonacci* we must do to calculate F_n.

In this cached version, we calculate a value in the Fibonacci sequence once, and do not recurse unless we must. We remove the recursion completely with this iterative version that includes an elegant use of simultaneous assignment:

```
def iterative_fibonacci(n):
    if not isinstance(n, int) or n < 0:
        raise ValueError("Argument must be a non-negative integer.")

    if n < 2:
        return n
```

```
    a, b, m = 0, 1, 2

    while m < n:
        a, b, m = b, a + b, m + 1

    return a + b

for j in range(16):
    print(f"{iterative_fibonacci(j)} ", end="")

0 1 1 2 3 5 8 13 21 34 55 89 144 233 377 610
```

Exercise 6.23

Insert print(f"{a = } {b = } {m = }") into the while loop. Explain the values of a, b, and m when you compute iterative_fibonacci(5).

Exercise 6.24

For n = 0 to 9, compute how many calls C_n to *iterative_fibonacci* we must do to calculate F_n.

Use recursion in Python carefully and with a lot of thought. In this section, we saw three different algorithms to accomplish the same task, and they have very different execution times and usage of computer resources.

6.12.4 Execution timing

Let's define a function to run and time our implementations of Fibonacci.

```
import time

def time_a_fib(n, fib_function, fib_function_name):
    # time_a_fib computes a fibonacci value for n.
    # fib_function is the function object and
    # fib_function_name is a string with the function's name
```

```
    start_time = time.time()
    fib_function(n)
    elapsed_time = round((time.time() - start_time), 9)
    print(f"{fib_function_name}({n}) "
          f"took {elapsed_time} seconds")


time_a_fib(30, recursive_fibonacci, "recursive_fibonacci")
time_a_fib(30, cached_fibonacci, "cached_fibonacci")
time_a_fib(30, iterative_fibonacci, "iterative_fibonacci")

recursive_fibonacci(30) took 0.351046801 seconds
cached_fibonacci(30) took 0.0 seconds
iterative_fibonacci(30) took 0.0 seconds
```

We see that the recursive version is much slower than the other two, even for 30. A time of 0.0 means that the elapsed time is extremely short when we round it.

We need to go right up against Python's recursion limit to start to see a difference between our cached and iterative versions.

```
time_a_fib(999, cached_fibonacci, "cached_fibonacci")
time_a_fib(999, iterative_fibonacci, "iterative_fibonacci")

cached_fibonacci(999) took 0.000996828 seconds
iterative_fibonacci(999) took 0.0 seconds
```

The iterative version is the only one that can go further.

```
time_a_fib(20000, iterative_fibonacci, "iterative_fibonacci")

iterative_fibonacci(20000) took 0.013954163 seconds
```

The elapsed time you see when you run this depends on your computer hardware and operating system.

When you are solving a Python coding problem, it may help to develop a recursive solution. You take a large problem and break it down into one or more smaller problems of the same form. Once they get small enough, you solve them and then roll up to the original problem's solution. When you have that working, transform your code into an iteration if performance is an issue.

For another example of recursion, see section 7.10.6, where we implement the repeated squaring algorithm for exponentiation.

6.13 Summary

Functions allow you to save sequences of Python expressions and statements and reuse them. You can parameterize a function, which means that its behavior might change based on the arguments you give it.

There is rarely only one way to accomplish a task, and we saw that the choice of algorithm could strongly influence a function's execution time. Python is well-tuned for iterative operations, and you should use recursion carefully.

All our functions have taken arguments and returned objects that were built into Python by its developers. They have operated on numbers, strings, tuples, lists, sets, and dictionaries. In the next chapter, I introduce classes and the objects we construct with them. A function we define within a class goes by a new name: *method.*

7

Organizing Objects into Classes

The creative mind plays with the objects it loves.

—Carl Jung, *Psychological Types*

Although we previously grouped several expressions into collections, such as lists and dictionaries, they don't have the natural feel or use of built-in types like integers or strings. Are you limited to only those core objects that come with Python? Can you construct new kinds of things complete with operations you can perform on them? Can you hide the data you use and the way you operate on it? If you want to extend an object with additional data or operations, can you do it?

Of course you can, or this chapter would not exist! In what follows, I describe how to create classes and the objects, or *instances*, that belong to them. A function that operates on class instances gets a new name: *method*. By careful definition of a class, you can hide its internal information and how you manipulate it. We call this concealment *encapsulation*.

If you have an existing class, you extend it through *inheritance*, though some of the rules to do what you want are tricky. This chapter explains how Python implements the class behavior we first saw in section 1.10.

Topics covered in this chapter

7.1 Objects

As we saw in section 5.6, Python does not have a built-in object that implements polynomials like

$$p(x) = (x - 3)(x - 1)(x + 2) = x^3 - 2x^2 - 5x + 6.$$

You can import them from **sympy** or other packages, but we implement a simple version of them in this chapter.

In this example polynomial, x is an *indeterminate*, and it is the only indeterminate that appears in the polynomial. We call such a polynomial *univariate* ("one variable").

We evaluate a polynomial by substituting a value for x and performing the algebra. For this reason, we can think of a polynomial as a function.

```
def p(x):
    return x**3 - 5*x**2 + 7*x + 1

[p(x) for x in range(-2, 4)]

[-41, -12, 1, 4, 3, 4]
```

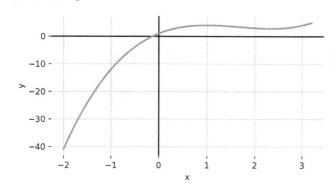

Figure 7.1: Plot of the univariate polynomial $x^3 - 5x^2 + 7x + 1$

In our definition of p, there are four *coefficients*: 1, −5, 7, and 1. These are the explicit or implicit numbers before x^3, x^2, x, and the invisible $x^0 = 1$, respectively. Every one of our coefficients must be an **int** in our implementation.

An individual coefficient-indeterminate-exponent expression is a *term*. A term is zero if and only if its coefficient is zero. A polynomial with one term is a *monomial*.

For any given exponent, there is at most one term with that exponent. That is, we always simplify anything like $3x^2 - 5x^2$ to $-2x^2$.

Every exponent must be an **int** greater than or equal to 0. The largest exponent belonging to a non-zero term is the *degree* of the polynomial. The zero polynomial has degree 0. A polynomial of degree 0 is called a *constant polynomial*.

If we have two polynomials

$$p = x^2 + 7x + 1 \quad \text{and} \quad q = 4x^2 - 7x + 3$$

then

$$2p = 2x^2 + 14x + 2 \quad \text{and} \quad p + q = 5x^2 + 4.$$

Note that the addition of two polynomials sums the coefficients of corresponding terms with equal exponents.

Our implementation of polynomials should be as natural as possible for users. In this chapter, we incrementally define:

- a Python class for univariate polynomial objects,
- algebraic operators like addition and multiplication for polynomials,
- utility methods and properties for polynomials like *degree*, and
- the derivative of a polynomial.

Regarding the derivative, don't worry if you have not studied or do not remember calculus.

Exercise 7.1

If you have studied calculus, what restriction have we placed on our polynomials that prevents our implementing integration?

Now that we understand polynomial objects and how they should behave, it's time to start building them in Python!

7.2 Classes, methods, and variables

A class is a Python object that creates other objects, known as *instances* of the class. We also say that a class *instantiates* new objects of its type. Its creation syntax is similar to that of a function definition, except we use class instead of def. A class can be "parameterized," and we discuss this in section 7.13 when we consider inheritance.

We begin by defining a class, **UniPoly**, that creates a univariate monomial from a coefficient, indeterminate, and exponent. We implement a preliminary way to display these polynomials, but we do not yet allow any algebraic operations. The code for the final version of **UniPoly** that we develop in this chapter is in *Appendix C*.

```
1  class UniPoly:
2      # UniPoly creates univariate polynomials with integer coefficients
3      # Version 1
4
5      def __init__(self, coefficient, indeterminate, exponent):
6          # Create a UniPoly monomial.
7
```

```
 8          # Validate that the coefficient is an int.
 9          if not isinstance(coefficient, int):
10              raise ValueError(
11                  "The coefficient for a UniPoly must "
12                  "be an int.")
13
14          # Validate that the exponent is a non-negative int.
15          if not isinstance(exponent, int) or exponent < 0:
16              raise ValueError(
17                  "The exponent for a UniPoly must "
18                  "be a non-negative int.")
19
20          # Validate that the indeterminate is an alphabetic
21          # string of length 1.
22          if (not isinstance(indeterminate, str) or
23                  len(indeterminate) != 1 or
24                  not indeterminate[0].isalpha()):
25              raise ValueError(
26                  "The indeterminate for a UniPoly must "
27                  "be an alphabetic str of length 1.")
28
29          self.coefficient = coefficient
30          self.indeterminate = indeterminate
31          self.exponent = exponent
32
33      def __repr__(self):
34          # create the object representation of the polynomial
35          return f"UniPoly({self.coefficient}, " + \
36              f"{repr(self.indeterminate)}, {self.exponent})"
37
38      def __str__(self):
39          # create the displayable string form of the polynomial
40          return f"{self.coefficient}*" + \
41              f"{self.indeterminate}**{self.exponent}"
```

Our initial definition for **UniPoly** begins with the class keyword, followed by the name of the class. The first line ends with a colon ":". I then define three instance methods using def.

These are *instance methods* because they are functions I define within the class that operate upon instances that the class creates. The first parameter for each is required, and it is named "self". self represents the instance to which the method is applied.

Not all methods have names that begin and end with double underscores, but these are special methods with predefined roles.

- *__init__* on line 5 is a class instance initializer, and you can have only one in a class.

- *__repr__* on line 33 produces a string that is the instance input **representation**. Python calls *__repr__* when a user of your code calls *repr*. In theory, you could type this string into Python, and it would reproduce the object. You can also use it to display information useful to you as the coder to distinguish one object from another. I recommend you always code it, even if it has the same definition as *__str__*.

- *__str__* on line 38 creates a "human-readable" string describing the instance. Python calls *__str__* when a user of your code calls *str*. For our class, **UniPoly**, the implementation of *__str__* will eventually produce strings like "x", "y + 10", and "3*z**2 + z - 4". The first example already works, but only because our **UniPoly** is trivial so far.

__init__ has two main code segments: argument validation and initialization. In the first segment beginning on line 8, we check that the value passed in as indeterminate is a string of length one that contains an alphabetic character. If not, we raise a **ValueError** exception with an error message.

If the indeterminate argument is valid, on line 30, we store the parameter value in the *instance variable* self.indeterminate. Each object has its own associated and separate instance variables. Think of these as the core data that defines an instance. We also assign the instance variables self.coefficient and self.exponent.

Unlike some other programming languages, you do not allocate system memory for the instance object, nor must you release that memory when you are done with the instance. Python implements *garbage collection* and handles low-level memory allocation and release for you.

Now we create a **UniPoly** instance and assign it to x. *__repr__* creates the string that Python displays for the polynomial.

```
x = UniPoly(1, 'x', 1)
x

UniPoly(1, 'x', 1)

isinstance(x, UniPoly)

True
```

We use __str__ to make a displayable form of the polynomial, though we need to improve its style to look more mathematical. When you use the *str* function on your object, Python invokes the __str__ method. Python can also call __str__ in situations where it expects a **str** object, such as in f-strings.

```
str(x)
```

```
'1*x**1'
```

```
print(x)
```

```
1*x**1
```

Exercise 7.2

How would you define **UniPoly** so you can call it with no arguments and default values for the three arguments? What should those values be?

The class definition so far produces useless polynomials.

```
2 * UniPoly(1, 'J', 1)
```

```
TypeError: unsupported operand type(s) for *:
          'int' and 'UniPoly'
```

```
UniPoly(2, 'α', 3) + 1
```

```
TypeError: unsupported operand type(s) for +:
          'UniPoly' and 'int'
```

Exercise 7.3

Are you surprised that 'α' returned True for the *isalpha* test? Research what Python accepts as alphabetic, starting with "letters" in Unicode. [UNI]

Now let's introduce coefficients, exponents, and terms.

7.3 Object representation

As we saw previously, each polynomial has its indeterminate stored as a string with a single alphabetic character. Remember that one polynomial might have an indeterminate of `'x'` while another might use `'z'`. We don't need to worry about that now, but we will when it comes time to perform binary operations like addition and multiplication.

We need to design how we represent multiple terms. Since all terms share the one indeterminate, each term needs only **int** values for its coefficient and exponent. I could put these in a list of lists and say that those are the terms. Our polynomials

$$p = x^2 + 7x + 1 \quad \text{and} \quad q = 4x^3 - 7x^2 + x - 9$$

could be represented as

```
[[1, 2], [7, 1], [1, 0]]
```
```
[[1, 2], [7, 1], [1, 0]]
```

and

```
[[4, 3], [-7, 2], [1, 1], [-9, 0]]
```
```
[[4, 3], [-7, 2], [1, 1], [-9, 0]]
```

where each inner list is a coefficient-exponent pair.

There are three rules if we use this representation:

1. No two different terms have the same exponent.
2. Never include a term with zero coefficient.
3. Therefore, there is a unique way for us to represent a zero polynomial, which is by an empty list.

We represent the polynomial

$$3x^6 - 5x^4 + x$$

as

```
[[3, 6], [-5, 4], [1, 1]]
```
```
[[3, 6], [-5, 4], [1, 1]]
```

in this design.

The polynomial 0 is [ ].

When I display a polynomial like $3x^6 - 5x^4 + x$ in mathematical format, I do so in descending order of exponents. I could also do it in ascending order, but one or the other is needed to make it easy to read the polynomial. Not coincidentally, I used descending sort order for the inner lists based on the second item, the exponent, in the Python representation. Is this necessary? Do I need to keep the list of terms sorted?

If I do, every time I multiply two polynomials and create new terms for the product, I have to either insert the new terms in the correct sequence or sort the terms at the end.

```
unsorted_terms = [[-5, 4], [1, 1], [3, 6]]
unsorted_terms
```

```
[[-5, 4], [1, 1], [3, 6]]
```

```
sorted(unsorted_terms, key=(lambda term: term[1]), reverse=True)
```

```
[[3, 6], [-5, 4], [1, 1]]
```

The optional keyword argument key for *sorted* takes a function that returns some value for comparison. In this case, it returns the second item in a term, the exponent. *sorted* uses "<" to compare the results of key. If you set reverse=True, *sorted* uses ">".

Exercise 7.4

The default value of key is None. What comparison values does *sorted* use then?

You may think that careful term insertion or repeated sorting is not onerous, but it can become a severe performance bottleneck when polynomials have tens or hundreds of terms. **I now decide** that I will not worry about the exponent order in the representation, and I will sort only if I need to display the polynomial.

Dropping the ordered sequence of exponents seems better, but what if I am adding two polynomials and I must look at the terms in each that have the same exponent? If I have a term for the first polynomial in hand, I must search each term in the second polynomial to try to locate the corresponding one with the same exponent, if it is there.

Exercise 7.5

Let p and q be two univariate polynomials with the same indeterminate and let us assume that the degree of each is n. This means that the largest exponent that appears in either of them is n. What's the maximum number of non-zero terms that each polynomial can have?

Exercise 7.6

Continuing with p and q, assume my Python representation is an unsorted list of coefficient-exponent pairs. What's the maximum number of comparisons you need to match a term in p with its corresponding term in q with the same exponent?

If I had kept the exponent ordering, I could have found corresponding terms much more efficiently. For example, I could use a binary search, which we discuss in section 11.1. This begs a more fundamental question: why am I still using a list, an ordered collection, for unordered data? I shouldn't be.

I will use a dictionary for the polynomial terms where the keys are the term exponents, and the values are the coefficients. Checking to see if a term exists is fast, as is accessing a term of a given exponent if it is present. When I print the polynomial, I sort the terms and display them in the natural display order.

```
1  def __init__(self, coefficient, indeterminate, exponent):
2      # Create a UniPoly monomial.
3
4      # Validate that the coefficient is an int.
5      if not isinstance(coefficient, int):
6          raise ValueError(
7              "The coefficient for a UniPoly must "
8              "be an int.")
9
10     # Validate that the exponent is a non-negative int.
11     if not isinstance(exponent, int) or exponent < 0:
12         raise ValueError(
13             "The exponent for a UniPoly must "
14             "be a non-negative int.")
```

```
15
16        # Validate that the indeterminate is an alphabetic
17        # string of length 1.
18        if (not isinstance(indeterminate, str) or
19                len(indeterminate) != 1 or
20                not indeterminate[0].isalpha()):
21            raise ValueError(
22                "The indeterminate for a UniPoly must "
23                "be an alphabetic str of length 1.")
24
25        # The 'variable' for the polynomial.
26        self.indeterminate = indeterminate
27
28        # The terms in the polynomial. Each key is an int
29        # exponent and each value is the int coefficient.
30
31        if coefficient != 0:
32            self.terms = {exponent: coefficient}
33        else:
34            self.terms = dict()
```

In this second version of *__init__*, I define a dictionary instance variable, `self.terms`, on lines 32 and 34. `self.terms` holds the exponent-coefficient key-value pairs.

Do you see how I handle the case where the coefficient is zero? A **UniPoly** instance is zero if and only if its terms dictionary is empty. In our instance methods, we will often check for this special case.

I am deferring re-implementing *__repr__* and *__str__* until we can create polynomials with more than one term.

7.4 Magic methods

You do not call the three methods *__init__*, *__repr__*, and *__str__* directly; Python "magically" calls them for you when it needs to initialize an instance, get an input representation, or convert to a string. You can define several dozen such "magic methods" with predefined uses. The algebraic and comparison methods are important to us for polynomials, so let's look at some examples.

7.4.1 Negation

I now define negation via __neg__, where I create a new polynomial by applying "-" to each coefficient. In this third version of **UniPoly**, I extend __init__ to accept a coefficient and exponent.

```
1   class UniPoly:
2       # UniPoly creates univariate polynomials with integer coefficients
3       # Version 3
4
5       def __init__(self,
6                    coefficient=1,
7                    indeterminate='x',
8                    exponent=0):
9
10          # Create a UniPoly object from an integer coefficient,
11          # string indeterminate, and non-negative integer
12          # exponent
13
14          if not isinstance(coefficient, int):
15              raise ValueError(
16                  "The coefficient for a UniPoly must "
17                  "be an int.")
18
19          if not isinstance(exponent, int) or exponent < 0:
20              raise ValueError(
21                  "The exponent for a UniPoly must "
22                  "be a non-negative int.")
23
24          if (not isinstance(indeterminate, str) or
25              len(indeterminate) != 1 or
26                  not indeterminate[0].isalpha()):
27              raise ValueError(
28                  "The indeterminate for a UniPoly must "
29                  "be an alphabetic str of length 1.")
30
31          # The 'variable' for the polynomial
32          self.indeterminate = indeterminate
33
34          # The terms in the polynomial. Each key is an int
35          # exponent and each value is the int coefficient.
36
37          if coefficient != 0:
38              self.terms = {exponent: coefficient}
```

```
39              else:
40                  self.terms = dict()
41
42      def __neg__(self):
43          # Negate the polynomial term-wise unless it is already 0.
44          if not self.terms:
45              return self
46
47          new_poly = UniPoly(0, self.indeterminate)
48          for exponent in self.terms:
49              new_poly.terms[exponent] = -self.terms[exponent]
50
51          return new_poly
52
53      def __repr__(self):
54          return str(self)
55
56      def __str__(self):
57          # Do a very rough and incomplete conversion to str.
58
59          if not self.terms:
60              return '0'
61
62          print_terms = [
63              f"{self.terms[exponent]}*{self.indeterminate}**{exponent}"
64              for exponent in sorted(self.terms, reverse=True)
65          ]
66
67          return " + ".join(print_terms)
```

Now we verify that we can negate a polynomial:

```
p = UniPoly(2, 'z', 5)

p, -p

(2*z**5, -2*z**5)

UniPoly(0), -UniPoly(0)

(0, 0)
```

In my definition of *__neg__* on line 43, I copied the instance data for self except that I negated the coefficients. **I did not change** the coefficients in the terms dictionary in self. If I had, any variable pointing to the original polynomial would point to a negated version.

As you can see, class definitions can get lengthy quickly. Validating data can take a lot of code and time, so you should only do it once. For internal methods that you do not call outside other instance methods, skip the validation or use `assert`. We discuss `assert` in section 10.1.2.

Exercise 7.7

Define an instance method, __*copy*__, for polynomials that creates a duplicate of a **UniPoly** instance. Begin by following the example of __*neg*__.

Exercise 7.8

The definition of __*str*__ is incomplete, even for a single-term polynomial.

```
UniPoly(1, 'a', 1)

1*a**1

-UniPoly(1, 'a', 1)

-1*a**1
```

Improve the method by not printing the "1" when the coefficient is "1" or "-1". Does your code work for UniPoly(1, 'x', 0)?

7.4.2 "Boolness"

We have seen that Python objects such as 0, 0.0, [ ], and empty sets and dictionaries behave like `False` in conditionals. Non-zero numbers and non-empty collections act like `True`. We implement this in new classes with the __*bool*__ method. For polynomials, we test if the term dictionary is empty.

I'm no longer going to repeat the full **UniPoly** class definition, but just show you the new methods.

```
1  def __bool__(self):
2      # Returns "True" if self is nonzero.
3
4      return bool(self.terms)
```

While I could have defined __*bool*__ via

```
def __bool__(self):
    # Returns "True" if self is nonzero.
    return True if self.terms else False
```

the version in the class is more concise.

Once we have done this, we can use tests like `if self:` instead of `if self.terms()`:

```
def poly_is_zero(p):
    print("not zero" if p else "zero")

poly_is_zero(UniPoly(0, 'x', 1))

zero

poly_is_zero(UniPoly(1, 'x', 1))

not zero
```

In your implementation of __*bool*__, explicitly return `True` or `False`.

7.4.3 Equality testing

The next magic method we implement is __*eq*__, and it tests whether two polynomials are equal when you call "==".

I've complicated equality testing because I want to return `True` if a constant polynomial is equal to an integer. Recall that a constant is a zero-degree polynomial.

My strategy for testing whether two polynomials are equal is to look at all the ways they might not be. Once I have exhausted those, the polynomials must be the same.

```
 1  def __eq__(self, other):
 2      # Tests for equality between a polynomial and an int
 3      # or between two polynomials.
 4
 5      if isinstance(other, int):
 6          # If we have a constant polynomial and the
 7          # coefficient is the int, return True.
 8
 9          if not self.terms:
10              # self is the zero polynomial.
11              return other == 0
```

```
12          if other == 0:
13              # self is not 0 but other is 0.
14              return False
15          if len(self.terms) > 1:
16              return False
17          if 0 not in self.terms:
18              return False
19          # self is a constant polynomial.
20          if self.terms[0] == other:
21              return True
22          return False
23
24      if isinstance(other, UniPoly):
25          # Both are polynomials.
26
27          if self.indeterminate != other.indeterminate:
28              # The indeterminates are different.
29              return False
30          if len(self.terms) != len(other.terms):
31              # They have different numbers of terms.
32              return False
33
34          for exponent in self.terms:
35              if exponent not in other.terms:
36                  # An exponent in self is missing from other.
37                  return False
38              if self.terms[exponent] != other.terms[exponent]:
39                  # The coefficients of terms of equal exponent
40                  # are different.
41                  return False
42
43          return True
44
45      return NotImplemented
```

We first check that testing with constants works:

```
UniPoly(0) == 0, UniPoly(3, 'w', 0) == 3
```

```
(True, True)
```

and now with some non-trivial polynomials:

```
p = UniPoly(3, 'w', 1)
p
```

```
3*w
```

```
p == 3
```

```
False
```

```
UniPoly(3, 'x', 7) == UniPoly(3, 'y', 7)
```

```
False
```

```
q = UniPoly(3, 'x', 7)
q == q
```

```
True
```

It may surprise you that "!=" now also works.

```
q != q
```

```
False
```

```
UniPoly(3, 'x', 7) != UniPoly(3, 'y', 7)
```

```
True
```

The magic method to implement "!=" is __ne__. If you do not implement it, Python computes not (a == b) instead of a != b.

I implemented the test to see if a polynomial p is equal to an integer n: p == n. I did not explicitly implement n == p. Python reverses the arguments for us if necessary and performs the test.

```
-5 == UniPoly(-5, 'w', 0)
```

```
True
```

Though this is appropriate here, it would not be for an operation like subtraction.

Note the very last line of __eq__: return NotImplemented. Within the definition, I check equality between a polynomial and an integer and between two polynomials. My returning NotImplemented means, "I've done all I can here. Look around to see if there is another method you can use to test for equality." For example, I may later write another mathematical class and test for equality between its instances and polynomials.

7.4.4 "is" versus "=="

Suppose I have the three polynomial definitions

```
p = UniPoly(-3, 'z', 10)

q = p

r = UniPoly(-3, 'z', 10)
```

They all look the same:

```
p, q, r
(-3*z**10, -3*z**10, -3*z**10)
```

and they all test the same using "==":

```
p == q, p == r, q == r
(True, True, True)
```

However, we get different answers when we use is.

```
p is q, p is r, q is r
(True, False, False)
```

We use is when we want to know if two objects are literally the same within Python. We're not testing if they look the same; we're asking, "if I go to the location in my computer's system memory where Python stores p, will this be the same location where Python stores q?".

Now let's change the indeterminate in the polynomial p.

```
p
-3*z**10

p.indeterminate = 'M'

p, q, r
(-3*M**10, -3*M**10, -3*z**10)
```

As you can see, I changed p, but I also changed q. They are the same Python objects addressed by two different names. As an analogy, you can call me "Bob" or "Robert", but I'm still me.

You do not need to do anything special to implement is.

7.5 Attributes and properties

Each polynomial created by **UniPoly** has an attribute named `indeterminate` and another for terms. These are the instance variables we defined. **UniPoly** also has attributes for each instance method that you define or that Python creates for you. If `obj` is a Python object, then `dir(object)` returns a list of attribute names.

```
"indeterminate" in dir(p)

True
```

```
"terms" in dir(p)

True
```

```
"__init__" in dir(p)

True
```

An instance variable is a read-write attribute, meaning that you can both get and set its value with ".." notation.

In your class design, you may wish to allow your users direct and public access to an instance variable. If you do not, prefix the variable's name with two underscores "__". Python will rename the variable internally and make it more difficult, but not impossible, to get to it. We call this renaming "mangling."

Let's go ahead and do this in our class. Here is what *__init__* looks like now:

```
 1  def __init__(self,
 2                  coefficient=1,
 3                  indeterminate='x',
 4                  exponent=0):
 5
 6      # Create a UniPoly object from an integer coefficient,
 7      # string indeterminate, and non-negative integer
 8      # exponent
 9
10      if not isinstance(coefficient, int):
11          raise ValueError(
12              "The coefficient for a UniPoly must "
13              "be an int.")
14
15      if not isinstance(exponent, int) or exponent < 0:
16          raise ValueError(
17              "The exponent for a UniPoly must "
18              "be a non-negative int.")
```

```
19
20        if (not isinstance(indeterminate, str) or
21                len(indeterminate) != 1 or
22                not indeterminate[0].isalpha()):
23            raise ValueError(
24                "The indeterminate for a UniPoly must "
25                "be an alphabetic str of length 1.")
26
27        # The 'variable' for the polynomial.
28        self.__indeterminate = indeterminate
29
30        # The terms in the polynomial. Each key is an int
31        # exponent and each value is the int coefficient.
32
33        if coefficient != 0:
34            self.__terms = {exponent: coefficient}
35        else:
36            self.__terms = dict()
```

See the changes on lines 28, 34, and 36?

```
p = UniPoly(2, 'w', 5)
p
```

```
2*w**5
```

```
p.indeterminate
```

```
AttributeError: 'UniPoly' object has no attribute 'indeterminate'
```

```
p.__indeterminate
```

```
AttributeError: 'UniPoly' object has no attribute '__indeterminate'
```

So __indeterminate__ is safely hidden away. But what if we do want to access it? It's certainly reasonable to want to know the variable in a polynomial. We can define a *getter* for it using a @property "decorator". A *decorator* is a statement starting with "@" that causes Python to handle a definition in some special way.

```
1  @property
2  def indeterminate(self):
3      return self.__indeterminate
```

We define the property as a method with only `self` as a parameter, but we refer to it as a variable.

```
p.indeterminate
```

```
'w'
```

This gives us the indeterminate's value, but it does not let us set it to something new.

```
p.indeterminate = 'y'
```

```
AttributeError: can't set attribute
```

I now define an attribute *setter* so we can alter the indeterminate.

```
1  @indeterminate.setter
2  def indeterminate(self, value):
3      if (not isinstance(value, str) or
4              len(value) != 1 or
5              not value[0].isalpha()):
6          raise ValueError(
7              "The indeterminate for a UniPoly must "
8              "be an alphabetic str of length 1.")
9      self.__indeterminate = value
```

```
p.indeterminate = 'y'
p
```

```
2*y**5
```

Our evolving definition of **UniPoly** creates polynomials of one variable. When you design a new class, you must decide whether your objects are mutable. If so, what can change them and when? I allow the user of my code to reset the indeterminate after initialization, but I likely would not do that in practice. Methods inside my class could change a polynomial as they built it, but I would not allow later alterations once the polynomial was returned to the user by a public method.

Exercise 7.9

Define and test read-only properties `degree` and `minimum_degree` for **UniPoly** that return the largest and smallest exponents in a polynomial. Remember that the degree of a zero or other constant polynomial is 0. If you created a list of exponents in either implementation, rewrite the code without the list.

7.6 Naming conventions and encapsulation

In section 6.3, we covered naming conventions for functions. Let's complete the story for classes and their contents.

Although many built-in classes like `int` and `float` have lowercase names, your class names should begin with a capital letter. If the name has multiple "words", start those with capital letters as well. Don't use underscores. Examples are `BreadRecipe`, `ElectricGuitar`, and `Matrix`.

Instance variables should follow the same naming conventions as other variables, and methods should follow the conventions for functions. All letters should be lowercase, and you may use underscores.

End a name with an underscore to avoid confusion with a Python keyword.

```
class_ = "Quantum Computing 101"
class_
```

```
'Quantum Computing 101'
```

Magic methods (section 7.4) begin and end with two underscores.

Python does not thoroughly enforce hiding methods and variables from outside the class. Begin a name with an underscore to tell others and remind yourself that the variable or method is "private" to the class. For example, you may have a helper method called only by other instance methods.

If you start a name with two underscores and end it with no more than one underscore, Python "mangles" the name to make it hard for you to access outside the class. Of course, you can look up how Python does the mangling and then get to the variable or methods. Behave yourself and don't do this! We say that a method or variable that does not start with an underscore is *public*.

If you are familiar with C++, Python has no corresponding `public`, `private`, and `protected` concepts or keywords.

Defining a class with contained data and methods, along with some way of limiting access to them, is called *encapsulation*. Think of it as putting a protective "capsule" around your class and its instances. Python's capsules are relatively porous compared to the C++, Java, and PHP programming languages.

Encapsulation helps hide your class data and implementation from its users. As the class author, this allows you to change the representations and implementations without breaking code that relies on your code's functionality. As a class user, **do not** reach into the class and use variables that start with an underscore, "_".

7.7 Commenting Python code

A Python *comment* begins with a "#" and continues to the end of the line. This symbol is called a *hash sign*. Your comments help those who *read* your code understand why and how you implemented the functionality the way you did. Your *documentation* helps those who *use* your code understand the functionality it provides. We discuss commenting Python code in this section. In the next section, we explore how you can best document your source code.

Although I have used comments for this purpose so far, use documentation techniques to describe your functions, methods, and classes. Use these to describe parameters, their types, and their roles. Starting in the next section, I will do this.

Ideally, you should beautifully write your code and carefully choose your names. This way, anyone reading your Python source can instantly understand what it does. At all times, aim for elegance in design and implementation. Where there can be any confusion, misunderstanding, or non-obvious behavior, add comments. If anyone else looks at your code and cannot tell what it does, add comments. It's their opinion that counts, not yours. (There is a strong parallel here with suggestions from editors when you are writing a book!)

You may place a comment on its own line or after the Python code on a line. Try not to do the latter frequently, but if you do, place two spaces before the "#" to improve readability. For full-line comments, use proper grammar and punctuation. Don't be too wordy. Put at least one space after the "#".

Some languages like C++ and Java distinguish between single- and multi-line comments. They use // instead of Python's "#" for a single line, but they also have /* ... */ for multiple lines. In Python, make a multi-line block comment with a "#" at the start of each line.

You can see the three styles of comments in this function we wrote in section 5.2.2.

```
1   def sieve_of_eratosthenes(n):
2       # Return a list of the primes less than or equal to n.
3       # First check that n is an integer greater than 1.
4
5       if not isinstance(n, int) or not n > 1:
6           print(
7               "The argument to sieve_of_eratosthenes "
8               "must be an integer greater than 1.")
9           return []
10
11      # Make a list holding the integers from 0 to n.
12
13      potential_primes = list(range(n + 1))
14
```

```
15        # If index is not prime, set potential_primes[index] to 0.
16        # We start with 0 and 1
17
18        potential_primes[0] = 0
19        potential_primes[1] = 0
20
21        p = 2  # 2 is prime, so start with that.
22
23        while p <= n:
24            # If at an index whose value in potential_primes
25            # is not 0, it is prime.
26
27            if potential_primes[p]:
28                i = 2 * p
29
30                # Mark p+p, p+p+p, etc. as not prime.
31                while i <= n:
32                    if i != p:
33                        potential_primes[i] = 0
34                    i += p
35            p += 1
36
37        # The only non-zero integers left in potential_primes
38        # are primes. Return a list of those.
39
40        return [prime for prime in potential_primes if prime]
```

While you are developing code, use comments together with `pass` statements to create and fill out the skeleton of your implementation. A comment that starts `#  TODO:` is easy to find, as is one beginning with `#  BUG:`.

My final advice on comments is to phrase them so your code is not "write-only." Write-only code means that you knew perfectly well what it did when you created it, but now neither you nor anyone else can decipher it.

7.8 Documenting Python code

A *docstring* is a free-standing string surrounded by three double quotes that describes what your code does.

Developers read the docstrings when browsing your code. By their location, Python associates docstrings with your modules, classes, functions, and methods. Python collects the docstrings and makes them available to your code's users via functions like *help*. Third-party tools like **Sphinx** can process your code to create beautiful web and PDF documents for your users' reference. [SPH]

One of the most popular Python documentation conventions is part of the *Google Python Style Guide*. [GSG, Section 3.8] To add lightweight markup for formatting, many coders use "reStructuredText" within their docstrings. [RST] The NumPy and SciPy package developers combined and tweaked these into the "numpydoc docstring guide." I use their conventions from this point forward in this book. [NDG]

7.8.1 Structure

For a function or method, the documentation consists of a brief one-line description, an optional longer multi-line description, and several sections with headings like

- **Parameters**
- **Raises**
- **Yields**
- **Notes**
- **Examples**
- **References**
- **Returns**

Each heading is "underlined" with dashes.

As we saw in section 7.4.2, __bool__ implements the behavior of "falseness" for polynomials equal to zero. It needs minimal documentation: a brief functional description and the statement of what it returns.

```
1  def __bool__(self):
2      """Returns True if the polynomial is non-zero, False otherwise.
3
4      Returns
5      -------
6      bool
7      """
8
9      return bool(self.__terms)
```

This is how the Python *help* function renders that information:

```
help(UniPoly.__bool__)

Help on function __bool__ in module __main__:

__bool__(self)
    Returns True if the polynomial is non-zero, False otherwise.

    Returns
    -------
    bool
```

My documentation for __*add*__ has more sections.

```
help(UniPoly.__add__)

Help on function __add__ in module __main__:

__add__(self, other)
    Adds a polynomial to a polynomial or an integer.

    Parameters
    ----------
    other : :class:`UniPoly` or int
        A polynomial or integer.

    Notes
    ------
    If the second argument is not a polynomial or integer,
    or if the indeterminates in non-constant polynomials
    are different, returns NotImplemented.

    Returns
    -------
    :class:`UniPoly` or NotImplemented
        The sum of self and other, or NotImplemented.
```

We don't need to include a description of self. I explain the meaning of :class:`UniPoly` in the next section.

The brief descriptions for your functions and methods should start with words like "Computes," "Calculates," "Extracts," and "Copies." As a last resort, use "Returns."

If you run `help(UniPoly)` in Python, you will see the class's complete documentation, including its methods. To see just the top-level class documentation, examine the *__doc__* attribute.

```
print(UniPoly.__doc__)

Polynomial with one variable and integer coefficients.

    UniPoly creates a univariate polynomial with a single term
    and an integer coefficient. Polynomials may use different
    variables, here called 'indeterminates', but their names
    must each be a single alphabetic character. These polynomials
    are immutable.

    Parameters
    ----------
    coefficient : int, optional
        The coefficient of the term. Default value is 1.
    indeterminate : str, optional
        The "variable." Must be a single alphabetic character.
        Default value is 'x'.
    exponent : int, optional
        The exponent of the term. Default value is 1.

    Raises
    ------
    ValueError
        If the coefficient is not an int.
    ValueError
        If the exponent is not a non-negative int.
    ValueError
        If the indeterminate is not a single
        alphabetic character.
```

My convention for the brief description of the class is that it should be the second part of the sentence "This creates a ...". The class parameters are those for *__init__*, and the attributes are the instance and class variables.

I'll point out additional documentation features as we encounter them.

7.8.2 Formatting

Python code tools such as **Sphinx** understand how to translate reStructuredText markup to create web and print documentation from your code.

Surround text with double asterisks to embolden it: "**This is so bold**" yields **This is so bold**. Single asterisks give you italic text: "*I am italic*" yields *I am italic.* For a monospace font, use double backticks: "``This looks like code``" produces `This looks like code`.

We separate paragraphs with a blank line. For bulleted lists, start each line with an asterisk:

```
* The first item
* The second item
* The third item
```

The markup produces

* The first item

* The second item

* The third item

Use "#" instead of "*" to get a numbered list.

Your Python code and what you see from *help* and looking at *__doc__* will contain this formatting. It is only once a tool like **Sphinx** processes your code that you will see it beautifully displayed.

The documentation for *__add__* contains markup like `:class:`UniPoly``. This is semantic markup that says, "`UniPoly` is a Python class, and you should format it in whatever fancy way you display classes." It also allows **Sphinx** to cross-reference the class and provide an automatic link to its definition.

`:class:` is called a *role*. Others roles include

* `:attr:` (data attribute / instance variable),

* `:const:` (module-level constant),

* `:data:` (module-level non-constant variable),

* `:exc:` (exception),

* `:func:` (function),

* `:meth:` (method), and

* `:mod:` (module).

If you do not want to use this semantic markup, surround the names with single backticks.

To include mathematical markup, use `:math:` and LaTeX within the backticks. For example, `:math:`2x^3-5z^2`` would display as $2x^3 - 5z^2$.

If you wish, you can use formatting and semantic markup for every Python element in your documentation. This will create lovely web and printed documents but will make your code hard to read. Strike a balance between readability and just enough markup to make your code documentation attractive.

It can be tedious to create docstrings, but tools like the Python Docstring Generator for **Visual Studio Code** may speed up the work.

7.9 Enumerations

Bob's Bakery Bistro makes and sells four kinds of cookies: chocolate, sugar, oatmeal, and shortbread. Bob hires ace coder Will to write a new cookie inventory application for the bistro using Python.

How should Will represent the cookies? One way is to use global constants.

```
CHOCOLATE_COOKIE = 0
SUGAR_COOKIE = 1
OATMEAL_COOKIE = 2
SHORTBREAD_COOKIE = 3

COOKIE_NAMES = [
    "chocolate",
    "sugar",
    "oatmeal",
    "shortbread"
]
```

It's easy to get the name if you have the cookie constant:

```
COOKIE_NAMES[SUGAR_COOKIE]
```

```
'sugar'
```

It's less obvious how to go from the name to the cookie constant unless we make some assumptions about positions and indices. You can easily introduce bugs into your code if you start adding more kinds of cookies.

It would be nice to have a simpler and class-oriented way of doing this in Python. This functionality is the role of *enumerations*.

```
from enum import Enum

class Cookie(Enum):
    CHOCOLATE = "chocolate"
    SUGAR = "sugar"
    OATMEAL = "oatmeal"
    SHORTBREAD = "shortbread"
```

There is no comma at the end of a line assigning an enumeration value to its name. Each name-value pair is called an *enumeration member*.

We access a member using "." syntax:

```
Cookie.SHORTBREAD
```

```
<Cookie.SHORTBREAD: 'shortbread'>
```

```
Cookie.SUGAR
```

```
<Cookie.SUGAR: 'sugar'>
```

Exercise 7.10

What is the type of an enumeration member?

If you have a member, you can retrieve its name and value.

```
cookie = Cookie.OATMEAL
cookie.name
```

```
'OATMEAL'
```

```
cookie.value
```

```
'oatmeal'
```

Given the value, you can get the member.

```
Cookie("sugar")
```

```
<Cookie.SUGAR: 'sugar'>
```

You can also get the member from the name.

```
Cookie["SUGAR"]
```

```
<Cookie.SUGAR: 'sugar'>
```

To add a new cookie, add a new member definition. You make one change, and everything works.

Exercise 7.11

What happens if you use a name or value more than once? How does this change if you use the @unique decorator on the line before the enumeration class definition?

You may use enumeration members in sets and as dictionary keys. Use is to compare enumeration members instead of "==".

If you want to use enumeration members as if they were integers, subclass from **IntEnum** instead of **Enum**. **IntEnum** itself is a subclass of **int**.

```
from enum import IntEnum

class CatAge(IntEnum):
    GUS = 3
    FERDINAND = 6
    FRANZ = 6
    GEORGIA = 9
```

```
CatAge.GUS.name, CatAge.GUS.value
```

```
('GUS', 3)
```

```
CatAge.FERDINAND >= CatAge.FRANZ
```

```
True
```

```
CatAge.GEORGIA < 10
```

```
True
```

```
2 * (CatAge.GUS + CatAge.GEORGIA) - 1
```

```
23
```

7.10 More polynomial magic

Now that we've laid the groundwork for polynomials, it's time to implement operations like
addition, subtraction, and multiplication. Since we are restricting ourselves to integer
coefficients, we do not implement division.

7.10.1 A better __str__

We'll start to see more than one term, and we need a more robust implementation of *__str__*.

```python
def __str__(self):
    """Creates a human-readable string representation.

    This returns forms that look like 2x**6 + x**3 - 1.

    Returns
    -------
    str
        Mathematical human-readable form of the polynomial.
    """
    if not self.__terms:
        return '0'

    def format_term(coefficient, exponent):
        """Format a single term in the polynomial.

        This function formats a term and handles the special
        cases when the coefficient is +/- 1 or the exponent is
        0 or 1.

        Parameters
        ----------
        coefficient : int
            The coefficient of the term.
        exponent : int
            The exponent of the term.
```

```
28            Returns
29            -------
30            str
31                The human-readable representation of a polynomial term.
32            """
33            coefficient = abs(coefficient)
34
35            if exponent == 0:
36                return str(coefficient)
37
38            if exponent == 1:
39                if coefficient == 1:
40                    return self.__indeterminate
41                return f"{coefficient}*{self.__indeterminate}"
42
43            if coefficient == 1:
44                return f"{self.__indeterminate}**{exponent}"
45
46            return f"{coefficient}*{self.__indeterminate}**{exponent}"
47
48        # Collect the formatted and sorted terms in result, taking
49        # case to handle leading or middle negative signs.
50
51        result = ""
52
53        for exponent in sorted(self.__terms, reverse=True):
54            coefficient = self.__terms[exponent]
55            term = format_term(coefficient, exponent)
56
57            if result:
58                result += f" - {term}" \
59                    if coefficient < 0 else f" + {term}"
60            else:
61                result = f"-{term}" \
62                    if coefficient < 0 else term
63
64        return result
```

See where and how I handled negative coefficients and those equal to ±1? Why is *format_term* a nested function and not an instance method?

```
UniPoly(-3), UniPoly(-1), UniPoly(0), UniPoly(1), UniPoly(3)

(-3, -1, 0, 1, 3)
```

```
UniPoly(-1, 'z', 1)
```

```
-z
```

```
UniPoly(3, 'x', 7)
```

```
3*x**7
```

7.10.2 Constantly thinking

The *constant coefficient* is the coefficient of the term that has the exponent 0. I've implemented *constant_coefficient* as a Python property because it is an intrinsic part of a polynomial.

```
1  @property
2  def constant_coefficient(self):
3      """The coefficient of the term with exponent 0."""
4      if self.__terms and 0 in self.__terms:
5          return self.__terms[0]
6      return 0
```

```
UniPoly(3, 'x', 0).constant_coefficient
```

```
3
```

Similarly, *is_constant* is a Boolean-valued property that tells us if there are no terms with an exponent greater than 0. I defer showing you its implementation until the end of section 7.14 because I need to introduce other Python features first.

```
UniPoly(3, 'x', 7).is_constant, UniPoly(2, 'y', 0).is_constant
```

```
(False, True)
```

7.10.3 Addition

We define addition with the *__add__* instance method.

```
1  def __add__(self, other):
2      """Adds a polynomial to a polynomial or an integer.
3
4      Parameters
5      ----------
6      other : :class:`UniPoly` or int
7          A polynomial or integer.
8
```

```
 9      Notes
10      ------
11      If the second argument is not a polynomial or integer,
12      or if the indeterminates in non-constant polynomials
13      are different, returns NotImplemented.
14
15      Returns
16      -------
17      :class:`UniPoly` or NotImplemented
18          The sum of self and other, or NotImplemented.
19      """
20
21      if not self.__terms:
22          return other
23
24      # if other is an integer, create a constant polynomial
25      if isinstance(other, int):
26          other = UniPoly(other, self.__indeterminate, 0)
27
28      # both objects are polynomials
29
30      if isinstance(other, UniPoly):
31          # if other == 0, return self
32          if not other.__terms:
33              return self
34
35          if self.__indeterminate != other.__indeterminate:
36              if self.is_constant and other.is_constant:
37                  return UniPoly(self.constant_coefficient +
38                                 other.constant_coefficient,
39                                 self.__indeterminate,
40                                 0)
41              return NotImplemented
42
43          new_poly = UniPoly(0, self.__indeterminate, 0)
44
45          for exponent, coefficient in self.__terms.items():
46              if exponent not in other.__terms:
47                  new_poly.__terms[exponent] = coefficient
48              else:
49                  sum = self.__terms[exponent] \
50                      + other.__terms[exponent]
51                  if sum:
52                      new_poly.__terms[exponent] = sum
53
```

```
54                for exponent, coefficient in other.__terms.items():
55                    if exponent not in self.__terms:
56                        new_poly.__terms[exponent] = coefficient
57
58            return new_poly
59
60        return NotImplemented
```

I avoided extra code for the case when `other` is an **int** by turning it into a constant polynomial on line 26.

```
w = UniPoly(1, 'w', 1)
w, w + w, w + -w

(w, 2*w, 0)

UniPoly(3, 'x', 7) + UniPoly(5, 'x', 3) + 1

3*x**7 + 5*x**3 + 1

UniPoly(3, 'x', 7) + UniPoly(-5, 'x', 3) + 1

3*x**7 - 5*x**3 + 1

UniPoly(3, 'x', 7) + 2

3*x**7 + 2
```

Did we fully define addition? *__add__* handles adding two polynomials. It also can compute the sum of a polynomial and an integer, **in that order**. It does not handle adding an integer and a polynomial. For that, we need to implement *__radd__*.

```
1  def __radd__(self, other):
2      """Compute other + self.
3
4      Returns
5      -------
6      :class:`UniPoly`
7          The sum of other and self.
8      """
9
10     # Addition is commutative, so we call __add__.
11
12     return self.__add__(other)
```

```
2 + UniPoly(3, 'x', 7)

3*x**7 + 2
```

This implementation works because addition is commutative for our polynomials: $p + q = q + p$ and $p + n = n + p$ for **UniPoly** p and q, and **int** n.

Exercise 7.13

Does the "r" in __*radd*__ stand for "right-side" or "reverse"?

Exercise 7.14

In Python we use "+" for string concatenation. Is this commutative?

7.10.4 Subtraction

The definition of __*sub*__ for subtraction is almost the same as that for __*add*__. Some of the comments should differ, and a few "+" operations should change into "-".

Exercise 7.15

Implement __*sub*__.

```
w = UniPoly(1, 'w', 1)
w, w - w, w - -w
(w, 0, 2*w)

UniPoly(3, 'x', 7) - UniPoly(5, 'x', 3) + 1
3*x**7 - 5*x**3 + 1

UniPoly(3, 'x', 7) - UniPoly(-5, 'x', 3) + 1
3*x**7 + 5*x**3 + 1

UniPoly(3, 'x', 7) - 2
3*x**7 - 2
```

Exercise 7.16

While it is mathematically correct that $p - q = p + -q$, why is it inefficient to implement __*sub*__ with __*neg*__ and __*add*__? What if q had 10,000 terms?

Exercise 7.17

Is this implementation of __*rsub*__:

```
1  def __rsub__(self, other):
2      """Computes other - self.
3
4      Returns
5      -------
6      :class:`UniPoly`
7          The difference of other and self.
8      """
9
10     # other - self == -(self - other)
11     # For efficiency, this should be computed like __sub__
12     # so we create fewer intermediate polynomials.
13
14     return self.__sub__(other).__neg__()
```

as efficient as it could be?

```
2 - UniPoly(3, 'x', 7)

-3*x**7 + 2
```

What if you wrote and used an internal-only instance method __*destructive_negate*__ that negated the coefficients of self in place?

7.10.5 Multiplication

To implement multiplication "*", we code the __*mul*__ and __*rmul*__ instance methods. Since our polynomial multiplication is commutative, __*rmul*__ is straightforward:

```
1  def __rmul__(self, other):
2      """Computes other * self.
3
4      Returns
5      -------
6      :class:`UniPoly`
7          The product of other and self.
8      """
9
```

```
10        # Multiplication is commutative, so we call __mul__.
11
12        return self.__mul__(other)
```

To multiply polynomials *p* and *q*, multiply each term of *p* by each term of *q* and sum the results.

```
 1  def __mul__(self, other):
 2      """Multiplies a polynomial to a polynomial or an integer.
 3
 4      Parameters
 5      ----------
 6      other : :class:`UniPoly` or int
 7          A polynomial or integer.
 8
 9      Notes
10      ------
11      If the second argument is not a polynomial or integer,
12      or if the indeterminates in non-constant polynomials
13      are different, returns NotImplemented.
14
15      Returns
16      -------
17      :class:`UniPoly` or NotImplemented
18          The product of self and other, or NotImplemented.
19      """
20
21      # if self == 0, return it
22      if not self.__terms:
23          return self
24
25      # if other is an integer, create a constant polynomial
26      if isinstance(other, int):
27          other = UniPoly(other, self.__indeterminate, 0)
28
29      # both objects are polynomials
30
31      if isinstance(other, UniPoly):
32          # if other == 0, return other
33          if not other.__terms:
34              return other
35
36          # We now have two non-zero polynomials.
```

```
37
38              if self.__indeterminate != other.__indeterminate:
39                  if self.is_constant and other.is_constant:
40                      return UniPoly(self.constant_coefficient *
41                                     other.constant_coefficient,
42                                     self.__indeterminate,
43                                     0)
44                  return NotImplemented
45
46              new_poly = UniPoly(0, self.__indeterminate, 0)
47
48              for exponent_1, coefficient_1 in self.__terms.items():
49                  for exponent_2, coefficient_2 in other.__terms.items():
50                      new_poly += UniPoly(coefficient_1 * coefficient_2,
51                                          self.__indeterminate,
52                                          exponent_1 + exponent_2)
53              return new_poly
54
55          return NotImplemented
```

```
z = UniPoly(1, 'z', 1)
(z + 1) * (z - 1)

z**2 - 1

(z + 1) * (z + 1)

z**2 + 2*z + 1

(2*z + 1) * (-3*z*z*z + 4)

-6*z**4 - 3*z**3 + 8*z + 4
```

Exercise 7.18

On the line with "new_poly +=", we are doing immutable addition. This creates and copies many terms. Define an internal-only instance method, *__destructive_add*, that changes self in place to be the sum of itself and other. Use this in an improved version of *__mul__*.

Our math is looking good! We have one more basic algebraic operation to code.

7.10.6 Exponentiation

To implement exponentiation by a non-negative integer, we code the __pow__ instance method.

If p is a polynomial, then p^4 is $p \times p \times p \times p$. Generally, p^n requires $n - 1$ multiplication operations if we do it this way. The algorithm is O($n - 1$).

Note, though, that p^4 is also $q \times q$ if $q = p \times p$. We only need to do two multiplications instead of three! For p^5, we compute $p \times q \times q$. If $n = 8$,

$$p^8 = p^4 \times p^4$$

and we do 3 multiplications instead of 7. We are using the *repeated squaring algorithm for exponentiation*. It is recursive.

Exercise 7.19

Are you suspicious that $3 = \log_2(8)$? Do an O() analysis for repeated squaring.

To compute p^n via repeated squaring,

- Return 1 if n is 0.

- Return p if n is 1.

- Return $p \times p$ if n is 2.

- Recursively compute $q = p^{n \,//\, 2}$. (Remember that "//" is integer division, so that 7 // 2 == 3, for example.)

- Set q equal to $q \times q$.

- If n is even, return q.

- Return $p \times q$.

```
1  def __pow__(self, exponent):
2      """Raises a polynomial to a non-negative integer exponent.
3
4      `__pow__` used by both `**` and `pow`.
5
6      Parameters
7      ----------
8      exponent : int
9          The non-negative exponent to which self is raised.
```

```
10
11      Notes
12      -----
13      If the exponent is not a non-negative integer, this method
14      returns NotImplemented. This allows another class to implement
15      quotients of polynomials and expressions like `p**(-1)`.
16
17      Returns
18      -------
19      :class:`UniPoly` or NotImplemented
20          The polynomial `self` raised to `exponent`, or NotImplemented.
21      """
22
23      # Validate the exponent.
24      if not isinstance(exponent, int) or exponent < 0:
25          return NotImplemented
26
27      # Handle the simple cases
28      if exponent == 0:
29          return UniPoly(1, self.__indeterminate, 0)
30
31      if exponent == 1:
32          return self
33
34      if exponent == 2:
35          return self * self
36
37      # Compute the exponent by recursive repeated squaring.
38      power = self**(exponent // 2)
39      power *= power
40      if exponent % 2 == 1:
41          power *= self
42
43      return power
```

The simple cases each work.

```
(z + 1)**0, (z + 1)**1, (z + 1)**2
```

```
(1, z + 1, z**2 + 2*z + 1)
```

Larger examples work as well.

```
(z + 1)**6
```

```
z**6 + 6*z**5 + 15*z**4 + 20*z**3 + 15*z**2 + 6*z + 1
```

```
(z + 1)**25
```

```
z**25 + 25*z**24 + 300*z**23 + 2300*z**22 + 12650*z**21 + 53130*z**20 +
177100*z**19 + 480700*z**18 + 1081575*z**17 + 2042975*z**16 +
3268760*z**15 + 4457400*z**14 + 5200300*z**13 + 5200300*z**12 +
4457400*z**11 + 3268760*z**10 + 2042975*z**9 + 1081575*z**8 +
480700*z**7 + 177100*z**6 + 53130*z**5 + 12650*z**4 + 2300*z**3 +
300*z**2 + 25*z + 1
```

See how small expressions can generate large polynomials? For this reason, it's important for you to use memory-efficient and high-performance algorithms, though you may sometimes need to trade off one for the other.

When you are creating Python classes, focus on good design and correctness first. Then iterate until your architecture is elegant and your code performs very well.

7.10.7 The derivative

The *derivative* from calculus gives us the instantaneous rate of change of a function at a point. If

$$p = a_n x^n + a_{n-1} x^{n-1} + \cdots + a_2 x^2 + a_1 x + a_0$$

is a polynomial with integer coefficients and exponents, the derivative of p is

$$p' = n\, a_n x^{n-1} + (n-1)\, a_{n-1} x^{n-2} + \cdots + 2\, a_2 x + a_1$$

The implementation is straightforward from the definition:

```
1   def derivative(self):
2       """Computes the derivative of a polynomial.
3
4       Returns
5       -------
6       :class:`UniPoly`
7           The derivative.
8       """
9
10      # Initialize the result to the zero polynomial
11      # with the correct indeterminate.
12      result = UniPoly(0, self.__indeterminate, 0)
```

```
13
14          # Compute the term-by-term derivative, and sum.
15          if self.__terms:
16              for exponent, coefficient in self.__terms.items():
17                  if exponent > 0:
18                      result += UniPoly(exponent * coefficient,
19                                        self.__indeterminate,
20                                        exponent - 1)
21          return result
```

If

```
p = 3*z**6 - 2*z**5 + 7*z**2 + z - 12
p
```

```
3*z**6 - 2*z**5 + 7*z**2 + z - 12
```

then

```
p.derivative()
```

```
18*z**5 - 10*z**4 + 14*z + 1
```

This is another example where the addition of polynomials can get expensive when the polynomials get large.

Exercise 7.20

In our implementation of *derivative*, we always return a polynomial, even for constants.

```
q_prime = UniPoly(4, 'y', 1).derivative()
print(q_prime)
print(type(q_prime))

4
<class '__main__.UniPoly'>
```

What are the pros and cons of returning an **int** instead of a **UniPoly** for constant polynomials?

7.10.8 Objects as functions

At the beginning of this chapter, I showed how we could think of polynomials as mathematical functions. We implement function behavior for instances with ___call__.

```
1  def __call__(self, x):
2      """Evaluates a polynomial by substituting `x` for the indeterminate.
3
4      Python calls this when it sees a polynomial being used
5      in function application. If `p` is a polynomial, then
6      `p(3.4)` substitutes 3.4 for the indeterminate and
7      does the math as expressed by the polynomial.
8
9      Parameters
10     ----------
11     x : any object that supports -, +, *, **
12         The value to be substituted for the indeterminate.
13
14     Returns
15     -------
16     Probably `type(x)`
17     """
18     result = 0
19
20     for exponent, coefficient in self.__terms.items():
21         result += coefficient * x**exponent
22
23     return result
```

Now we define a polynomial and show that it produces the same values as before.

```
x = UniPoly(1, 'x', 1)
p = x**3 - 5*x**2 + 7*x + 1

[p(x) for x in range(-2, 4)]

[-41, -12, 1, 4, 3, 4]
```

We can evaluate p for any type that has the basic algebraic operations.

```
p(2.75)        # float

3.234375
```

```
p(1 + 2j)     # complex

(12-8j)

p(x + 1)      # polynomial

x**3 - 2*x**2 + 4
```

I defined __*call*__ with one parameter after `self`:

$$\text{def \_\_call\_\_(self, x)}$$

More generally, you can use

$$\text{def \_\_call\_\_(self, *args, **kwargs)}$$

to implement full function argument capability.

Exercise 7.21

Just as we saw that repeated squaring does exponentiation with fewer multiplications, we can also evaluate polynomials more efficiently.

$$x^3 - 5x^2 + 7x + 1 = x(x^2 - 5x + 7) + 1$$

We can continue to reduce the number of multiplications within the parentheses. Implement __*call*__ more efficiently using this technique.

7.10.9 Square brackets

For a list, I use an index and square brackets to get an item.

```
small_primes = [2, 3, 5, 7, 11]
small_primes[4]
```

```
11
```

Similarly, I can use a key and square brackets with a dictionary to get the associated value.

```
books = {"math": 4, "cooking": 7, "fishing": 2}
books["fishing"]
```

```
2
```

I want to use square bracket notation to retrieve the coefficient of a given exponent. I implement the __*getitem*__ magic method to do that.

```
1  def __getitem__(self, exponent):
2      """Retrieves the coefficient of the term with the exponent.
3
4      Supports square bracket access to coefficients of terms in
5      the polynomial. If there is no term with the given exponent,
6      returns 0.
7
8      Parameters
9      ----------
10     exponent : int
11         A non-negative integer representing an exponent
12         in a polynomial term.
13
14     Returns
15     -------
16     int or NotImplemented
17         The coefficient of the term with the given exponent.
18
19     Raises
20     ------
21     ValueError
22         Raised if the exponent is an integer but less than zero.
23     """
24     if not isinstance(exponent, int):
25         return NotImplemented
26     if exponent < 0:
27         raise ValueError("polynomial exponents are integers >= 0")
28
29     if not self.__terms:
30         return 0
31     return self.__terms.get(exponent, 0)
```

```
p = (x**2 - 3)**3
p
```

```
x**6 - 9*x**4 + 27*x**2 - 27
```

```
p[0]     # the constant term
```

```
-27
```

```
p[3]      # this exponent does not appear

0

p[6]      # the exponent of highest degree

1
```

Exercise 7.22

The __getitem__ method is very flexible in what it can take as its argument. Although I only use it for the exponent, you could accept a slice object.

Issue help(slice) within Python to learn about **slice** objects.

Augment __getitem__ to accept a slice in addition to an integer.

The first number in the slice is the starting exponent, the second is the exponent limit, and the third is the step. Produce a new polynomial from __getitem__ that includes only those exponents represented in the slice.

7.10.10 Type conversions

To implement conversion to **int**, we implement __int__.

```
1  def __int__(self):
2      """Returns an `int` if the polynomial is constant.
3
4      Returns
5      -------
6      int
7          The constant term of the constant polynomial.
8
9      Raises
10      ------
11      ValueError
12          Raised if self is not constant.
13      """
14      if not self.__terms:
15          return 0
16      if len(self.__terms) == 1 and 0 in self.__terms:
17          return self.__terms[0]
18      raise ValueError("non-constant polynomial for int()")
```

```
int(UniPoly(3, 'x', 0))
```

```
3
```

```
int(UniPoly(3, 'x', 2))
```

```
ValueError: non-constant polynomial for int()
```

Exercise 7.23

Implement and test conversions of polynomials to floating-point and complex numbers via *__float__* and *__complex__*.

7.10.11 Additional exercises

There are many ways in which you can move beyond our basic definition of **UniPoly**, though it already has significant functionality. Consider doing these additional exercises by yourself or with a coding partner. Focus on efficiency, and document and comment your code well.

Exercise 7.24

Extend **UniPoly** in *Appendix C* to support both **int** and **float** coefficients.

Implement *divmod* to return a quotient and remainder tuple from dividing two extended polynomials.

Exercise 7.25

Extend **UniPoly** to support both **int** and **Fraction** coefficients.

Implement *divmod* to return a quotient and remainder tuple from dividing two extended polynomials.

Exercise 7.26

Extend **UniPoly** to support **int**, **float**, **complex**, and **Fraction** coefficients.

Exercise 7.27

Code the **BiPoly** class that creates polynomials of two specified and different invariants. These are *bivariate polynomials*. For example, BiPoly(4, 'x', 6, 'y', 2).

Exercise 7.28

Create an abstract base class **Poly**. Make **UniPoly** and **BiPoly** concrete child classes of **Poly**.

Exercise 7.29

Allow addition, subtraction, and multiplication of a univariate polynomial and a bivariate polynomial. The indeterminate in the first must be one of the indeterminates in the second. For example, allow UniPoly(2, 'x', 3) + BiPoly(4, 'x', 6, 'y', 2):

$$2x^3 + 4x^6 y^2$$

7.11 Class variables

An instance variable is unique to each object created by a class. A class variable is shared, accessible, and changeable by all objects of a class. There are similarities with the global variables we saw in section 6.9.2. Coders frequently use class variables instead of top level global variables. Restricting their use within a class is more acceptable, especially if their names start with "__".

As we have seen, if obj is the name of an instance, you reference an instance variable with a name prefixed by "obj.". obj is often self.

You define a class variable at the top level of a class, along with the instance methods.

```
1  class Guitar:
2      number_of_guitars = 0
3
4      def __init__(self, brand):
5          Guitar.number_of_guitars += 1
```

```
6          plural = "s" if Guitar.number_of_guitars != 1 else ""
7          print(f"We have {Guitar.number_of_guitars} guitar{plural}")
8          self.brand = brand
9
10     def __repr__(self):
11         return f"Guitar({repr(self.brand)})"
12
13     def __str__(self):
14         return self.brand
```

```
Guitar("Fender")
```

```
We have 1 guitar
Guitar('Fender')
```

```
Guitar("Taylor")
```

```
We have 2 guitars
Guitar('Taylor')
```

```
g = Guitar("Gibson")
g
```

```
We have 3 guitars
Guitar('Gibson')
```

Reference a class variable either by the class name followed by ".", followed by the variable name, or an instance name followed by "." followed by the variable name. I prefer the first version, where the class name is part of this prefix. It makes it clear where the variable is defined and does not require you to create an instance.

```
Guitar.number_of_guitars
```

```
3
```

```
g.number_of_guitars
```

```
3
```

Exercise 7.30

Why do I call *repr* on the instance variables within the definition of __repr__?

7.12 Class and static methods

Class methods and *static methods* are both defined within a class. Unlike an instance method, Python does not bind a *class method* to a particular instance. It is bound to and knows about the entire class. A static method is a function that sits within a class but knows nothing special about the class or its instances.

You create a class method by placing the @classmethod decorator on the line before the method's definition. Instead of using self for the first parameter as you would in an instance method, use "cls". At runtime, Python passes the class object to the method as the first argument.

```
1  @classmethod
2  def get_count_class(cls):
3      return cls.number_of_guitars
```

If you refer to a class variable within a class method, prefix the variable name with cls and a ".". Do not hardcode the name of the class into the call, as this can interfere with inheritance.

If you call the class method from the Python command line, you may use either the class name or an instance name as the prefix.

```
Guitar.get_count_class(), g.get_count_class()
```

```
(3, 3)
```

We define a static method with the @staticmethod decorator instead of @classmethod, and there is no special first parameter. You can access objects inside the class using the class's hardcoded name, but you do not have the class object in hand.

```
1  @staticmethod
2  def get_count_static():
3      return Guitar.number_of_guitars
```

```
Guitar.get_count_class(), g.get_count_static()
```

```
(3, 3)
```

The differences between class and static methods seem slight and subtle.

- If you want to package related functions into a class and don't plan to instantiate class objects, use class methods if you have class variables and static methods otherwise.

- If you write a non-instance method that creates new instances of the class, use a class method.

- If you plan to redefine the method in a subclass (next section), use a class method.

7.13 Inheritance

Starting from one class, we can derive other classes that inherit and modify the original's behavior and functionality. Inheritance allows us to specialize instances while reusing code from the class hierarchy.

In section 1.10, we focused on Bay leaves, but here we return to guitars and their various types. Let's begin with a basic definition of the **Guitar** class. The full and documented code that we develop here is in *Appendix D*.

```
1   class Guitar:
2       def __init__(self, brand, model, year_built, strings=6):
3           self._brand = brand
4           self._model = model
5           self._year_built = year_built
6           self._strings = strings
7
8       def __str__(self):
9           return f"{self._strings}-string " \
10              + f"{self._year_built} " \
11              + f"{self._brand} {self._model}"
```

```
print(Guitar("Fender", "Stratocaster", "2012"))
```

```
6-string 2012 Fender Stratocaster
```

In the **Guitar** class, we have

- instance methods *__init__* and *__str__*, and
- instance variables _brand, _model, _year_built, and _strings.

The instance variable names start with an underscore, and this indicates to us that the variables should only be used privately by other coders within this and any derived classes. Python does not enforce this, but it is a convention.

7.13.1 Parent and child classes

`Guitar` is a standalone class, but if we look at the brand, model, and build year attributes, they could apply to any commercial musical instrument. So let's define `MusicalInstrument`:

```
1   class MusicalInstrument:
2       def __init__(self, brand, model, year_built):
3           self._brand = brand
4           self._model = model
5           self._year_built = year_built
6
7       def __str__(self):
8           return f"{self._year_built} " \
9               + f"{self._brand} {self._model}"
```

In `MusicalInstrument`, we have

- instance methods __init__ and __str__, and
- instance variables _brand, _model, and _year_built.

Now we derive `Guitar` from `MusicalInstrument`.

```
1   class Guitar(MusicalInstrument):
2       def __init__(self, brand, model, year_built, strings=6):
3           super().__init__(brand, model, year_built)
4           self._strings = strings
5
6       def __str__(self):
7           return f"{self._strings}-string " \
8               + super().__str__()
```

`Guitar` adds the instance variable _strings.

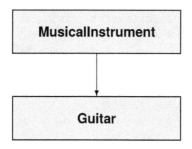

Figure 7.2: MusicalInstrument class hierarchy with Guitar

We call **MusicalInstrument** the *parent class, superclass,* or *base class* of **Guitar. Guitar** is a *child class* or *subclass* of **MusicalInstrument**.

The **Guitar** __*init*__ method has an optional argument, strings, with a default value of 6. It calls the **MusicalInstrument** initialization code via

$$\texttt{super().\_\_init\_\_(brand, model, year\_built)}$$

super() means, "think of self not as an instance of this class, but as an instance of its immediate parent class." The __*init*__ in **Guitar** then sets the new instance variable, _strings.

__*str*__ in **Guitar** overrides the implementation in **MusicalInstrument**. It reuses and extends the definition from the parent class. The methods can freely use the parent class instance variables.

Let's create some instances and see what they can do.

```
instrument = MusicalInstrument("Gibson", "Les Paul", "1997")
print(instrument)
```

```
1997 Gibson Les Paul
```

```
guitar = Guitar("Fender", "Stratocaster", "2012")
print(guitar)
```

```
6-string 2012 Fender Stratocaster
```

A guitar is a musical instrument, but not every musical instrument is a guitar.

```
isinstance(instrument, MusicalInstrument), isinstance(instrument,
Guitar)
```

```
(True, False)
```

```
isinstance(guitar, MusicalInstrument), isinstance(guitar, Guitar)
```

```
(True, True)
```

Exercise 7.31

Add a _number_of_guitars class variable to **Guitar**, along with a class function, *get_count*, that returns the class variable. Increment _number_of_guitars in *__init__* every time we initialize a new instance.

We can derive a child class without doing any additional work if we only want to distinguish it via *isinstance* from the parent class.

```
1  class Clarinet(MusicalInstrument):
2      pass
```

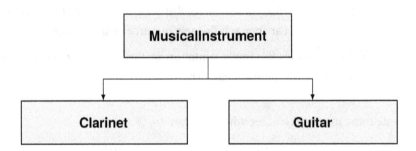

Figure 7.3: MusicalInstrument class hierarchy with Guitar and Clarinet

```
help(Clarinet)
```

```
Help on class Clarinet in module src.code.guitar_04:
```

```
class Clarinet(MusicalInstrument)
 |  Clarinet(brand, model, year_built)
 |
 |  Method resolution order:
 |      Clarinet
 |      MusicalInstrument
 |      builtins.object
 |
 |  Methods inherited from MusicalInstrument:
 |
```

```
 |  __init__(self, brand, model, year_built)
 |      Initialize self.  See help(type(self)) for accurate signature.
 |
 |  __str__(self)
 |      Return str(self).
 |
 |
 |  ----------------------------------------------------------------
 |  Data descriptors inherited from MusicalInstrument:
 |
 |  __dict__
 |      dictionary for instance variables (if defined)
 |
 |  __weakref__
 |      list of weak references to the object (if defined)
```

```
clarinet = Clarinet("Yamaha", "YCL-650", "2021")
print(clarinet)
```

```
2021 Yamaha YCL-650
```

```
isinstance(clarinet, MusicalInstrument), isinstance(clarinet, Clarinet)
```

```
(True, True)
```

Exercise 7.32

Implement read-only properties brand, model, and year_built in **MusicalInstrument**. Demonstrate that they work for instances of **Guitar** and **Clarinet**.

Now let's create the child class **ElectricGuitar** from **Guitar**. It adds a new enumeration instance variable, _pickup, of type **Pickup**.

```
1  class Pickup(Enum):
2      UNKNOWN = "unknown"
3      SINGLECOIL = "single coil"
4      HUMBUCKER = "Humbucker"
5      ACTIVE = "active"
6      GOLDFOIL = "gold foil"
7      TOASTER = "toaster"
```

The user can both set and get the `pickup` property after initialization.

```
1  class ElectricGuitar(Guitar):
2      def __init__(self, brand, model, year_built, strings=6):
3          super().__init__(brand, model, year_built, strings)
4          self._pickup = Pickup.UNKNOWN
5
6      def __str__(self):
7          return f"{self._strings}-string " \
8              + f"{self._year_built} " \
9              + f"{self._brand} {self._model} " \
10             + f"with {self._pickup.value} pickup"
11
12     @property
13     def pickup(self):
14         return self._pickup
15
16     @pickup.setter
17     def pickup(self, value):
18         if not isinstance(value, Pickup):
19             raise ValueError("The pickup must be of "
20                              "enumeration type Pickup.")
21         self._pickup = value
```

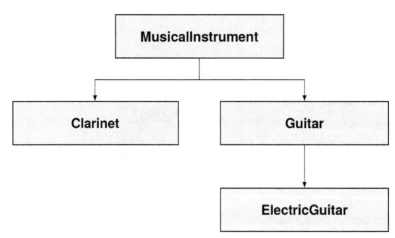

Figure 7.4: MusicalInstrument class hierarchy with ElectricGuitar added

```
electric_guitar = ElectricGuitar("Fender", "Stratocaster", "2012")
print(electric_guitar)

6-string 2012 Fender Stratocaster with unknown pickup
```

```
electric_guitar.pickup = Pickup.SINGLECOIL
print(electric_guitar)
```

```
6-string 2012 Fender Stratocaster with single coil pickup
```

7.13.2 Abstract base classes

Why do we let users create instances of **MusicalInstrument**? We only need it to store generic information about any instrument. An *abstract base class* is a class from which you can inherit but not create instances. You can state which methods developers must override in child classes.

```
from abc import ABC, abstractmethod
```

```
1  class MusicalInstrument(ABC):
2      def __init__(self, brand, model, year_built):
3          self._brand = brand
4          self._model = model
5          self._year_built = year_built
6
7      @abstractmethod
8      def __str__(self):
9          return NotImplemented
```

The @abstractmethod decorator indicates that the coder of a child class must override the following method, __str__. Nothing will ever call this method in **MusicalInstrument**, which is why I return NotImplemented. Alternatively, return None or use pass.

Guitar works as before.

```
print(Guitar("Fender", "Stratocaster", "2012"))
```

```
6-string 2012 Fender Stratocaster
```

We cannot create a **MusicalInstrument** instance.

```
print(MusicalInstrument("Gibson", "Les Paul", "1997"))
```

```
TypeError: Can't instantiate abstract class
          MusicalInstrument with abstract methods __str__
```

We cannot create a **Clarinet** because we did not implement __*str*__ and so it is still abstract.

```
print(Clarinet("Yamaha", "YCL-650", "2021"))

 TypeError: Can't instantiate abstract class
             Clarinet with abstract methods __str__
```

Guitar is a *concrete class* because it implements all the required abstract methods. In the diagrams, concrete classes are in boxes with solid lines, and abstract base classes have dotted lines.

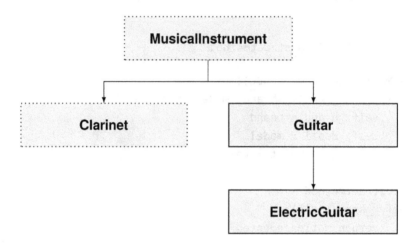

Figure 7.5: MusicalInstrument as an abstract base class

Your abstract base classes will be larger in practice and have more methods that developers must implement in derived concrete classes.

7.13.3 Multiple inheritance

I now complete my guitar class hierarchy by adding **AcousticGuitar**. Though it is not strictly correct, I am assuming that all acoustic guitars are wooden. In particular, I'm talking about the soundboard, which is the top of the guitar.

Figure 7.6: An acoustic guitar

For **AcousticGuitar**, I inherit both from **Guitar** but also from **Wooden**.

```
1  class Wooden:
2      def __init__(self, wood_name):
3          self._wood_name = wood_name
4
5      def __str__(self):
6          return self._wood_name
```

When I inherit from more than one class in a `class` definition, I am implementing *multiple inheritance.*

```
1   class AcousticGuitar(Guitar, Wooden):
2       def __init__(self, brand, model, year_built,
3                       soundboard_wood, strings=6):
4           super().__init__(brand, model, year_built, strings)
5           Wooden.__init__(self, soundboard_wood)
6
7       def __str__(self):
8           return f"{self._strings}-string " \
9               + f"{self._year_built} " \
10              + f"{self._brand} {self._model} " \
11              + f"with {self._wood_name} soundboard"
```

```
guitar = AcousticGuitar("Taylor", "AD27e", "2021", "mahogany")
print(guitar)
```

```
6-string 2021 Taylor AD27e with mahogany soundboard
```

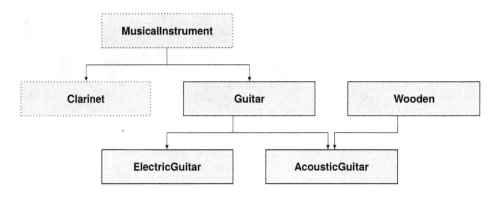

Figure 7.7: AcousticGuitar with multiple inheritance

Note how I coded __*init*__ to call the **Guitar** and **Wooden** initialization methods.

If I call a method on an instance of **AcousticGuitar**, Python looks first in **AcousticGuitar** itself, then **Guitar**, then **MusicalInstrument**, and finally in **Wooden**. This is called the *method resolution order*.

If there were a third class from which **AcousticGuitar** inherited, Python would search it and its parent's class after **Wooden**.

Exercise 7.33

For each class in our hierarchy, define a *whoami* method that returns a string like "I am an ElectricGuitar object". Call this on instances of each concrete type to ensure it works correctly.

When you have multiple inheritance from classes that themselves have a common ancestor, you have *diamond inheritance*. It is an excellent source of bugs if you can even get your code to work at all. In Figure 7.8, I show the diamond using double lines.

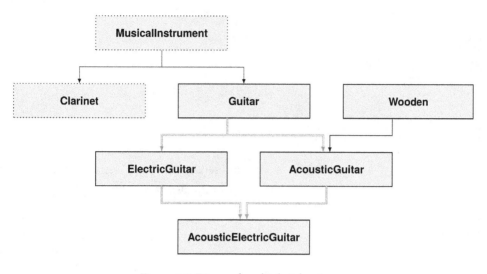

Figure 7.8: Diamond multiple inheritance

You can often fix the "wrong method being called" problem with diamond inheritance by reordering the names of the parents in the class definition. For example, if class **A** inherits from both **B** and **C**, you can try either

$$\texttt{class A(B, C):}$$

or

$$\texttt{class A(C, B):}$$

Exercise 7.34

Define a new class, **AcousticElectricGuitar**, that inherits from both **ElectricGuitar** and **Acoustic** guitar. Do not implement *whoami* in **AcousticElectricGuitar**. When you call *whoami* on an **AcousticElectricGuitar** instance, what does it return? What happens if you reverse the order of classes from which **AcousticElectricGuitar** inherits?

7.14 Iterators

What really happens when Python iterates over a list?

```python
for i in [2, 3, 5]:
    print(i)

2
3
5
```

Python keeps track of your position in the list and returns the "next" item every time you ask for it. When you reach the end of the list, the loop terminates. It behaves something like this:

```python
the_list = [2, 3, 5]
the_index = 0

while True:
    if the_index < len(the_list):
        print(the_list[the_index])
        the_index += 1
    else:
        break

2
3
5
```

This iteration looks like normal progression through the list from the beginning to the end. What "iteration" means is up to you: you can define it to do anything you wish.

```python
the_list = [2, 3, 5]
the_index = len(the_list) - 1

while True:
    if the_index >= 0:
        print(the_list[the_index])
        the_index -= 1
    else:
        break

5
3
2
```

In this case, the iteration framework we have set up selects the items from the last to the first. The "next" item is before the one we last accessed in the list.

An *iterator* object has a __next__ magic method defined for it by its class. The first time you call __next__, it returns the "initial" object. The Python code remembers your position and subsequent calls to __next__ work their way through all the items.

An *iterable* object has a __iter__ magic method that returns an iterator object. An object may be both an iterable and an iterator. In this case, __iter__ often just returns self.

An iterable returns an iterator via __iter__.

Exercise 7.35

Is an object returned by *range* an iterable, an iterator, or both?

Python calls __iter__ and __next__ automatically in a for loop via *iter* and *next*. You can call them yourself.

```
my_iter = iter([2, 3, 5])
print(next(my_iter))
```

```
2
```

```
print(next(my_iter))
```

```
3
```

```
print(next(my_iter))
```

```
5
```

If you call it one more time, *next* raises the **StopIteration** exception.

```
print(next(my_iter))
```

```
StopIteration
```

You should raise this exception in your __next__ code so that for works correctly. Look for and catch **StopIteration** in a try block.

If your object changes while you are iterating over it, you may get errors or at least unexpected results. For example, you likely do not want to insert or remove items from a list while working with it in a loop.

Exercise 7.36

Let numbers be the dictionary:

    numbers = {1: "one", 2: "two", 3: "three"}

Explain what is happening in each of the following:

    next(iter(numbers))

    1

    next(iter(numbers.keys()))

    1

    next(iter(numbers.values()))

    'one'

    next(iter(numbers.items()))

    (1, 'one')

What are the iterators and what are the iterables?

For **UniPoly** instances, it's natural for us to use the term dictionary for the iteration.

```
1  def __iter__(self):
2      """Iterator returning exponent-coefficient tuple pairs.
3
4      Notes
5      -----
6      The order of the pairs is not guaranteed to be sorted.
7
8      Returns
9      -------
10     tuple:
11         (exponent, coefficient) in polynomial.
12     """
13
14     return iter(self.__terms.items())
```

```
s = UniPoly(1, 's', 1)

for exp_coef in (s - 1)**4:
    print(exp_coef)
```

```
(4, 1)
(3, -4)
(2, 6)
(1, -4)
(0, 1)
```

I used a **Notes** section within the documentation for __*iter*__.

If an object is both an iterable and an iterator, the object probably stores the iteration state within itself. That's why __*iter*__ might return `self`. As we have seen, though, there may be multiple ways you want to iterate across an object. In this case, you would create separate iterables as Python does with dictionaries.

Exercise 7.37

In our **UniPoly** class, I define a property `is_constant` that tells us whether a polynomial is constant. Remember that a constant polynomial has no term with the indeterminate exponent greater than 0.

```
1  @property
2  def is_constant(self):
3      """True if the polynomial only contains a term of
       exponent 0."""
5
6      if not self.__terms:
7          return True
8      if len(self.__terms) != 1:
9          return False
10      return next(iter(self.__terms)) == 0
```

What does the last line do?

7.15 Generators

Closely related to an iterator is the Python concept of a *generator*. A generator *g* is a special function with this behavior:

- The first time you call *g*, it initializes itself, passes back a value, and freezes its state.
- The next time you call *g*, it continues processing until it passes back a value and then freezes its state again.
- These actions continue until the function terminates, and then it returns no additional values.

We use `yield` to tell Python to pass back a value and freeze everything until the next call to the generator.

```python
def create_odd_number_generator():
    n = 1
    while True:
        yield n
        n += 2

odd_number_generator = create_odd_number_generator()
odd_number_generator
```

```
<generator object create_odd_number_generator at 0x0000024A93102660>
```

```python
next(odd_number_generator)
```

```
1
```

```python
next(odd_number_generator)
```

```
3
```

```python
next(odd_number_generator)
```

```
5
```

```python
for n in odd_number_generator:
    print(f"{n} ", end="")
    if n > 20:
        break
```

```
7 9 11 13 15 17 19 21
```

Generators are an example of using *lazy evaluation*. Rather than producing a big list of odd numbers and iterating over them, we only compute the next odd number when the user requests it. In this way, we don't waste memory holding a large collection, and we don't waste computation time producing values that no one might use.

Exercise 7.38

Using the iterative Fibonacci function from section 6.12, write a generator that returns the next Fibonacci number upon request.

Exercise 7.39

In section 5.2.2, we defined a function that implemented the Sieve of Eratosthenes to create lists of primes. Write a generator that returns the next prime number upon request.

Generators don't need to yield values forever. In this implementation of a generator, you can specify the maximum number you want it to yield. Once the generator reaches that, we return out of the function.

```
def create_even_number_generator(maximum_n=None):
    n = 0
    while True:
        if maximum_n is not None and n > maximum_n:
            return
        yield n
        n += 2

even_number_generator = create_even_number_generator(20)
even_number_generator
```

```
<generator object create_even_number_generator at 0x0000024A9310D040>
```

```
for n in even_number_generator:
    print(f"{n} ", end="")
```

```
0 2 4 6 8 10 12 14 16 18 20
```

Exercise 7.40

If you call `next(even_number_generator)` now, what happens? Did I code that explicitly?

Exercise 7.41

If you have studied calculus: you can compute approximations to the exponential function e^x via the Maclaurin series

$$e^x = \frac{x^0}{0!} + \frac{x^1}{1!} + \frac{x^2}{2!} + \frac{x^3}{3!} + \cdots$$

Create a generator that yields the next term in the Maclaurin series for a given x. For example, x equals to 2.0. Code this generator in a `my_exp(x, n)` function that calculates the exponential function for x using n terms of the Maclaurin series. How do its results compare with those from *math.exp*?

7.16 Objects in collections

If you write a new class and plan to put its instances in a list, there are no special requirements. The list will hold the items in the order in which you place them. An operation like *reverse*, which only changes the list order in place, works fine.

```
polys = [UniPoly(1, 'x', 2), UniPoly(2, 'x', 3), UniPoly(3, 'x', 4)]
polys

[x**2, 2*x**3, 3*x**4]

polys.reverse()
polys

[3*x**4, 2*x**3, x**2]
```

Generally, you do want to tell when one instance is equal to another, so you should implement __eq__. For completeness, I often define __ne__ as well. Python uses these for "==" and "!=", respectively. If __ne__ is missing, Python negates the result of __eq__.

If you plan to sort the list, you need a way to compare items to know which object is "less than" another. For the *sorted* function and the in-place *sort* method, Python looks for __lt__ to implement "<". If it is not present, you get an error message.

```
polys.sort()

TypeError: '<' not supported between instances
            of 'UniPoly' and 'UniPoly'
```

When working with mathematical objects, you must be careful in mixing algebraic meanings of operations with structural interpretations. Mathematically, you cannot say when one **UniPoly** polynomial is less than another. For some subsets such as constants, you can decide this, but it does not generally work. Structurally, we could compare the polynomials lexicographically by tuples of their exponents and coefficients.

Given

$$p = x^2 + 7x + 1 \quad \text{and} \quad q = 4x^2 - 7x + 3$$

look at the tuples of the highest degree. For p, this is (2, 1), and for q, it is (2, 4). If one exponent were smaller, that polynomial would be "less than" the other. In this case, the exponents are equal, so we compare the coefficients. Since 1 < 4, p is "less than" q. Had they been equal, we would have moved to the terms of the next lower degree.

Exercise 7.42

sort and *sorted* each accept a key optional function argument that returns objects the functions use to compare with "<". Write a method for **UniPoly** to use with key that returns a suitable tuple of tuples.

Exercise 7.43

In section 7.9, we defined the **Cookie** enumeration. There is a problem: we can't sort lists of cookies.

```
cookies = [Cookie.SUGAR, Cookie.SHORTBREAD,
    Cookie.OATMEAL, Cookie.CHOCOLATE]
sorted(cookies)

TypeError: '<' not supported between instances of
            'Cookie' and 'Cookie'
```

Write an anonymous function for key so that we can sort the list by enumeration value.

Exercise 7.44

There is no comparison problem if you sort a list of **CatAge** enumeration members. Why?

Dictionary keys and set members must be immutable objects, as we saw in sections 3.9.1 and 3.10. Since you can't change them, Python can compute fixed information with the *hash* function that speeds up equality checking and access time. For example, if two objects have different *hash values,* they cannot be equal.

When Python sees hash(object), it looks for an instance method, *__hash__*, to call on object.

```
hash(-0.8367)
```

```
-1929298845809097728
```

```
hash(735678)
```

```
735678
```

```
hash("A solitary string")
```

```
5073272553319032547
```

```
hash((4, 3, 2, 1))
```

1111739957947434390

___hash___ must return an integer, and your implementation must adhere to one rule: if a == b for instances a and b, then hash(a) == hash(b). Ideally, the opposite is true, but that does not need to be the case.

Your ___hash___ method should be non-trivial. Don't, for example, always return 0. A common method that coders use is to make a tuple of the instance variables and then call *hash* on that. You could also sum the hash values for the instance variables.

The full and final code for **UniPoly** is in *Appendix C*, and it creates immutable instances. I define ___hash___ in this way:

```
1  def __hash__(self):
2      """Computes a hash code for a polynomial.
3
4      If p and q are equal polynomials, then their hash codes
5      are equal.
6
7      Returns
8      -------
9      int
10         An integer hash code.
11     """
12     h = hash(self.__indeterminate)
13     for exponent_coefficient in self.__terms:
14         h += hash(exponent_coefficient)
15     return h
```

```
hash(UniPoly(10, 'z', 100))
```

-483907656708034775

```
hash(UniPoly(10, 'w', 100))
```

1477461016404213130

Exercise 7.45

Why didn't I compute hash(self.__terms)? Why didn't I build a tuple of the exponent-coefficient pairs and call *hash* on that?

Python lists are mutable, so there is no hash function for them. You cannot use them as dictionary keys or put them in sets.

```
title = ["For", "Whom", "the", "Bell", "Tolls"]

hash(title)

TypeError: unhashable type: 'list'

{title: "Ernest Hemingway"}

TypeError: unhashable type: 'list'

books = set()
books.add(title)

TypeError: unhashable type: 'list'
```

If I were thinking about collections of objects and saw this error message, I likely would not understand why Python was complaining. Strive to make your error messages clear in the contexts within which your users see them.

7.17 Creating modules

We have seen many expressions, variable assignments, and function and class definitions. When you put those together in a file, you get a Python *module*. For example, in *Appendix D*, I group the class definitions that we developed for guitars in section 7.13 on inheritance.

The name of your module file should be a valid Python name followed by ".py". My module with **Guitar** and other classes is **guitar**. In particular, do not use hyphens in your module name because Python will think you are subtracting.

The "top level" of a module is code that exists outside any function or class definition. It starts in column 1. Everything we have seen about constants, global variables, and name mangling applies to top-level code.

Bring the top-level contents of your module into other modules or your interactive environment by using import and from, as we discussed in section 5.1. For example:

```
import guitar
```

or

```
from guitar import Guitar, Pickup, ElectricGuitar
```

The **guitar** module itself imports from two other modules.

```
from abc import ABC, abstractmethod
from enum import Enum
```

Put the `import` statements at the top of the module file after the documentation. List the standard modules first, then the third-party modules, followed by your own modules. Sort the `import` statements by module name within each group.

You should start a module with a one-line summary docstring, though I recommend you include an extended summary after that. Also, consider including **Notes**, **References**, and **Examples** sections.

7.18 Summary

In this chapter, we learned how to create new kinds of objects with Python `class` definitions. A class brings together data in the form of instance and class variables with functions that operate on them. These functions can be instance methods operating on objects created by the class or class and static methods that work across all instances.

Class inheritance is a crucial feature of Python's support for object-oriented programming. Inheritance gives us the ability to reuse code within a logical hierarchy of objects while specializing those objects with their own data and operations.

Commenting and documenting Python code is a critical activity so that other developers can understand and use what you created. It also allows you to go back and fix or improve the code when you may not remember how it works.

We finish the first part of this book in the next chapter, where we learn how to read and write information to and from files.

8

Working with Files

Follow your own path, and let people talk.

—Dante Alighieri

Files provide persistent storage of information. If you want data to be available for longer than the length of time it takes your app to run, perhaps even years, you store that data in files. A physical storage device might be a hard disk drive (HDD), solid-state drive (SSD), or a flash drive in today's phones, laptops, and servers.

This book doesn't cover how the physical devices work or low-level software access to the hardware. Instead, in this chapter, we look at the abstraction of storing and retrieving information with files. We group files into *directories* or *folders*, which may be nested. The files and the folders constitute the *file system.* The operating system manages the file system and is one of the OS's most essential roles.

Topics covered in this chapter

8.1 Paths and the file system

The text string that describes how to get from the top of the file system to the exact file we want is called the *pathname*. Think of it as a set of directions to navigate to the information we want.

The tree in Figure 8.1 shows a subset of the folders and files I used to create this book.

Figure 8.1: A directory tree with folders and files

The full path to the file `Working-with-Files.html` on Linux and macOS has the pathname

`/src/part-I/Working-with-Files.html`

On Microsoft Windows, we would have a drive letter like "C:", and we use backslashes instead of slashes. The pathname's case is not significant on Windows, but it is on the other operating systems.

`C:\src\part-I\Working-with-Files.html`

I will use slashes, and most Python routines that work with pathnames will work with either on Windows. You may find it convenient to use a raw string as in section.4.1 with a Windows pathname. Raw strings reduce the number of characters you must escape.

```
"C:\\src\\part-I\\Working-with-Files.html"
```

```
'C:\\src\\part-I\\Working-with-Files.html'
```

```
r"C:\src\part-I\Working-with-Files.html"
```

```
'C:\\src\\part-I\\Working-with-Files.html'
```

The **/** directory is the *root directory*.

We use the **os** package and the **os.path** module to manipulate pathnames. "**os**" stands for "**o**perating **s**ystem."

```
import os
```

```
pn = "/src/part-I/Working-with-Files.html"
```

```
os.path.split(pn)
```

```
('/src/part-I', 'Working-with-Files.html')
```

The second item in the tuple is the *basename*.

```
os.path.basename(pn)
```

```
'Working-with-Files.html'
```

Get the extension by splitting the filepath or the basename with *splitext*.

```
os.path.splitext(pn)
```

```
('/src/part-I/Working-with-Files', '.html')
```

```
os.path.splitext(os.path.basename(pn))
```

```
('Working-with-Files', '.html')
```

On Windows, use *splitdrive* to get the drive letter and the following colon.

```
os.path.splitdrive("C:\\src\\part-I\\Working-with-Files.html")
```

```
('C:', '\\src\\part-I\\Working-with-Files.html')
```

Operating systems keep track of a lot of data about files, and **os.path** provides ways of getting that. Figure 8.2 shows the file information that Windows 10 had for the source file for this chapter at one point during in its development:

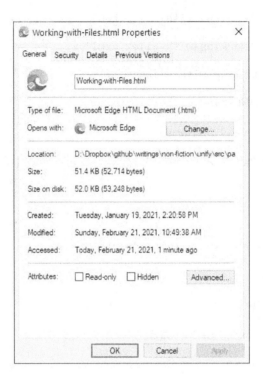

Figure 8.2: Windows Explorer file information

The *getctime*, *getmtime*, and *getatime* functions give you the file creation time, last modification time, and last access time.

```
mod_time = os.path.getmtime("src/part-I/Working-with-Files.html")
mod_time
```

```
1628111739.0
```

This result is the time someone last modified the file, and it is the number of seconds since the *epoch*.

The epoch is a peculiar computer system concept, with second 0 being midnight on January 1, 1970, for many operating systems. We can make more sense of mod_time if we use *gmtime* and *asctime*.

```
import time

time.gmtime(mod_time)

time.struct_time(tm_year=2021, tm_mon=8, tm_mday=4, tm_hour=21,
tm_min=15, tm_sec=39, tm_wday=2, tm_yday=216, tm_isdst=0)

time.asctime(time.gmtime(mod_time))

'Wed Aug  4 21:15:39 2021'
```

The file creation, modification, and access times, along with permission data stating who can do what to the file, are collectively part of the file's *metadata*.

To get the number of bytes in the file, use *getsize*.

```
f'{os.path.getsize("src/part-I/Working-with-Files.html"):,} bytes'

'53,449 bytes'
```

I used an f-string to format the output nicely.

Most of the **os.path** functions raise **OSError** if you try to access a file or path that does not exist.

```
os.path.getsize("src/part-I/Working-with-Flies.html")

FileNotFoundError: The system cannot find the file specified:
                  'src/part-I/Working-with-Flies.html'
```

Use *exists* to test if a file or directory is present.

```
os.path.exists("src/part-I/Working-with-Files.html")

True

os.path.exists("src/part-I")

True
```

isfile tests if a file exists and is not a directory.

```
os.path.isfile("src/part-I/Working-with-Files.html")

True

os.path.isfile("src/part-I/Working-with-Flies.html")

False
```

```
os.path.isfile("src/part-I")
```

```
False
```

To see if a path points to a directory or folder, use *isdir*.

```
os.path.isdir("src/part-I/Working-with-Files.html")
```

```
False
```

```
os.path.isdir("src/part-I")
```

```
True
```

Exercise 8.1

Write a function that takes two file pathnames, validates that they point to existing files, and prints the name of the most recently modified file.

8.2 Moving around the file system

If you have used a command line in Windows, macOS, or Linux, you know about the idea of the "current working directory" and how to move to another directory. Python has the same concept and functionality.

Figure 8.3 is another partial view of the directory tree for the files that make up this book.

Figure 8.3: A directory tree showing the current directory

If I am in the **src** directory, I can issue pwd on the operating system command line and see something like

```
Path
----
C:\src
```

on Windows, or

    /src

on macOS or Linux. "pwd" is an abbreviation for "**p**rint **w**orking **d**irectory." The corresponding Python function is *getcwd*, which is short for "**get** the **c**urrent **w**orking **d**irectory,"

```
import os

os.getcwd()
```

    'C:\\src'

I issue cd code on the command line to change to the code subdirectory. In Python, the function is *chdir*.

```
os.chdir("code")
os.getcwd()
```

    'C:\\src\\code'

```
📁 /
├── 📁 bin
├── 📁 final
│   └── 📁 images
└── 📁 src
    ├── 📁 │code│
    ├── 📁 images
    └── 📁 frontmatter
```

Figure 8.4: A directory tree showing the new current directory

A shorthand for the current directory is ".". Use ".." to move to the directory above the current one, the *parent directory*. Combine this with other directory names to "go up, over, and down." Here, we use a *relative directory path*.

```
os.chdir("../../final/images")
os.getcwd()
```

    'C:\\final\\images'

Use an *absolute path* to go to a fully specified directory.

```
os.chdir("/src/frontmatter")
os.getcwd()
```

    'C:\\src\\frontmatter'

8.3 Creating and removing directories

mkdir creates a single directory below an existing parent. If I am in **src**, then

```
os.mkdir("cookie-recipes")
```

creates the **src/cookie-recipes** directory. Use *rmdir* to remove it.

```
os.rmdir("cookie-recipes")
```

Suppose we want to create two new directories under **src**, first the **recipes** subdirectory, and then **cookies** within that. You cannot do it with one call to *mkdir*:

```
os.mkdir("recipes/cookies")

FileNotFoundError: The system cannot find the path specified:
                  'recipes/cookies'
```

To use *mkdir*, you could make two calls:

```
os.mkdir("recipes")
os.mkdir("recipes/cookies")
```

Alternatively, use *makedirs*. It creates any necessary intermediate directories.

```
os.makedirs("recipes/cookies")
```

Exercise 8.2

Now that you have created **recipes/cookies**, what does

```
os.rmdir("recipes/cookies")
```

do?

8.4 Lists of files and folders

If I have the five file names

guitar_01.py,

guitar_02.py,

unipoly_01.py,

unipoly_02.py, and

unipoly_10.py,

then I can use *wildcard characters* to match some of the names. Here are some examples:

- The pattern "***.py**" matches all of them since the pattern means "the name may start with any characters but must end in ".**py**"."

- The pattern "***01***" matches **guitar_01.py** and **unipoly_01.py**.

- "**unipoly_0?.py**" matches **unipoly_01.py** and **unipoly_02.py**. The "**?**" is a single character wildcard and matches anything in that one position.

Using patterns like these to match file names is called "globbing" and goes back to the early days of UNIX in the late 1960s and early 1970s. The **glob** module implements globbing in Python.

The module's two most important functions are *glob*, which returns a list of matches, and *iglob*, which produces an iterator that gives you the matches on demand. Unless you want a list, I recommend you use *iglob* in loops.

This is how I list all source HTML files in Part I of this book:

```
import glob

for html_file in glob.iglob("src/part-I/*.html"):
    html_file = os.path.normcase(html_file)
    print(html_file)
```

```
src\part-i\collecting-things-together.html
src\part-i\computing-and-calculating.html
src\part-i\defining-and-using-functions.html
src\part-i\organizing-objects-into-classes.html
```

```
src\part-i\part-i.html
src\part-i\stringing-you-along.html
src\part-i\working-with-expressions.html
src\part-i\working-with-files.html
```

normcase is a convenience function mainly useful on Windows. It makes the pathname lowercase and converts any forward slashes to backslashes on Windows. The function does nothing under other operating systems.

glob and *iglob* return files as well as directories. You can test which is which with *isfile* and *isdir*.

```
directories = []

for name in glob.iglob("src/**", recursive=True):
    if os.path.isdir(name):
        dir = os.path.normcase(name)
        directories.append(dir)

print(directories)
```

```
['src\\', 'src\\appendices', 'src\\art', 'src\\backmatter', 'src\
\code', 'src\\code\\linting', 'src\\code\\__pycache__', 'src\\css',
'src\\examples', 'src\\frontmatter', 'src\\images', 'src\\matplotlib',
'src\\matplotlib\\__pycache__', 'src\\part-i', 'src\\part-ii', 'src\
\part-iii', 'src\\preface', 'src\\preliminary']
```

The combination of the "**" wildcard and the recursive=True keyword argument tells *iglob* to descend into the directory tree to find matches.

Exercise 8.3

Code a comprehension that produces the same result as directories.

Exercise 8.4

Instead of using "**" and recursive=True, write a recursive function that finds all the directories beginning at some initial directory.

If you want to list the files in a directory, use *listdir* from **os**. This code counts the files in the **src/part-I** directory:

```
len(os.listdir("src/part-I"))
```

8

Exercise 8.5

Write a **DirectoryTree** class that accepts a directory pathname argument to *__init__* and holds the full subdirectory tree within the instance data. Use *os.walk* in your implementation. Learn about how to call the function by issuing `help(os.walk)`.

Exercise 8.6

How would you change **DirectoryTree** to lazily discover the directory tree instead of building it in *__init__*? Implement the *iter* method to allow users of your code to iterate over the tree.

Exercise 8.7

For the *__str__* method of **DirectoryTree**, draw a tree using the "+", "-", and "|" characters. Example:

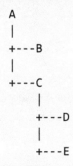

```
A
|
+---B
|
+---C
    |
    +---D
    |
    +---E
```

Check that this works on system directories like `C:\Program Files` on Windows or `/etc` on other operating systems.

8.5 Names and locations

If you wish to change the name of a directory, call *rename*.

```
os.rename("recipes/cookies", "recipes/moms-cookies")
os.rename("recipes", "moms-recipes")
```

We can also *rename* files. The file stays in the same directory but has a new name.

```
os.rename("test.py", "chapter-test.py")

os.rename("recipes/cookies/shortbread.docx",
          "recipes/cookies/moms-shortbread.docx")
```

To make a copy of a file's contents, you can duplicate it using a new name in its current or another directory, or use its current name in a different directory. The **shutil** module provides several functions to copy files. They take the same arguments.

```
import shutil

shutil.copyfile("t.py", "test-01.py")        # same dir, new name

'test-01.py'

shutil.copy("t.py", "/tests")                # different dir, same name

'tests\\t.py'

shutil.copy2("t.py", "/tests/test-01.py")  # different dir, different
name

'tests/test-01.py'
```

copyfile preserves no metadata, *copy* preserves the permission metadata, and *copy2* preserves the permission and time metadata. Each function returns the new pathname if the copy operation was successful. I usually use *copy2* unless I have a good reason to use one of the other two.

Exercise 8.8

What exception does Python raise if you try to copy a file onto itself?

Exercise 8.9

What does *copytree* from **shutil** do? Use `help(shutil.copytree)` after importing **shutil**.

One way to "move" a file is to copy it and then delete the original. To do it in one step, use *move* from **shutil** with the same arguments as the copy functions. You can rename the file as you move it.

move also works with directories if both arguments are directories. The directory in the first argument is moved to be a child of the directory in the second argument.

To delete a directory, its files, its subdirectories, and their content, use *shutil.rmtree*. To delete an empty directory, call *os.rmdir*. As we saw above, to delete a file, use *os.remove*.

Before you try to delete a file or directory, you should know that it exists or use a `try` / `except` block.

8.6 Types of files

Files fall into two main types: *text* and *binary*. We can think of a text file as something that is humanly readable. It could be plain text with words, numbers, and punctuation. It could be an HTML file with all those, plus markup that tells us what the information means and how we should display it. The text file could contain code that you read, create, update, and extend in a programming editor. The information is likely in a Unicode encoding.

We read and write text files as strings, with newline characters separating individual lines. We might read and write the file as a list of strings or one long string.

Examples of binary files include photos, images, and videos. We don't represent the data as characters in strings but as bytes. Each byte contains eight 0s and 1s. Instead of asking for the next line as you would for a text file, you ask for the next *n* bytes, for some $n > 0$. Once you have those bytes, you would process them appropriately. If you were looking for something and found it, you would be done. Otherwise, you would request more bytes.

Here's one way to remember the difference between the file types: we would use the string character `"2"` for the digit two for text. For binary, we would use the byte 00000010 for the number two.

We *open* a file to start using it and *close* it when we are finished. When we open the file, we tell Python the access mode we need. The modes are:

- **r** – open as a text file to read from it. This is the default mode.
- **rb** – open as a binary file to read from it.
- **w** – open as a text file to write to it.
- **wb** – open as a binary file to write to it.
- **r+** – open as a text file to read from and write to it.
- **rb+** – open as a binary file to read from and write to it.

We begin working at the beginning of the file when using the preceding modes.

If you open a file in the following modes, Python will position you at the end of the file's contents.

- **a** – open as a text file to append to it.
- **ab** – open as a binary file to append to it.
- **a+** – open as a text file to read from and append to it.
- **ab+** – open as a binary file to read from and append to it.

Python creates a new file for the modes that write or append if the specified file does not already exist. I mostly work with text files, so I frequently use **w** and **r**, and occasionally **a**.

8.7 Reading and writing files

The file **src/examples/sonnet-18.txt** contains the text of Shakespeare's Sonnet 18. The following code sequence opens the file for reading as text, reads the file into content, closes the file, and prints the string content.

```
input_file = open("src/examples/sonnet-18.txt", "r")
content = input_file.read()
input_file.close()

print(content)

Shall I compare thee to a summer's day?
Thou art more lovely and more temperate.
Rough winds do shake the darling buds of May,
And summer's lease hath all too short a date.
Sometime too hot the eye of heaven shines,
```

```
And often is his gold complexion dimmed;
And every fair from fair sometime declines,
By chance, or nature's changing course, untrimmed;
But thy eternal summer shall not fade,
Nor lose possession of that fair thou ow'st,
Nor shall death brag thou wand'rest in his shade,
When in eternal lines to Time thou grow'st.
So long as men can breathe, or eyes can see,
So long lives this, and this gives life to thee.
```

Close the file as soon as you are done reading it! This is not a coding requirement but helps you avoid errors with files you have processed but not closed.

Remember that the string is Unicode-encoded in Python and the file. We can process the string as we would any other. This code counts the number of times the letter 'a' appears in the sonnet.

```
count = 0
for c in content:
    if c == 'a':
        count += 1

count
```

37

Exercise 8.10

Write code that counts how many lines begin with the letter 'S'.

Exercise 8.11

Use *split* to break contents into a list of strings without trailing newlines.

Use *readlines* to read the file contents into a list of strings with trailing newlines.

```
input_file = open("src/examples/sonnet-18.txt", "r")
content = input_file.readlines()
input_file.close()
```

```
print(content)
```

["Shall I compare thee to a summer's day?\n", 'Thou art more lovely and more temperate.\n', 'Rough winds do shake the darling buds of May,\n', "And summer's lease hath all too short a date.\n", 'Sometime too hot the eye of heaven shines,\n', 'And often is his gold complexion dimmed;\n', 'And every fair from fair sometime declines,\n', "By chance, or nature's changing course, untrimmed;\n", 'But thy eternal summer shall not fade,\n', "Nor lose possession of that fair thou ow'st,\n", "Nor shall death brag thou wand'rest in his shade,\n", "When in eternal lines to Time thou grow'st.\n", 'So long as men can breathe, or eyes can see,\n', 'So long lives this, and this gives life to thee.']

The *readline* function reads the text file one line at a time. The length of each line successfully read from the file is at least one, since the line has a trailing newline or is the last non-empty line in the file. If the line is empty, we have reached the file's end.

The following code reads the file line by line and then prints the line in uppercase, preceded by its line number.

```python
input_file = open("src/examples/sonnet-18.txt", "r")

line_number = 0
line = input_file.readline()

while line:
    line_number += 1
    print(f"{line_number:>2}  {line.upper()}", end="")
    line = input_file.readline()  # Don't forget this!

input_file.close()
```

```
 1   SHALL I COMPARE THEE TO A SUMMER'S DAY?
 2   THOU ART MORE LOVELY AND MORE TEMPERATE.
 3   ROUGH WINDS DO SHAKE THE DARLING BUDS OF MAY,
 4   AND SUMMER'S LEASE HATH ALL TOO SHORT A DATE.
 5   SOMETIME TOO HOT THE EYE OF HEAVEN SHINES,
 6   AND OFTEN IS HIS GOLD COMPLEXION DIMMED;
 7   AND EVERY FAIR FROM FAIR SOMETIME DECLINES,
```

```
 8  BY CHANCE, OR NATURE'S CHANGING COURSE, UNTRIMMED;
 9  BUT THY ETERNAL SUMMER SHALL NOT FADE,
10  NOR LOSE POSSESSION OF THAT FAIR THOU OW'ST,
11  NOR SHALL DEATH BRAG THOU WAND'REST IN HIS SHADE,
12  WHEN IN ETERNAL LINES TO TIME THOU GROW'ST.
13  SO LONG AS MEN CAN BREATHE, OR EYES CAN SEE,
14  SO LONG LIVES THIS, AND THIS GIVES LIFE TO THEE.
```

Exercise 8.12

Explain what {line_number:>2} does in the call to *print*. How would the printing behavior change if I omitted end=""?

Exercise 8.13

You know that you can use import to read a Python module and make its contents available to you. You can also *read* a text file of code and call *exec* to run the string result. Research the *exec* function and compare it with import. Are there any security concerns?

The behavior of *write* for a string and *writelines* for a list of strings complements that of their *read* counterparts. Neither function automatically appends a newline character '\n' to a string. Remember to use **a** when you open the file if you want to write to the end of it.

Exercise 8.14

Why don't we need a *writeline* function?

Exercise 8.15

Code a function, *my_file_copy*, with two file pathname parameters: source_file and destination_file. After checking that source_file exists, copy its contents into destination_file using *read* and *write*. Be sure to check for exceptions.

In addition to the `sep` and `end` optional arguments we saw in section 6.5, the *print* function has another called `file`. You can use it instead of *write* for text files. It will automatically append a newline unless you change that to something else with `end`.

```python
output_file = open("src/examples/print-test.txt", "w")

print("First line", file=output_file)
print("Second line", file=output_file)

output_file.close()
```

We also use *read* for binary files. If we include a positive **int** argument for *read*, it is the number of bytes to read from a binary file. For a text file, it is the number of characters to read. Remember that a Unicode character can use more than one byte!

```python
input_file = open("src/examples/sonnet-18.txt", "rb")
content = input_file.read(39)
input_file.close()

content
```

```
b"Shall I compare thee to a summer's day?"
```

The b preceding the value of `content` indicates that we have a byte string, not a **str**.

```python
type(content)
```

```
bytes
```

Python tries to print the byte string using ASCII. If you look at an individual byte, Python shows it as an integer.

```python
content[0]
```

```
83
```

```python
bin(content[0])   # binary
```

```
'0b1010011'
```

```python
hex(content[0])   # hexadecimal
```

```
'0x53'
```

Given a Unicode string, we can encode it as a binary string. We can then use the binary string and *write* it to a binary file.

```
s = "Sîne klâwen æðelen ᛉᛘ ᚺᛈ ᚼᚱ ᚲᚻ ᛘ"

t = s.encode("utf-8")
t
```

```
b'S\xc3\x83\xc2\xaene kl\xc3\x83\xc2\xa2wen
\xc3\x83\xc2\xa6\xc3\x83\xc2\xb0elen
\xc3\xa1\xc5\xa1\xc2\xb7\xc3\xa1\xe2\x80\xba\xe2\x80\x93
\xc3\xa1\xc5\xa1\xc2\xbb\xc3\xa1\xc5\xa1\xc2\xb9
\xc3\xa1\xe2\x80\xba\xc2\xa6\xc3\xa1\xe2\x80\xba\xc5\xa1
\xc3\xa1\xc5\xa1\xc2\xb3\xc3\xa1\xc5\xa1\xc2\xa2
\xc3\xa1\xe2\x80\xba\xe2\x80\x94'
```

To go in the opposite direction from a byte string to a string, use *decode*.

```
t.decode("utf-8")
```

```
'Sîne klâwen æðelen ᛉᛘ ᚺᛈ ᚼᚱ ᚲᚻ ᛘ'
```

Sîne klâwen æðelen
ᛉᛘ ᚺᛈ ᚼᚱ ᚲᚻ ᛘ

Figure 8.5: Example of Unicode text

A string and its encoded byte string are not equal objects.

```
s == t
```

```
False
```

To summarize the above, the two primary functions for getting information from and putting information into files are *read* and *write*. For text files, we have the additional *readline*, *readlines*, and *writelines* functions.

Python provides a simplified syntax for opening and closing files.

```
with open("src/examples/sonnet-18.txt", "r") as input_file:
    print(input_file.readline())
```

```
Shall I compare thee to a summer's day?
```

Within the body of the `with` statement, we use `input_file` as we normally would. Python automatically closes the file. Note how `with` is paired with the keyword `as`. If an exception occurs in the `with` code block, Python will automatically close the file.

Using `with` / *open* / `as` is considered more Pythonic than *open* / *close*.

Exercise 8.16

Code a function, *my_other_file_copy*, with two file pathname parameters: `source_file` and `destination_file`. After checking that `source_file` exists, copy its contents into `destination_file` using *read* and *write*. Use nested `with` / `as` blocks. Be sure to check for exceptions, though test what happens if you don't.

8.8 Saving and restoring data

We save the code we write in text file modules and later `import` it into other modules or our environment. How can we save data and reload it at another time? There are two straightforward options, one binary and one text.

8.8.1 Pickling

Pickling is a Python-specific way of writing some Python objects and collections to binary files. You can later read them back in and use them. The Python documentation lists the kinds of data we have covered here that can be pickled: [**PYL**]

- `None`, `True`, and `False`,
- integers, floating-point numbers, and complex numbers,
- strings and byte strings, and
- tuples, lists, sets, and dictionaries containing only "picklable" objects.

If you can pickle something, it is *picklable*. (Don't blame me, I didn't coin the term.)

Let's reinstate our guitar dictionary from section 3.9.3.

```
guitars = {
    "Fender Stratocaster": 1954,
    "Fender Telecaster": 1950,
    "Gibson Les Paul": 1952,
    "Gibson Flying V": 1958,
```

```
    "Ibanez RG": 1987
    }
```

Now we import the **pickle** module, open a binary file for writing, and *dump* the dictionary.

```
import pickle

with open("work/pickle_example", "wb") as pickle_file:
    pickle.dump(guitars, pickle_file)
```

Remember that with / as automatically closes `pickle_file`.

At this point, we can go about our business. A few months later, we want to restore the guitar data. We open the binary file for reading and *load* the dictionary.

```
with open("work/pickle_example", "rb") as pickle_file:
    new_guitars = pickle.load(pickle_file)
```

The data we loaded looks correct:

```
print(new_guitars)
```

```
{'Fender Stratocaster': 1954, 'Fender Telecaster': 1950, 'Gibson Les
Paul': 1952, 'Gibson Flying V': 1958, 'Ibanez RG': 1987}
```

For this example, I still have `guitars`, and we see that the old dictionary is the same as the new.

```
guitars == new_guitars
```

```
True
```

Pickling works for the basic types. Here we pickled a dictionary, strings, and integers.

What about more complicated types or instances of classes we build ourselves? We cannot use **pickle** if there are class variables. If the instance variables are picklable, Python can automatically *dump* and *load* class instances. Hence the class instances are picklable.

Exercise 8.17

Show that you can pickle fractions made with **Fraction**.

Let's show this with a polynomial from the last chapter. We don't need to write any new code to make **pickle** work.

```
t = UniPoly(1, 't', 1)
p = (t**3 - 7)**5
p

t**15 - 35*t**12 + 490*t**9 - 3430*t**6 + 12005*t**3 - 16807
```

Now we *dump* and *load* the polynomials.

```
with open("work/pickle_example", "wb") as pickle_file:
    pickle.dump(p, pickle_file)

with open("work/pickle_example", "rb") as pickle_file:
    new_p = pickle.load(pickle_file)
```

They are equal:

```
p == new_p

True
```

The **dill** module extends **pickle** and handles class variables. [DIL] Like **pickle**, **dill** is Python-specific. The files they create may not work with Python versions significantly different from your own.

8.8.2 JSON

JSON is a standardized text-based interchange format created as part of the JavaScript programming language. "JSON" stands for "JavaScript Object Notation." There is nothing special that ties it to JavaScript, and Python has good support for the format. [JSN]

JSON matches up well with Python data types, and Python has four primary functions for working with it.

The **src/examples/example.json** file contains an extended subset of my current Visual Studio Code settings file. JSON files should have the extension ".json". Let's read the example file with *load* and see what Python does with it.

```
import json

with open("src/examples/example.json", 'r') as input_file:
    json_ = json.load(input_file)
```

The first thing to note is that Python created a dictionary.

```
type(json_)
```

```
dict
```

```
json_
```

```
{'editor.columnSelection': False,
 'editor.detectIndentation': True,
 'editor.autoClosingQuotes': 'never',
 'nothing': None,
 'pi.approximation': 3.14159,
 'python.formatting.autopep8Args': ['--max-line-length', '100'],
 'githubIssues.queries': [{'label': 'My Issues', 'query': 'default'}],
 'debug.console.fontSize': 12,
 'files.associations': {'.ncx': 'xml', '.opf': 'xml'}}
```

The keys are unsorted, and the values are True, False, None, integers, floating-point numbers, strings, lists, and dictionaries. The values that are collections can also hold objects of all these types.

The *load* function does two things: it reads the file into a string and decodes it into a Python object. To do the second yourself, use *loads*. The "s" is for "**string**."

If you have a Python object, use *dumps* to create a JSON-formatted string.

```
json_string = json.dumps(json_, indent=4, sort_keys=True)
print(json_string)
```

```
{
    "debug.console.fontSize": 12,
    "editor.autoClosingQuotes": "never",
    "editor.columnSelection": false,
    "editor.detectIndentation": true,
    "files.associations": {
        ".ncx": "xml",
        ".opf": "xml"
    },
    "githubIssues.queries": [
        {
            "label": "My Issues",
            "query": "default"
        }
    ],
```

```
    "nothing": null,
    "pi.approximation": 3.14159,
    "python.formatting.autopep8Args": [
        "--max-line-length",
        "100"
    ]
}
```

I used the indent and sort_keys optional arguments to make the JSON expression more readable.

Exercise 8.18

Python True, False, and None correspond to which JSON values?

Now that I have the data in JSON format, I can *write* it to a text file in the usual way. I can also *dump* it directly into a file.

```
with open("work/new_example.json", 'w') as output_file:
    json.dump(json_, output_file, indent=4, sort_keys=True)
```

8.9 Summary

We use files to store data in information for later reuse. In this chapter, we looked at Python functions for moving around and examining the file system controlled by the computer's operating system. We learned that files come in two varieties, text and binary. We saw how to access the files and read and write to them for each variety.

Other than pickling and using JSON, you can use a database to save and retrieve information, though we do not cover that here. We explore working with datasets with **pandas** in *Chapter 14, Analyzing Data*.

We have now completed Part I and the fundamentals of Python. We now examine and implement classical and quantum algorithms and circuits in Part II.

PART II

Algorithms and Circuits

9

Understanding
Gates and Circuits

*Logical consequences are the scarecrows of fools and
the beacons of wise men.*

—Thomas Huxley, *Science and Culture and Other Essays*

Classical computers use logical *gates* to manipulate bits. Using them, we assemble *circuits* to implement more complicated processes like addition and multiplication. Eventually, we get all the software that runs on computers everywhere.

Quantum computers use qubits to significantly extend the power of bits, as we saw in section 1.11. We assemble these into quantum circuits to implement algorithms.

There is a strong connection between classical and quantum computing, and a *quantum computing system* is a classical computing system extended with one or more quantum devices. These devices are the physical implementations of qubits and the software and hardware that control them.

This chapter examines bits and qubits, gates that operate upon them, and how we assemble them into circuits.

Topics covered in this chapter

9.1 The software stack

Software runs on hardware, and you use both to build applications and solve problems. This is true whether you are working entirely on a laptop or access computing resources on the cloud. Software is not monolithic. It begins with a layer very close to the hardware. On top of that, we implement abstractions and higher-level features to make it easier to code and develop solutions.

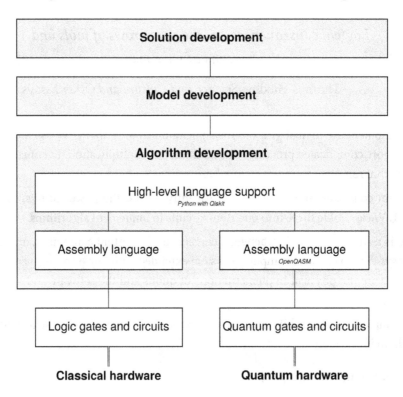

Figure 9.1: Classical and quantum software stack

We have gates and circuits within the hardware and the software layers. These are the most primitive operations, and we make compositions of them into useful processes.

Assembly languages allow us to code gates into custom circuits. They let us control the underlying hardware and, in the classical case, access and use system memory. OpenQASM3 is an example of a quantum computing assembly language developed by IBM and others.

Coders and researchers have created dozens of high-level programming languages above classical computing assembly languages. These include Python, C, C++, Swift, Java, JavaScript, PHP, Go, Rust, and R. Should we develop a new language for quantum computing or take advantage of existing languages?

Though it may seem geeky, it is *fun* to write new languages. It's less fun to support them and convince thousands of programmers to learn yet another way of creating code. With **qiskit**, you take everything you know from Python and use the open source code developed by hundreds of quantum experts. That is the approach we take in this book.

With this hardware and software foundation, we design and implement algorithms. Domain-expert model developers create the computational building blocks for solutions and applications using algorithms. From there, we construct apps for consumers and workflows for industry use cases.

Classical software development is decades ahead of its quantum counterpart. In the next section, I show how to code higher-level functions such as addition using logic gates and circuits. In Python and other languages, addition itself is low-level compared to other language features and the code in thousands of libraries. Nevertheless, we can then analogously examine quantum gates and circuits and begin our trek toward algorithms.

9.2 Boolean operations and bit logic gates

As we saw in section 2.5, there is a direct correspondence between Booleans and bits, equating True to 1 and False to 0. I first learned about Booleans and their operations in a logic course in my college Philosophy department. [MOL] You can also take a mathematical approach and look at Boolean algebra. Bits and circuits are fundamental to electrical engineering and computer science. Perceived distinctions among these approaches can be misleading because we talk about the same ideas using different words.

In this section, I work with bits and use their associated gate language.

I cover logic gates and circuits in Chapter 2 of *Dancing with Qubits.* [D.W.Q]

9.2.1 1-bit gates

The simplest 1-bit gate is **id**, which leaves each of 0 and 1 alone. It is the *identity* gate. A much more useful gate is **not**, which maps 0 to 1 and 1 to 0.

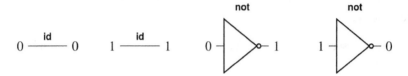

Figure 9.2: 1-bit logical "id" and "not" gates

The implementation of **id** is trivial because there is nothing to do. **not** is more subtle in Python.

```
not 0

True

not 1

False
```

Though we started with objects that looked like bits, we ended up with Booleans. This happens because of Python's extended notion of "trueness" and "falseness." We saw this when we discussed __bool__ in section 7.4.2. We can fix this by defining

```
def bit_not(bit):
    return 0 if bit else 1

bit_not(0)

1

bit_not(1)

0
```

Python does provide a bitwise **not** operation "~" that flips every bit in an **int**.

```
~0

-1
```

```
~1

-2
```

You probably did not expect those results! Python uses a "two's complement" representation for negative integers. When I write a binary number like 101101100, the "high-order bit" is the left-most. When I use two's complement, the number is negative if the high-order bit is 1, non-negative otherwise.

Because of Python's behavior, when I mention **not** for a bit, I mean the function we implemented as *bit_not*. We call this gate a "bit flip" because it "flips" 0 to 1 and vice versa.

Exercise 9.1

By examining these results:

```
[~5, ~2, ~1, ~0, ~1, ~2, ~5]

[-6, -3, -2, -1, -2, -3, -6]
```

explain how Python represents small integers. What is a formula for ~x, for x an `int`?

We define two additional gates for a single bit: **0**, which takes both 0 and 1 to 0; and **1**, which takes both 0 and 1 to 1.

Exercise 9.2

Show that **id**, **not**, **0**, and **1** are the only possible 1-bit gates.

9.2.2 2-bit gates

In section 2.5.2, we saw the Boolean "truth tables" for **and**, **or**, and exclusive-or "^". A truth table displays all the possible Boolean or bit values and the results of applying an operation or gate to them.

To see the 2-bit gate truth tables, let's write a function with some simple formatting. The top line has the gate's name and the second bit's values. The first column shows the values of the first bit.

```
def print_truth_table(op, name):
    print(f"{name:4} | 0    1")
```

```
    print("-----+--------")
    for bit in (0, 1):
        print(f"{bit}    | {op(bit, 0)}    {op(bit, 1)}")
```

The truth table for **and** is:

```
print_truth_table(lambda x,y: x and y, "and")

and  | 0    1
-----+-------
0    | 0    0
1    | 0    1
```

This table says, for example, that for inputs of 0 and 0, the output of 0 and 0 is 0. Electrical engineers use a standardized set of symbols to represent gates. Figure 9.3 shows the "American" symbols for **and** gates with their two inputs and single outputs:

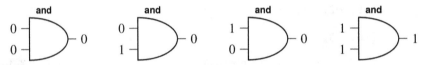

Figure 9.3: 2-bit logical "and" gate

The table for **or** is:

```
print_truth_table(lambda x,y: x or y, "or")

or   | 0    1
-----+-------
0    | 0    1
1    | 1    1
```

Figure 9.4: 2-bit logical "or" gate

We frequently write exclusive-or "^" as **xor**.

```
print_truth_table(lambda x,y: x ^ y, "xor")

xor  | 0    1
-----+-------
0    | 0    1
1    | 1    0
```

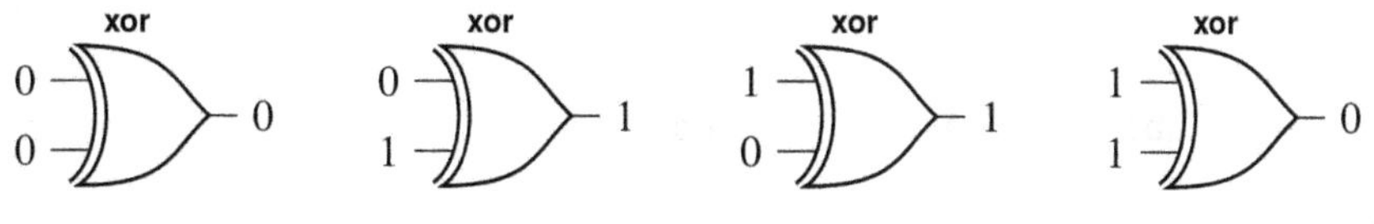

Figure 9.5: 2-bit logical "xor" gate

In addition to these three primary 2-bit gates, we define secondary gates by applying **not** to them. **nand** is **not and**:

```
print_truth_table(lambda x,y: bit_not(x and y), "nand")
```

```
nand | 0   1
-----+-------
0    | 1   1
1    | 1   0
```

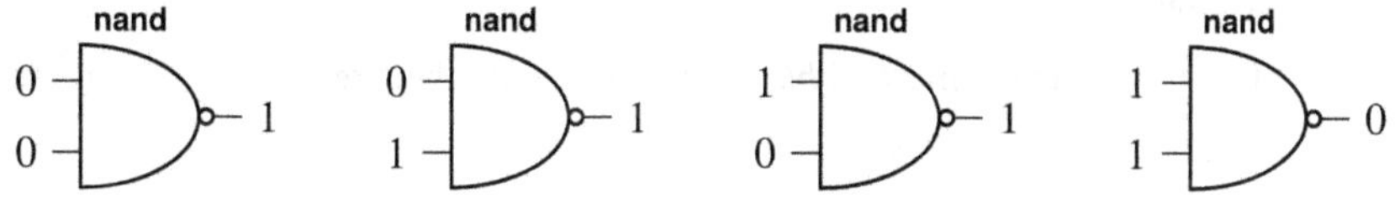

Figure 9.6: 2-bit logical "nand" gate

Exercise 9.3

Verify that the function

```
def bit_nand(bit_0, bit_1):
    return bit_not(bit_0 and bit_1)
```

has the same truth table as the one above.

nor is **not or**:

```
print_truth_table(lambda x,y: bit_not(x or y), "nor")
```

```
nor  | 0   1
-----+-------
0    | 1   0
1    | 0   0
```

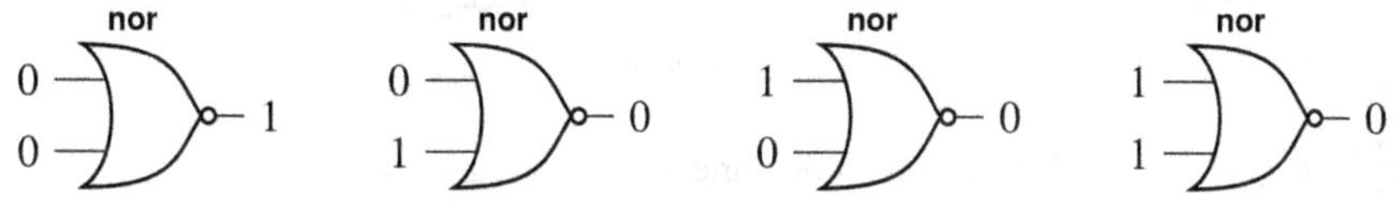

Figure 9.7: 2-bit logical "nor" gate

xnor is **not xor**:

```
print_truth_table(lambda x,y: bit_not(x ^ y), "xnor")

xnor | 0    1
-----+--------
0    | 1    0
1    | 0    1
```

Figure 9.8: 2-bit logical "xnor" gate

Exercise 9.4

Define functions *bit_nor* and *bit_xnor* and validate their results.

Exercise 9.5

Define the function

```
def mystery_gate(bit_0, bit_1):
    x = bit_nand(bit_0, bit_1)
    y = bit_nand(bit_0, x)
    z = bit_nand(bit_1, x)
    return bit_nand(y, z)
```

Display its truth table and determine which 2-bit gate this is. Note that we built it from four **nand** gates, which means that the *mystery_gate* is superfluous.

Exercise 9.6

Define the function

```
def another_mystery_gate(bit_0, bit_1):
    return abs(bit_0 - bit_1)
```

Display its truth table and determine which 2-bit gate this is.

The 2-bit controlled-**not** (**CNOT**) gate has the following characteristics:

- The first bit is the *control bit*, and the second bit is the *target bit*.

- Based on the control bit's value, we either apply **not** to the target bit or leave it alone. The first bit controls whether we apply **not** to the target bit.

- Specifically, if the control bit is 1, we apply **not** to the target bit. If the control bit is 0, we do nothing to the target bit.

- The gate has two output bits, the control bit and whatever we did to the target bit.

CNOT is the first 2-bit gate we have seen with two inputs and two outputs.

Exercise 9.7

Write the function *controlled_not* that implements the **CNOT** gate. It should return a tuple of two bits. What is its truth table if we use the second bit of the returned tuple as the result? Have we seen this gate before?

Exercise 9.8

What do we get if we use the output bits from *controlled_not* as the arguments to another call to *controlled_not*? That is, what is

```
controlled_not(*controlled_not(x, y))
```

for any bits x and y?

9.2.3 Reversible gates

A *reversible gate* **F** has two essential properties:

- The number of inputs to **F** is the same as the number of outputs.

- There is a gate **G** with the same number of inputs and outputs as **F** such that if you apply **G** to the outputs of **F**, you get the inputs to **F**.

In mathematical notation, $G = F^{-1}$ and $G \circ F = id$. We get **G** if we "run **F** backward." **G** is the *inverse* of **F**.

Exercise 9.9

Is **G** reversible? What is its inverse?

For the 1-bit gates, **id** and **not** are reversible and are their own inverses. **0** and **1** are not reversible.

None of the single output 2-bit gates are reversible. **CNOT** is reversible and is its own inverse.

To better show reversible gates, I now introduce a new way of drawing them in diagrams.

$$0 \longrightarrow \boxed{\textbf{not}} \longrightarrow 1$$

$$1 \longrightarrow \boxed{\textbf{not}} \longrightarrow 0$$

When the input to the **not** gate is 0, the result is 1. When the input is 1, the result is 0.

The **CNOT** gate has a control bit, shown with a black dot, and a target bit, shown with a plus sign in a circle. We connect them with a line.

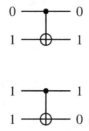

The example illustrates how a 0 or 1 in the control bit affects a 1 in the target bit. We can implement **CNOT** in Python with the function

```python
def bit_cnot(bit_0, bit_1):
    return (bit_0, bit_not(bit_1) if bit_0 else bit_1)
```

The **SWAP** gate interchanges the values of the bits. We draw the gate with two Xs connected by a line. For example,

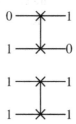

In the second example, the bit values are the same, and the swap is not apparent.

Exercise 9.10

Implement the function *bit_swap* in Python using a simultaneous assignment.

9.2.4 Gates on more than two bits

We can extend many of the 2-bit gates to three or more inputs. The 3-bit **and** returns 1 if all its inputs are 1, and 0 otherwise. For example,

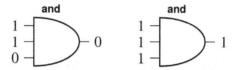

Though these are useful in electrical engineering, they are less so in coding.

The Toffoli or **CCNOT** gate is a reversible 3-bit gate. The first two bits are control bits, and the third is the target gate. If both the control bits are 1, we apply **not** to the target. Otherwise, we leave the target bit alone. We draw it as shown on the right.

Exercise 9.11

Show that **CCX** is reversible.

 The Fredkin or **CSWAP** gate is another reversible 3-bit gate. The first bit is the control bit, and its value determines whether we swap the values of the second and third bits. If the control bit is 1, we do the swap. Otherwise, we leave bits 2 and 3 alone. We draw it as shown on the left.

Exercise 9.12

Show that **CSWAP** is reversible.

9.3 Logic circuits

When you compose gates, you get a circuit. Figure 9.9 is the addition circuit from *Chapter 1, Doing the Things That Coders Do,* that adds two bits, *a* and *b*. The sum is the 2-bit number *cs,* with *c* being the *carry bit.* If *a* = *b* = 1, then *cs* is 10.

When I add 16 and 9 in base 10, I say, "add 6 to 9 to get 5, and *carry* a 1." The "carry bit" is the analogy in base 2.

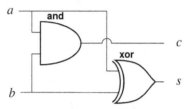

Figure 9.9: A simple logical addition circuit

Though the circuit has two inputs and two outputs, it is not reversible. If *cs* is 01, then *a* could have been 0 and *b* could have been 1, or the other way around. You can see this in our Python implementation of *bit_add*:

```
def bit_add(bit_0, bit_1):
    return (bit_0 and bit_1, bit_0 ^ bit_1)

for bit_0 in (0, 1):
    for bit_1 in (0, 1):
        sum_ = bit_add(bit_0, bit_1)
        print(f"{bit_0} + {bit_1} = {sum_[0]}{sum_[1]}")
```

```
0 + 0 = 00
0 + 1 = 01
1 + 0 = 01
1 + 1 = 10
```

I can use the same input bit for several gates in logic circuits. This is equivalent to using a function parameter more than once in the function definition.

We need to add more than single bits, so we should somehow combine several 1-bit adders and build larger circuits for addition. Let's begin by expressing our 1-bit adder in a circuit where **and** and **xor** are encapsulated in a new gate.

If a has n bits and b has m bits, the sum will have at most **maximum(n, m) + 1** bits, where the extra bit is for the carry. The carry bit might be 0. Expanding our diagram for a and b each having 2 bits, we get

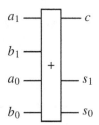

a_0 and b_0 are the right-most, low-order bits. For example, if a is 01, then a_0 is 1. Our task is to understand what is inside that big box with the "+". Can we break it down into calls to the 1-bit adder? Do we need to modify the 1-bit adder so it can be a component in larger circuits?

Our plan for 2-bit addition is this:

1. Add a_0 and b_0 to get s_0 and a temporary carry bit c_0.

2. Add a_1, b_1, and c_0 to get s_1 and carry bit c.

We add single bits in each step, but there is a problem in step 2: we need to add three single bits, not two. We modify the 1-bit adder to have three inputs:

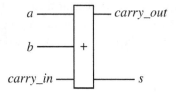

Now we can create our addition circuit for 2-bit numbers:

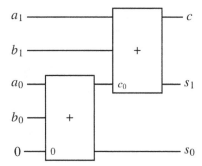

I labeled the carry bits.

If we write a 3-argument Python function *bit_add_with_carry* to implement the above, then we can define our earlier *bit_add* as

```
def bit_add(bit_0, bit_1):
    return bit_add_with_carry(bit_1, bit_2, 0)
```

Exercise 9.13

Implement *bit_add_with_carry* in Python using primitive 1-bit and 2-bit logic gates like **not**, **and**, and **xor**. **Do not** use "+".

Exercise 9.14

Implement the function *two_bit_add_with_carry* in Python to add two 2-bit numbers. You should use *bit_add_with_carry* and any primitive 1-bit and 2-bit logic gates you need. **Do not** use "+".

Exercise 9.15

Write functions in Python to implement multiplication of 1-bit and 2-bit numbers. Start by thinking about how you multiply integers by hand. Draw circuit diagrams as above. Discuss the role of the carry bit in your functions.

The next circuit has three **CNOT** gates, though we orient the middle one in the opposite direction to the outer two. It is a **reverse CNOT**. Since we make the circuit from reversible gates, it is also reversible.

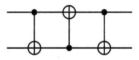

Exercise 9.16

Implement the function *bit_reverse_cnot* in Python.

Exercise 9.17

Using the Python functions *bit_cnot* and *bit_reverse_cnot* that you wrote, implement the circuit above.

Exercise 9.18

This circuit has the same behavior as a single 2-bit gate that we saw earlier. Which gate is it?

As you can see, we can implement logic circuits using Python functions.

9.4 Simplifying bit expressions

When we compose gates, we can often rearrange and simplify the combinations. If x and y are bits, then

- **not not** x = x
- x **and** 1 = 1
- x **and** x = x
- **0**(x) = x **and** (**not** x) = 0
- x **and** y = y **and** x
- x **or** 0 = x
- x **or** x = x
- **1**(x) = x **or** (**not** x) = 1
- x **or** y = y **or** x
- **not** (x **and** y) = (**not** x) **or** (**not** y)
- **not** (x **or** y) = (**not** x) **and** (**not** y)

The last two are known as *De Morgan's Laws* after British mathematician Augustus De Morgan. With a third bit z we also have:

- x **and** (y **and** z) = (x **and** y) **and** z
- x **or** (y **or** z) = (x **or** y) **or** z
- x **and** (y **or** z) = (x **and** y) **or** (x **and** z)
- x **or** (y **and** z) = (x **or** y) **and** (x **or** z)

Exercise 9.19

Which of these relationships hold for **xor, nand, nor, xnor**, and their combinations?

These rules are useful in Python conditionals. For example, the statement

```
if x != y and z != w then:
```

is the same as

```
if not (x == y or z == w) then:
```

Exercise 9.20

Replace x < y with an expression involving not. Do the same for "<=", ">", and ">=".

9.5 Universality for bit gates

Do we need all these gates? Is there some subset of the 1- and 2-bit gates that can generate all the others? Let's begin with a **nand** gate and see what we can build from there.

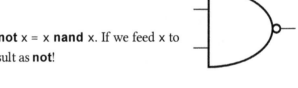

The first new construction is that **not** x = x **nand** x. If we feed x to both inputs of **nand**, we get the same result as **not**!

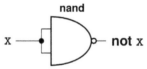

The black dot "●" reuses the bit's value so we can use it as two inputs.

This circuit shows that if we have **nand**, we don't need **not** as well. Though we may use **not** in circuits, it is superfluous.

The original set of 1- and 2-bit gates was

not and or xor nand nor xnor .

Since x **and** y = **not** (x **nand** y), the gates we still need are

or xor nand nor xnor .

We can construct **or** from three **nand** gates.

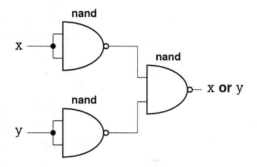

Since x **nor** y = **not** (x **nor** y), our list of required gates is now

xor nand xnor .

We need four **nand** gates to create an **xor** gate:

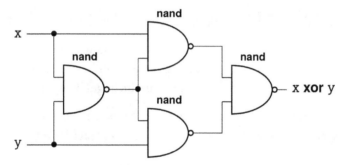

We are left with **nand**, from which we can generate all other logic gates. **nand** is therefore a *universal logic gate.* **nor** is also universal.

Exercise 9.21

The 1-bit adder circuit uses an **and** gate and an **xor** gate. Draw the circuit using only **nand** gates. How many **nand** gates do you need? How many do you need for a 2-bit adder circuit?

In practice, classical computer processors do not reduce all logic gates to millions of **nand** gates. Nevertheless, it gives you an appreciation for how much low-level computational machinery goes into computing something like 73569367 + 278356937563489! [ECA]

Exercise 9.22

Show how to construct the Fredkin **CSWAP** gate using only **and** and **xor**. Show how to build **and** and **not** (and therefore **nand**) from a **CSWAP**. Hint: let some of the inputs be 0 or 1.

Exercise 9.23

Show how to build **nand** from the Toffoli **CCNOT** gate.

These exercises show that the Fredkin **CSWAP** and the Toffoli **CCNOT** reversible logic gates are universal.

9.6 Quantum gates and operations

We've already encountered qubits twice in this book, first in the introductory section 1.11, and then again in section 5.8 where we explored quantum randomness. The remainder of this chapter expands those discussions and places them in the larger quantum computing context.

Though I'll discuss and use qubits here, I will not repeat the mathematical foundation of complex numbers, probability, and linear algebra from other texts. [DWQ]

9.6.1 Introduction to 1-qubit gates

We represent the value or *state* of a qubit as two complex numbers (a, b) such that

$$|a|^2 + |b|^2 = 1 .$$

For example, we might have $a = 1$ and $b = 0$, or the other way around. These are important cases, and we have special names for them. $(1, 0)$ is $|0\rangle$, pronounced "ket-0," and $(0, 1)$ is $|1\rangle$, pronounced "ket-1."

If you are familiar with linear algebra, $|0\rangle$ and $|1\rangle$ together are called the *computational basis* or *standard basis* for single qubits. In the linear algebra notation of column vectors, these are

$$|0\rangle = \begin{bmatrix} 1 \\ 0 \end{bmatrix} \qquad |1\rangle = \begin{bmatrix} 0 \\ 1 \end{bmatrix} .$$

Since a is a complex number, $a = a_1 + a_2 i$ with real part a_1 and imaginary part a_2. These are both real numbers. $|a| = \sqrt{(a_1{}^2 + a_2{}^2)}$, by definition. So, $|a|^2 = a_1{}^2 + a_2{}^2$. Similar statements hold for b.

We can write

$$(a, b) = a\,(1, 0) + b\,(0, 1) = a\,|0\rangle + b\,|1\rangle\,.$$

This equation for the quantum state is called *superposition*. a is the *amplitude* of $|0\rangle$, and b is the amplitude of $|1\rangle$. You will often see $a\,|0\rangle + b\,|1\rangle$ simplified to $|\psi\rangle$ in quantum computing texts and articles.

Though this notation may look strange to you, suppose I asked you to "walk 3 blocks east and 4 blocks north." Using ket notation, this would be:

$$3\,|east\rangle + 4\,|north\rangle\,.$$

As a result of this, you are "east and north simultaneously." This is the source of the statement people sometimes make when they say, "a qubit is 0 and 1 simultaneously!"

The amplitudes 3 and 4 don't work for a quantum state because $|3|^2 + |4|^3 \neq 1$.

Two quantum states $a_0\,|0\rangle + b_0\,|1\rangle$ and $a_1\,|0\rangle + b_1\,|1\rangle$ are *equivalent* if there exists a complex number c with $|c| = 1$ such that $a_0 = c\,a_1$ and $b_0 = c\,b_1$. We say that the quantum states "differ by the *global phase c*."

Now let's look at quantum gates on one qubit. I show how gates work with the notation

$$a\,|0\rangle + b\,|1\rangle \longmapsto c\,|0\rangle + d\,|1\rangle$$

The arrow "$\longmapsto$" means "maps to." For each gate that follows, verify that $|c|^2 + |d|^2 = 1$.

All quantum gates are reversible.

The id quantum gate

As in the bit case, the identity **id** gate does nothing. We use the symbol to the right for the **id** gate in circuit diagrams.

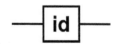

$$\textbf{id}: a\,|0\rangle + b\,|1\rangle \longmapsto a\,|0\rangle + b\,|1\rangle$$

The X quantum gate

The **X** gate generalizes the **not** bit gate to qubits.

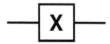

$$\textbf{X}: a\,|0\rangle + b\,|1\rangle \longmapsto b\,|0\rangle + a\,|1\rangle$$

We call **X** a bit flip because $\mathbf{X}(|0\rangle) = |1\rangle$ and $\mathbf{X}(|1\rangle) = |0\rangle$. It does more because quantum states are infinitely more varied than bit states. For example,

$$\mathbf{X}: (\sqrt{3}/2)\,|0\rangle - (1/2)\,|1\rangle \;\longmapsto\; (-1/2)\,|0\rangle + (\sqrt{3}/2)\,|1\rangle\,.$$

The last quantum state expression used the exact quantity $\sqrt{3}/2$. When we convert it to a **float** in Python, we get an approximation. Rounding errors will creep in as we use quantum gates and circuits.

```
import math

a = math.sqrt(3)/2
b = 1/2
(a, b)

(0.8660254037844386, 0.5)

abs(a)**2 + abs(b)**2

0.9999999999999999
```

You will learn to recognize several numbers that repeatedly occur in quantum calculations. In particular, the **float** and **complex** approximations to $\pm\sqrt{2}/2$ and $\pm\sqrt{2}/2\ i$ are

```
for x in (1, -1, 1j, -1j):
    print(x * math.sqrt(2)/2)

0.7071067811865476
-0.7071067811865476
0.7071067811865476j
-0.7071067811865476j
```

In the literature, you may sometimes see $\sqrt{2}/2$ written as the equivalent $1/\sqrt{2}$.

 I cover the **X** gate in section 7.6.1 of *Dancing with Qubits*. [DWQ]

The Z quantum gate

The **Z** gate changes the sign of the amplitude of $|1\rangle$.

$$\mathbf{Z}: a\,|0\rangle + b\,|1\rangle \;\longmapsto\; a\,|0\rangle - b\,|1\rangle$$

$$\boxed{\mathbf{Z}}$$

We call **Z** a *phase flip* for reasons coming from the physics of quantum mechanics.

When I write gates next to each other, it means that first I apply the gate on the right and then the gate on the left. For example,

$$\mathbf{XZ}(a\,|0\rangle + b\,|1\rangle) = \mathbf{X}(\mathbf{Z}(a\,|0\rangle + b\,|1\rangle)).$$

Exercise 9.24

Show that

$$\mathbf{XZ}(a\,|0\rangle + b\,|1\rangle) = -\,\mathbf{Z}(\mathbf{X}(a\,|0\rangle + b\,|1\rangle)).$$

I write this more simply as $\mathbf{XZ} = -\,\mathbf{ZX}$.

Exercise 9.25

Show that $\mathbf{XX} = \mathbf{ZZ} = \mathbf{id}$. The $\mathbf{X}$ and $\mathbf{Z}$ gates are each their own inverses.

We will use the gate and circuit facilities from **qiskit** in section 9.7.2, but in the meanwhile, we can define some Python functions to show the actions of gates on pairs of complex numbers a and b.

```
def qubit_id(a, b): return (a, b)

def qubit_X(a, b): return (b, a)

def qubit_Z(a, b): return (a, -b)
```

To show an application of $\mathbf{Z}$ after $\mathbf{X}$, I code the following:

```
ket_0 = (1, 0)
qubit_X(*ket_0)

(0, 1)

qubit_Z(*qubit_X(*ket_0))

(0, -1)
```

I encourage you to define similar functions for the other gates we see before using Qiskit.
[QIS]

I cover the **Z** gate in section 7.6.2 of *Dancing with Qubits*. [D.W.Q]

The Y quantum gate

The **Y** gate is the first 1-qubit gate we have seen that uses a non-real complex number.

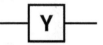

$$\mathbf{Y}: a\,|0\rangle + b\,|1\rangle \mapsto -\,b\,i\,|0\rangle + a\,i\,|1\rangle$$

Y is both a bit flip and a phase flip. We can rewrite this as

$$\mathbf{Y}: a\,|0\rangle + b\,|1\rangle \mapsto (-\,i)(b\,|0\rangle - a\,|1\rangle) = (-\,i)\,\mathbf{ZX}(a\,|0\rangle + b\,|1\rangle)$$

Since $|-\,i| = 1$, the quantum state we get by applying **Y** is equivalent to what we get by applying **X** and then **Z**. If we have **X** and **Z**, **Y** is superfluous.

 Exercise 9.26

Show that **XY** = − **YX**, **ZY** = − **YZ**, and **YY** = **id**.

 I cover the **Y** gate in section 7.6.3 of *Dancing with Qubits*. [D.W.Q]

The H quantum gate

The **H** gate, also known as the Hadamard gate after Jacques Hadamard, operates on $|0\rangle$ and $|1\rangle$ as

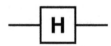

$$\mathbf{H}(|0\rangle) = (\sqrt{2}/2)(|0\rangle + |1\rangle) \quad \text{and} \quad \mathbf{H}(|1\rangle) = (\sqrt{2}/2)(|0\rangle - |1\rangle).$$

With a bit of algebra, we get

$$
\begin{aligned}
\mathbf{H}(a\,|0\rangle + b\,|1\rangle) \\
&= a\,\mathbf{H}(|0\rangle) + b\,\mathbf{H}(|1\rangle) \\
&= a\,(\sqrt{2}/2)\,(|0\rangle + |1\rangle) + b\,(\sqrt{2}/2)\,(|0\rangle - |1\rangle) \\
&= a\,(\sqrt{2}/2)\,|0\rangle + a\,(\sqrt{2}/2)\,|1\rangle + b\,(\sqrt{2}/2)\,|0\rangle - b\,(\sqrt{2}/2)\,|1\rangle \\
&= (\sqrt{2}/2)\,(a + b)\,|0\rangle + (\sqrt{2}/2)\,(a - b)\,|1\rangle.
\end{aligned}
$$

Exercise 9.27

A function f on a qubit is *linear* if $f(a\,|0\rangle + b\,|1\rangle) = a\,f(|0\rangle) + b\,f(|1\rangle)$ for complex numbers a and b. All 1-qubit quantum gates are linear.

Where did I use this above?

The square of the absolute value of the amplitude of $|0\rangle$ is

$$|(\sqrt{2}/2)\,(a + b)|^2 = (1/2)\,|a + b|^2$$

and that of $|1\rangle$ is

$$|(\sqrt{2}/2)\,(a - b)|^2 = (1/2)\,|a - b|^2 .$$

Exercise 9.28

Since $a = a_1 + a_2\,i$ and $b = b_1 + b_2\,i$,

$$|a + b|^2 = |(a_1 + b_1) + (a_2 + b_2)\,i|^2 .$$

This then equals $(a_1 + b_1)^2 + (a_2 + b_2)^2$. We also have $|a|^2 + |b|^2 = 1$.

Show that the sum of the squares of the amplitudes of $\mathsf{H}(a\,|0\rangle + b\,|1\rangle)$ equals 1.

I cover the **H** gate in section 7.6.5 of *Dancing with Qubits*. [D.W.Q]

9.6.2 Measurement

Many quantum calculations proceed in this way:

- Initialize each qubit to $|0\rangle$. You may think of this as setting it to the bit 0.
- If you want some qubits to start at "1", apply the **X** gate to them.
- Manipulate the qubits using 1- and 2-qubit gates. **You may not look at the qubits' quantum states while applying gates.**
- When you finish, *measure* the qubits. This is equivalent to "looking at" or "observing" them.
- Once measured, each qubit will be $|0\rangle$ or $|1\rangle$, which you may think of as 0 or 1.
- Interpret the 0s and 1s as numbers, bytes, strings, or whatever other data you are using.

As we saw in the last section with the **X**, **Y**, **Z**, and **H** gates, we can change a qubit's state to something else in a reversible way.

We can now interpret the a and b in the quantum state expression $a\,|0\rangle + b\,|1\rangle$. When we measure a qubit, the probability of getting $|0\rangle$ is $|a|^2$. The likelihood of getting $|1\rangle$ is $|a|^2$. While a and b are complex numbers, the squares of their absolute values are non-negative real numbers. There are only two options, $|0\rangle$ and $|1\rangle$, and the sum of the probabilities of our getting either must sum to 1. Though I refer to a or b as an amplitude, *probability amplitude* is the longer, more descriptive name.

If two quantum states $a_1\,|0\rangle + b_1\,|1\rangle$ and $a_2\,|0\rangle + b_2\,|1\rangle$ differ by a global phase c, then $a_1 = c\,a_2$ and $b_1 = c\,b_2$. Since $|c| = 1$, by definition, $|a_1|^2 = |c\,a_2|^2 = |c|^2\,|a_2|^2 = |a_2|^2$. A similar statement holds for b_1 and b_2. This means that *equivalent quantum states are indistinguishable when we measure.* Their probabilities of getting $|0\rangle$ or $|1\rangle$ are identical.

Measurement is not a gate and is not reversible. It is a one-way operation that "collapses" a quantum state to $|0\rangle$ or $|1\rangle$. We use the symbol to the right for measurement in circuit diagrams.

If a and b are the complex number amplitudes of a quantum state, we can roughly simulate measurement with pseudo-random numbers.

```
import random

random.seed(23)

def fake_measure(a, b):
    a_abs_squared = abs(a)**2

    # handle the endpoints
    if a_abs_squared == 0:
        return 1
    if a_abs_squared == 1:
        return 0

    # get a random number in a uniform distribution
    # that includes 0 and 1
    u = random.uniform(0.0, 1.0)

    return 0 if u <= a_abs_squared else 1
```

Now we "measure" the state $(\sqrt{3}/2)\,i\,|0\rangle - (1/2)\,|1\rangle$ 10,000 times.

```
a = complex(0, math.sqrt(3)/2)
a
```

```
0.8660254037844386j
```

```
b = -0.5
```

```
counts = [0, 0]
```

```
for _ in range(10000):
    counts[fake_measure(a, b)] += 1
```

```
for bit in (0, 1):
    print(f"Number of times we saw {bit}: {counts[bit]}")
```

```
Number of times we saw 0: 7485
Number of times we saw 1: 2515
```

Exercise 9.29

Did we see each value approximately the correct number of times?

Exercise 9.30

Why don't we use b in *fake_measure*? What validation tests should you code for the arguments? Why did I use `uniform(0.0, 1.0)` and not `random()`?

Please be very aware that this does not replace the real quantum randomness introduced in section 5.8.

We can express a superposition by writing a quantum state as $a\,|0\rangle + b\,|1\rangle$. If we say, "put the qubit in superposition," we usually mean "initialize the qubit to $|0\rangle$ or $|1\rangle$ and then apply the **H** gate." We call this a "balanced superposition" because the probabilities of getting $|0\rangle$ or $|1\rangle$ are equal.

9.6.3 This is getting complicated!

I now pause for a few moments to acknowledge that qubits and their quantum gates are more sophisticated than bits and their logic gates. Over time, both computing models will become second nature to you. You'll understand how to use them together to solve problems that classical techniques alone cannot tackle.

9.6.4 2-qubit gates

When we began talking about single qubits earlier in this chapter, we used 2-tuples to represent the pairs of complex numbers in the qubits' quantum states. We defined $(1, 0)$ to be $|0\rangle$ and $(0, 1)$ to be $|1\rangle$. We generalized this so that (a, b) was the sum $a\,|0\rangle + b\,|1\rangle$.

For two qubits with individual quantum states $a\,|0\rangle + b\,|1\rangle$ and $c\,|0\rangle + d\,|1\rangle$, we have a single quantum state when we work with them together. We apply the "tensor product" operation "$\otimes$":

$$(a\,|0\rangle + b\,|1\rangle) \otimes (c\,|0\rangle + d\,|1\rangle) = ac\,|00\rangle + ad\,|01\rangle + bc\,|10\rangle + bd\,|11\rangle$$

with

$$|00\rangle = (1, 0, 0, 0) = |0\rangle \otimes |0\rangle$$
$$|01\rangle = (0, 1, 0, 0) = |0\rangle \otimes |1\rangle$$
$$|10\rangle = (0, 0, 1, 0) = |1\rangle \otimes |0\rangle$$
$$|11\rangle = (0, 0, 0, 1) = |1\rangle \otimes |1\rangle$$

In the linear algebra notation of column vectors, these are

$$|00\rangle = \begin{bmatrix} 1 \\ 0 \\ 0 \\ 0 \end{bmatrix} \quad |01\rangle = \begin{bmatrix} 0 \\ 1 \\ 0 \\ 0 \end{bmatrix} \quad |10\rangle = \begin{bmatrix} 0 \\ 0 \\ 1 \\ 0 \end{bmatrix} \quad |11\rangle = \begin{bmatrix} 0 \\ 0 \\ 0 \\ 1 \end{bmatrix}.$$

Every time we use another qubit, we double the amount of information we can represent. For two qubits, this is $4 = 2 \times 2$.

Just as you knew about

adding, sum, and "+",
subtracting, difference, and "−",
multiplying, product, and "×",
dividing, quotient, and "/",

for numbers, you now know about

tensoring, tensor product, and "$\otimes$"

for qubits.

 I cover tensor products in section 8.1 of *Dancing with Qubits*. [DWQ]

The CNOT quantum gate

In the quantum case, **CNOT** is more correctly called the **controlled-X** or **CX** gate.

$$|0\rangle \longrightarrow\!\!\bullet\!\!\longrightarrow |0\rangle$$

$$a\,|0\rangle + b\,|1\rangle \longrightarrow\!\oplus\!\longrightarrow a\,|0\rangle + b\,|1\rangle$$

$$|1\rangle \longrightarrow\!\!\bullet\!\!\longrightarrow |1\rangle$$

$$a\,|0\rangle + b\,|1\rangle \longrightarrow\!\oplus\!\longrightarrow b\,|0\rangle + a\,|1\rangle$$

X is applied to the second qubit if the first qubit's quantum state is $|1\rangle$. If the state is $|0\rangle$, the second qubit's state is left unchanged.

The top gate application says

$$\textbf{CNOT}\,(|0\rangle \otimes (a\,|0\rangle + b\,|1\rangle)) = |0\rangle \otimes (a\,|0\rangle + b\,|1\rangle) = a\,|00\rangle + b\,|01\rangle\,.$$

The bottom says

$$\textbf{CNOT}\,(|1\rangle \otimes (a\,|0\rangle + b\,|1\rangle))$$
$$= |1\rangle \otimes \textbf{X}\,(a\,|0\rangle + b\,|1\rangle)$$
$$= |1\rangle \otimes (b\,|0\rangle + a\,|1\rangle)$$
$$= b\,|10\rangle + a\,|01\rangle\,.$$

 I cover the **CNOT** / **CX** gate in section 8.3.3 of *Dancing with Qubits*. [D.WQ]

The SWAP quantum gate

The **SWAP** interchanges the quantum states of two qubits.

$$a\,|0\rangle + b\,|1\rangle \longrightarrow\!\!\times\!\!\longrightarrow c\,|0\rangle + d\,|1\rangle$$

$$c\,|0\rangle + d\,|1\rangle \longrightarrow\!\!\times\!\!\longrightarrow a\,|0\rangle + b\,|1\rangle$$

It may surprise you, but you would violate the laws of physics if you could make a copy of a qubit's state. This is called "The No-cloning Theorem." You cannot do this in Python with actual qubits:

```
temp_state = qubit_1.quantum_state                # NO
qubit_1.quantum_state = qubit_2.quantum_state   # NO
qubit_2.quantum_state = temp_state                # NO
```

Instead, the **SWAP** gate interchanges the states simultaneously.

```
# This is not real Python code.

qubit_1.quantum_state, qubit_2.quantum_state \
    = qubit_2.quantum_state, qubit_1.quantum_state
```

If I want another qubit, wherever it is, to have my quantum state, I must transfer and destroy my copy simultaneously.

Exercise 9.31

Can you store a quantum state in classical memory or a database for later use?

Exercise 9.32

Show that

$$\text{SWAP} \left((a \,|0\rangle + b \,|1\rangle) \otimes (c \,|0\rangle + d \,|1\rangle) \right) = (c \,|0\rangle + d \,|1\rangle) \otimes (a \,|0\rangle + b \,|1\rangle).$$

I cover the **SWAP** gate in section 8.3.2 of *Dancing with Qubits*. [D.W.Q]

Other controlled gates

We can add a control qubit to the execution of any quantum gate, not just **X**. In the circuit in Figure 9.10, we have the state of the top qubit controlling whether we apply **X**, **Y**, **Z**, and **H** gates to other qubits. From left to right, we have **CX** (= **CNOT**), **CNOT** (= **CX**), **CY**, **CZ**, **CZ** (alternative form), and **CH**.

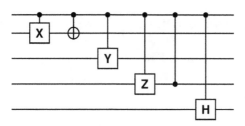

Figure 9.10: CX, CY, CZ, and CH quantum gates

The controls don't need to go from top to bottom. They can go in either direction and, as you can see, can jump across qubits.

Tensoring gates

Just as we can tensor two individual quantum states

$$|\varphi\rangle = a\,|0\rangle + b\,|1\rangle \text{ and } |\psi\rangle = c\,|0\rangle + d\,|1\rangle$$

to get a 2-qubit quantum state,

$$|\varphi\rangle \otimes |\psi\rangle = (a\,|0\rangle + b\,|1\rangle) \otimes (c\,|0\rangle + d\,|1\rangle) = ac\,|00\rangle + ad\,|01\rangle + bc\,|10\rangle + bd\,|11\rangle$$

we can tensor two 1-qubit gates.

If **A** and **B** are two 1-qubit gates, then

$$(\mathbf{A} \otimes \mathbf{B}(|\varphi\rangle \otimes |\psi\rangle) = (\mathbf{A}\,|\varphi\rangle) \otimes (\mathbf{B}\,|\psi\rangle)$$

is a 2-qubit gate.

For example, if **A** = **B** = **H**, the Hadamard gate, then

$$\mathbf{H}\,|0\rangle = (\sqrt{2}/2)\,|0\rangle + (\sqrt{2}/2)\,|1\rangle$$

and so

$$(H \otimes H)(|0\rangle \otimes |0\rangle) = (H |0\rangle) \otimes (H |0\rangle)$$
$$= ((\sqrt{2}/2) |0\rangle + (\sqrt{2}/2) |1\rangle) \otimes (\sqrt{2}/2) |0\rangle + (\sqrt{2}/2) |1\rangle$$
$$= (1/2) |00\rangle + (1/2) |01\rangle + (1/2) |10\rangle + (1/2) |11\rangle$$

Note that **H** ⊗ **H** is a 2-qubit gate while **HH** is a 1-qubit Hadamard gate applied twice.

We commonly write **H** ⊗ **H** as $H^{\otimes 2}$. When we tensor together n **H** gates for n qubits, we write it as $H^{\otimes n}$.

Exercise 9.33

For

$$|\varphi\rangle = a |0\rangle + b |1\rangle \text{ and } |\psi\rangle = c |0\rangle + d |1\rangle$$

compute **X** ⊗ **X**, **Z** ⊗ **Z**, and **id** ⊗ **H**.

9.6.5 Gates on more than two qubits

The Toffoli/**CCNOT**/**CCX** and the Fredkin/**CSWAP** 3-qubit gates are the direct analogs of their logic gate counterparts.

The Toffoli gate illustrates adding more than one control qubit to a target gate, which, in this case, is **CX.** We could continue building gates with three, four, or more control qubits.

The Fredkin gate shows that we can control 2-qubit gates and 1-qubit gates. Again, we can add additional control qubits to build, for example, a **CCCSWAP** gate.

Exercise 9.34

Draw diagrams for **CCCSWAP** and **CCCCCX** gates.

The standard basis for one qubit has two elements, $|0\rangle$ and $|1\rangle$. The standard basis for two qubits has four elements, $|00\rangle$, $|01\rangle$, $|10\rangle$, and $|11\rangle$. How many are in the standard basis for three qubits?

I said earlier that the amount of information in quantum states doubles each time you add a qubit. For three qubits, we can represent such a state as

$$a\,|000\rangle + b\,|001\rangle + c\,|010\rangle + d\,|011\rangle + e\,|100\rangle + f|101\rangle + g\,|110\rangle + h\,|111\rangle$$

with 8 elements in the basis. For four qubits, there are 16. For ten qubits, a quantum state holds 1,024 complex numbers. In each case, the sum of the squares of the absolute values of each amplitude must equal one.

Since we will quickly run out of letters for the amplitudes, you may see notation like

$$a_{000}\,|000\rangle + a_{001}\,|001\rangle + a_{010}\,|010\rangle + a_{011}\,|011\rangle +$$
$$a_{100}\,|100\rangle + a_{101}\,|101\rangle + a_{110}\,|110\rangle + a_{111}\,|111\rangle$$

instead of the above.

Exercise 9.35

How many basis elements (and therefore complex numbers representing data) are in a quantum state with 50 qubits? 300 qubits?

What are the estimates for the numbers of atoms in the Earth and the observable universe?

Exercise 9.36

What is $(a\,|0\rangle + b\,|1\rangle) \otimes (c\,|0\rangle + d\,|1\rangle) \otimes (e\,|0\rangle + f\,|1\rangle)$ expressed in terms of $|000\rangle$, $|001\rangle$, $|111\rangle$?

It may help to know that

$$(a|0\rangle + b|1\rangle) \otimes (c|0\rangle + d|1\rangle) \otimes (e|0\rangle + f|1\rangle)$$
$$= (a|0\rangle + b|1\rangle) \otimes \Big((c|0\rangle + d|1\rangle) \otimes (e|0\rangle + f|1\rangle)\Big).$$

9.6.6 1-qubit gates and the Bloch sphere

In section 1.11, I mentioned that we could show a qubit's quantum state on the surface of a sphere, the *Bloch sphere*, shown in Figure 9.11.

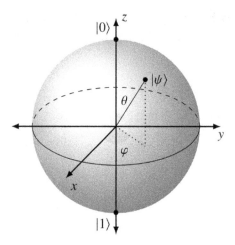

Figure 9.11: The Bloch sphere

The surface of a sphere is two-dimensional with real number coordinates, yet I've been showing quantum states as $a\,|0\rangle + b\,|1\rangle$. a and b are each complex numbers, so with the real and imaginary parts, we have four real dimensions. How do we go from four dimensions to two?

First, the condition $|a|^2 + |b|^2 = 1$ establishes a relationship between the possible values of a and b. This removes one degree of freedom, which we can loosely think of as a dimension.

Second, $a\,|0\rangle + b\,|1\rangle$ and $ac\,|0\rangle + bc\,|1\rangle$ are equivalent quantum states for complex c with $|c| = 1$. This removes another degree of freedom, and so a dimension.

Finally, we use trigonometry to move between $a\,|0\rangle + b\,|1\rangle$ and angles (θ, φ) on the surface of the sphere. φ is in the xy-plane and we measure from the positive x-axis. We measure θ from the positive z-axis. $0 \le \varphi < 2\pi$ and $0 \le \theta \le \pi$. [DWQ, Section 7.5]

$1\,|0\rangle + 0\,|1\rangle$ maps to the north pole, and $0\,|0\rangle + 1\,|1\rangle$ maps to the south pole. Any quantum state on the equator is in a balanced superposition.

A 1-qubit gate takes a quantum state on the Bloch sphere and moves it elsewhere on the sphere. This is true except for **id**, which does nothing.

Imagine holding a ball with your fingers at the opposite top and bottom points. If you spin the ball, you rotate it around the vertical axis connecting the poles. The **Z** gate rotates the Bloch sphere π radians (180°) around the vertical *z*-axis. This is halfway around the sphere, and θ is not changed. The angle φ is changed to an angle π radians away, which is the phase flip. If you rotate with **Z** again, you end up where you started, which is why **Z** is its own inverse.

The **X** gate rotates the Bloch sphere π radians around the horizontal *x*-axis. This flips |0⟩ and |1⟩.

The **Y** gate rotates the Bloch sphere π radians around the horizontal *y*-axis. This flips |0⟩ and |1⟩ and is a phase flip with φ.

The **H** gate requires some fancier geometric visualization on your part. If I draw the *xz*-plane, consider the diagonal line in the graph on the right.

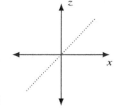

Now stand to the left of the sphere directly on the *y*-axis. Rotate the sphere π radians around the diagonal line. Equivalently, rotate π/2 radians around the *y*-axis and then π radians around the *z*-axis. That's what the **H** gate does.

The diagonal line is at a π/4 radians (45°) angle. From trigonometry, $\sin(\pi/4) = \cos(\pi/4) = \sqrt{2}/2$, which is why that number comes up when we look at the action of **H**.

Exercise 9.37

Where does **H** move |0⟩ on the sphere? |1⟩?

9.6.7 Additional 1-qubit gates

Now that we have connected 1-qubit gates to rotations on the Bloch sphere, I'll briefly introduce several additional gates.

The **RZ**$_\varphi$ gate changes a qubit's quantum state by a rotation of φ radians around the *z*-axis. φ can be any real number greater than or equal to 0 and less than 2π.

$$\mathbf{RZ}_\varphi\,(a\,|0\rangle + b\,|1\rangle) = a\,|0\rangle + (\cos(\varphi) + \sin(\varphi)\,i)\,b\,|1\rangle$$

Z = **RZ**$_\pi$. We have now gone from a few 1-qubit gates to **an infinite number of 1-qubit gates.**

Exercise 9.38

Show that $|a|^2 + |(\cos(\varphi) + \sin(\varphi)\, i)\, b|^2 = 1$.

The **S**, **S**†, **T**, and **T**† gates are the $\mathbf{RZ}_{\pi/2}$, $\mathbf{RZ}_{3\pi/2}$, $\mathbf{RZ}_{\pi/4}$, and $\mathbf{RZ}_{7\pi/4}$ gates.

We similarly define $\mathbf{RX}_\varphi$ and $\mathbf{RY}_\varphi$ as rotations of φ radians around the x-axis and y-axis, respectively. $\mathbf{X} = \mathbf{RX}_\pi$ and $\mathbf{Y} = \mathbf{RY}_\pi$.

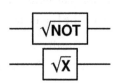

Exercise 9.39

The **H** gate is $\mathbf{RZ}_\psi\, \mathbf{RY}_\varphi$ for what values of ψ and φ? Here we apply $\mathbf{RY}_\varphi$ first and then $\mathbf{RZ}_\psi$. ("ψ" is the Greek letter "psi.")

The √**NOT** gate goes by several other names and symbols: √**X**, √¬, and **Sx**. Whatever you call it, when you apply the √**NOT** twice to a quantum state, you get a bit flip of the state.

√**NOT** operates on $|0\rangle$ and $|1\rangle$ by

$$|0\rangle \longmapsto (1/2)\left((1 + i)\,|0\rangle + (1 - i)\,|1\rangle\right)$$
$$|1\rangle \longmapsto (1/2)\left((1 - i)\,|0\rangle + (1 + i)\,|1\rangle\right)$$

Exercise 9.40

Using the formulas for the action of √**NOT** on $|0\rangle$ and $|1\rangle$, show that √**NOT**√**NOT** is a bit flip.

Exercise 9.41

Implement √**NOT** in Python

```
def qubit_SQRT_NOT(a, b):
    ...
```

for a quantum state $a\,|0\rangle + b\,|1\rangle$. It should return a 2-tuple.

I cover the √**NOT** gate in section 7.6.12 of *Dancing with Qubits*. [D.W.Q]

9.7 Quantum circuits

We have now seen all the quantum gates we will need in this book, and it's time to compose them into circuits.

The goal of a quantum algorithm as implemented in a circuit is to manipulate the qubits' states so that the 0s and 1s we get when we measure have a very high probability of being the correct answer.

9.7.1 A basic circuit

I've done this already in simple cases, but let's see what a basic circuit looks like from beginning to end.

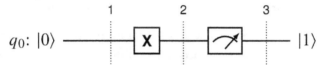

Figure 9.12: A basic quantum circuit

The circuit in Figure 9.12 contains a *quantum register* with one qubit. The initialization phase is to the left of the first vertical dotted line. In this phase, we initialize qubit q_0 to $|0\rangle$. If we do not explicitly display $|0\rangle$, you may assume that the qubit starts in that state. We call the horizontal line a *wire*.

Between the first and second dotted lines, we apply gates to qubits. From the second to the third lines, we measure the qubits. Afterward, we see if each measurement yields $|0\rangle$ or $|1\rangle$, which we interpret as the bits 0 or 1.

When quantum computers were first made available on the cloud by IBM in 2016, you could only put measurement at the end of a circuit. You can now measure mid-circuit and continue in the circuit with the $|0\rangle$ or $|1\rangle$ you saw.

The second operation is a *reset* and forces the quantum state to $|0\rangle$. It is not a gate, and it is not reversible. Nevertheless, it is a useful feature that allows you to reuse *ancillary* or "scratch" qubits in algorithms. We draw the reset operation as shown on the right.

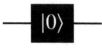

9.7.2 Physical and logical qubits

When we run the circuit, we expect to see that the **X** gate changed the quantum state from $|0\rangle$ to $|1\rangle$. The math tells us that the probability of getting $|1\rangle$ should be 100%, and this is true whether we run the circuit once or a thousand times.

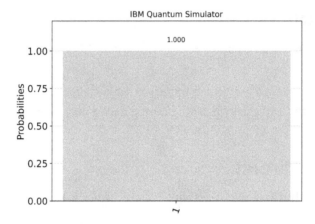

Figure 9.13: Results of circuit execution on a simulator

The histogram in Figure 9.13 shows that we get 1 for every execution of the circuit. We don't display a bar for 0 because it did not occur. The histogram is for a circuit simulated with pseudo-random numbers.

In the histogram in Figure 9.14, I show the result of running the circuit 8,192 times on actual quantum hardware through IBM Quantum on the IBM Cloud. We saw some 0s in addition to the 1s.

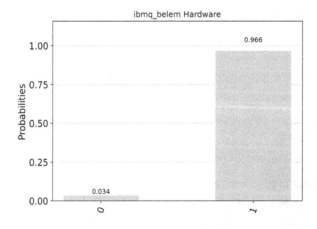

Figure 9.14: Results of circuit execution on quantum hardware

Why?

Since a quantum state for one qubit is $a\,|0\rangle + b\,|1\rangle$, we expect a mix of zeros and ones consistent with the probabilities $|a|^2$ and $|b|^2$. Since the amplitude of $|1\rangle$ is 1 after applying **X**, probability is not coming into play in this example.

Instead, we see the effect of noise and errors from the physical hardware and environment. These can occur in at least three places:

1. In the initialization phase, we may not be setting the starting state to exactly $|0\rangle$. To give an analogous example with real numbers, suppose we write the Python code x = 0.0, but the actual number assigned is 0.99999976. Though this would not happen in a Python assignment, we have seen floating-point rounding errors.

2. In the gate phase, the **X** gate's physical implementation might not provide full fidelity. This might indicate, for example, that instead of doing a perfect π rotation around the x-axis with **X**, it might only rotate 0.99999 π.

3. Finally, the hardware might have a *measurement error* and incorrectly read a 1 as a 0. Such an error could happen because the signal was distorted by environmental "static."

Another factor that can distort the result is *cross-talk* from nearby qubits. Though I only show one qubit in this example, the effects of our computing with nearby qubits could slightly change what we are doing with q_0.

The native hardware qubits affected by such noise are called *physical qubits*. Through hardware and software schemes, quantum computing providers implement mitigation schemes to reduce errors.

To remove errors, we must progress to fault tolerance. Quantum computer providers will eventually use error-correcting codes and hundreds of physical qubits to detect and remove errors. As I write this book, no one has implemented full quantum error correction.

9.7.3 Preparing to use Qiskit

Please see appendix section A.10 for information about installing **qiskit** and setting up your IBM Quantum account.

Let's import the classes and functions we need.

```
from qiskit import QuantumRegister, ClassicalRegister, QuantumCircuit
from qiskit import execute, transpile, Aer, IBMQ
from qiskit.visualization import plot_histogram
import math
import matplotlib
```

I define the following style dictionary to format the quantum circuits nicely for this book. You will see it in use as a **kwargs argument when I call the **QuantumCircuit** *draw* method.

```
draw_kwargs = {
    "output": "mpl",          # use matplotlib
    "cregbundle": False,      # separate classical register wires
    "initial_state": True,    # show |0> and 0
    "idle_wires": False,      # don't show unused wires

    "style": {
        "name": "bw",         # black-and-white for book
        "subfontsize": 9,     # font size of subscripts
        "dpi": 600            # image resolution
    }
}

histogram_color = "#82caaf"
```

9.7.4 Building the circuit

We begin by creating a circuit from its constituent parts: a quantum register and a classical measurement register. Each register has one qubit or bit, and we prefix each with "q" or "meas", respectively.

```
qreg_q = QuantumRegister(1, "q")
creg_c = ClassicalRegister(1, "meas")
```

We combine these into a circuit,

```
circuit = QuantumCircuit(qreg_q, creg_c)
```

and place an **X** gate on the qubit line.

```
circuit.x(qreg_q[0])
```

```
<qiskit.circuit.instructionset.InstructionSet at 0x1cf2bcef220>
```

```
circuit.draw(**draw_kwargs)
```

$$q\ |0> - \boxed{X} -$$

Finally, we add a measurement of the qubit state with the result going to the classical register.

```
circuit.measure(qreg_q[0], creg_c[0])
```

```
<qiskit.circuit.instructionset.InstructionSet at 0x1cf61a1bdf0>
```

```
circuit.draw(**draw_kwargs)
```

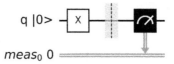

The code above is general and extends to more qubits and measurement bits. There is a simpler way to create this circuit using default values:

```
circuit = QuantumCircuit(1)
circuit.x(0)
circuit.measure_all()
```

```
circuit.draw(**draw_kwargs)
```

The vertical dashed line is a "barrier." Though it serves little purpose here, a barrier tells the Qiskit code optimizer to keep parts of a circuit separate.

9.7.5 Running the circuit

We now run this on a quantum simulator in Qiskit to see some of the steps and options. Qiskit can easily work on top of different hardware and kinds of simulators. We call these *backends*. We set simulator to the "aer_simulator" backend from the Qiskit **Aer** class. This runs on your laptop, desktop, or server where you installed Qiskit.

```
simulator = Aer.get_backend("aer_simulator")
```

Now we run the circuit 1,000 times and see how many 0s and 1s we get. The `counts` variable holds this data, which is a dictionary of the results we see and the number of times each appears. Each of the 1,000 executions is a *shot*.

```
result = execute(circuit, simulator, shots=1000).result()
counts = result.get_counts(circuit)
counts
```

```
{'1': 1000}
```

We only saw 1s, but let's display these in a histogram for later comparison.

```
histogram = plot_histogram(counts, color=histogram_color,
                           title="IBM Quantum Simulator")
```

```
<Figure size 504x360 with 1 Axes>
```

```
histogram
```

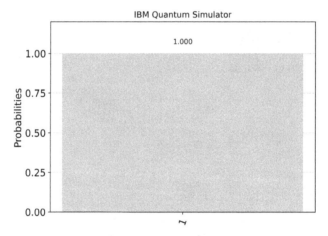

Now we use real hardware. While simulators are good for learning, debugging, and small experiments, the future is quantum hardware. Look at it this way: if you have only learned on a flight simulator and not a real airplane, there is no way I will let you fly me from New York to Tokyo. Don't expect to get a job as a pilot.

Start by loading your IBM Quantum account credentials.

```
IBMQ.load_account()
```

```
<AccountProvider for IBMQ(hub='ibm-q', group='open', project='main')>
```

If you saw the error

```
IBMQAccountCredentialsNotFound:
    'No IBM Quantum Experience credentials found.'
```

then see appendix section A.10 for information on setting up your account.

A *provider* offers one or more simulator or quantum computer backends. [QQP] You will likely use the "ibm-q" provider. I wrote some Python code to make it easier to see the system data.

```
IBMQ.providers()
provider = IBMQ.get_provider("ibm-q")

print("System Name            Number of Qubits")
print("-------------------------------------")
for backend in sorted(provider.backends(), key=lambda x: x.name()):
    config = backend.configuration()
    print(f"{config.backend_name:22}      {config.n_qubits:>3}")
```

```
System Name              Number of Qubits
-------------------------------------
ibmq_armonk                  1
ibmq_belem                   5
ibmq_bogota                  5
ibmq_lima                    5
ibmq_manila                  5
ibmq_qasm_simulator         32
ibmq_quito                   5
ibmq_santiago                5
simulator_extended_stabilizer    63
simulator_mps              100
simulator_stabilizer      5000
simulator_statevector       32
```

I choose the "ibmq_belem" quantum computing system as our backend, but you should select one to which you have access. The list of systems you see may be different from the previous list and "ibmq_belem" might not be in your list. To find which system is least busy, execute code like the following:

```
from qiskit.providers.ibmq import least_busy

device = least_busy(provider.backends(
    filters=lambda x: x.configuration().n_qubits >= 3 and
                      not x.configuration().simulator and
                      x.status().operational==True))
```

```
device
```

```
<IBMQBackend('ibmq_belem') from IBMQ(hub='ibm-q', group='open',
project='main')>
```

The code returns the least busy operational quantum hardware system with at least 3 qubits.

Please note that when you run a circuit on cloud hardware, there will be a delay while your job is queued and then executed. Depending on the system you selected, this can be fairly fast or many minutes. Please read on as you wait for your circuit execution to complete!

```
quantum_hw = provider.get_backend("ibmq_belem")
```

We run the circuit on the cloud on that IBM Quantum hardware.

```
result = execute(circuit, quantum_hw, shots=1000).result()
counts = result.get_counts(circuit)
```

```
counts
```

```
{'0': 34, '1': 966}
```

We did not see all 1s as we did with the simulator, but 96.6% of them were 1s.

```
histogram = plot_histogram(counts, color=histogram_color,
                           title="ibmq_belem Hardware")
```

```
<Figure size 504x360 with 1 Axes>
```

```
histogram
```

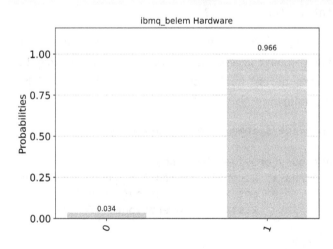

IBM provides the "open" backends at no charge. If you are a member of the IBM Quantum Network or belong to one of its hubs, you may see larger and more powerful quantum computers listed. IBM also upgrades and replaces systems over time. The names of the systems are cities, but generally, those are not where the hardware is located.

9.7.6 A 2-qubit circuit

Now let's build a 2-qubit circuit. Think of the gates as building blocks that we attach to the qubits, moving from left to right.

```
circuit = QuantumCircuit(2)
```

We begin by adding a **CNOT** and two Hadamard **H** gates.

```
circuit.cx(0, 1)
circuit.h(0)
circuit.h(1)
```

```
<qiskit.circuit.instructionset.InstructionSet at 0x1cf9c6749a0>
```

```
circuit.draw(**draw_kwargs)
```

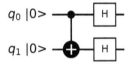

We repeat those gates. We could have coded a loop, but it would not have saved many lines.

```
circuit.cx(0, 1)
circuit.h(0)
circuit.h(1)
```

```
<qiskit.circuit.instructionset.InstructionSet at 0x1cf9cb8d550>
```

```
circuit.draw(**draw_kwargs)
```

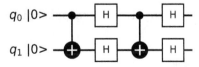

We insert a final **CNOT**.

```
circuit.cx(0, 1)
```

```
<qiskit.circuit.instructionset.InstructionSet at 0x1cf9e8be880>
```

```
circuit.draw(**draw_kwargs)
```

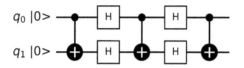

Exercise 9.42

The above circuit implements one of the 2-qubit gates we have seen before. Which is it and why?

The qubits q_0 and q_1 are initialized to $|0\rangle$ by the IBM Quantum hardware, but I want to set q_1 to $|1\rangle$. Here is a short circuit that is explicit about the values:

```
initialization_circuit = QuantumCircuit(2)
initialization_circuit.id(0)
initialization_circuit.x(1)
initialization_circuit.barrier()
```

```
<qiskit.circuit.instructionset.InstructionSet at 0x1cf9eaf6fd0>
```

$q_0\ |0\rangle$ — I —
$q_1\ |0\rangle$ — X —

I inserted the barrier to separate the initialization phase from what follows.

Now I "compose" the circuits together to make a new circuit.

```
new_circuit = initialization_circuit.compose(circuit)
new_circuit.measure_all()
```

```
new_circuit.draw(**draw_kwargs)
```

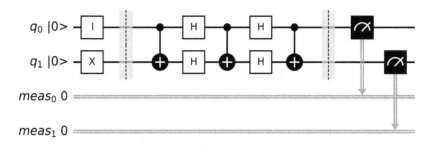

As we did before, we run the circuit 1,000 times with the simulator.

```
result = execute(new_circuit, simulator, shots=1000).result()
```

```
counts = result.get_counts(new_circuit)
counts
```

```
{'01': 1000}
```

Once again, we get the same result every time, but I must explain the '01'. Qiskit numbers its qubits from top to bottom in drawn circuits and writes out results *from right to left*. So '01' means that the result of measuring q_0 is 1, the right-most value. The measurement from q_1 is 0, the next value to the left. The gate and measurement phases of the circuit map $|0\rangle$ to 1 and map $|1\rangle$ to 0.

```
histogram = plot_histogram(counts, color=histogram_color,
                           title="IBM Quantum Simulator")
```

```
<Figure size 504x360 with 1 Axes>
```

```
histogram
```

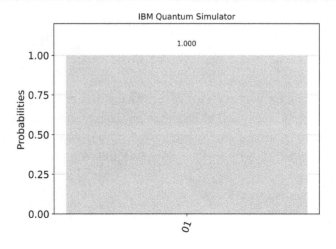

Let's see what we get on a quantum computing system. Remember, your results may vary because of probability and hardware errors.

```
result = execute(new_circuit, quantum_hw, shots=1000).result()
counts = result.get_counts(new_circuit)
```

```
counts
```

```
{'00': 49, '01': 923, '10': 12, '11': 16}
```

The results are mixed this time, but it's clear which value is the most likely.

```
histogram = plot_histogram(counts, color=histogram_color,
                           title="ibmq_belem Hardware")
```

```
<Figure size 504x360 with 1 Axes>
```

```
histogram
```

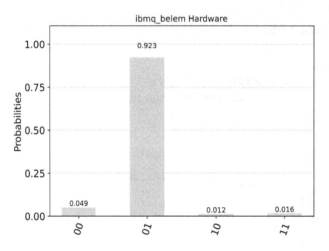

The gates you write are not necessarily the gates that run on the quantum hardware. As we will see in section 9.8, each kind of hardware provides a universal set of quantum gates specific to its physical qubit and control technology. Qiskit also optimizes circuits in both sophisticated and straightforward ways. For example, Qiskit can remove **HH** because it is the same as **id**.

The process of "translating and compiling" is called *transpiling*. (I did not coin that word.) Qiskit does this automatically, but you can call the *transpile* function to see how it changes your circuit. The optimization_level optional argument tells *transpile* how hard it should work to improve the circuit. It tries harder as optimization_level increases. The default value is 1, and Qiskit does no optimization when optimization_level is 0.

```
new_circuit_hw = transpile(new_circuit, quantum_hw,
    optimization_level=1)
```

```
new_circuit_hw.draw(**draw_kwargs)
```

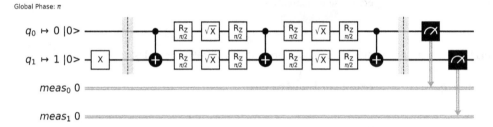

Do you see how the Hadamard **H** gates were replaced after transpiling?

9.7.7 Setting a specific quantum state

Given a quantum state $a |0\rangle + b |1\rangle$, the *initialize* method sets a qubit to that state in a circuit. Let $a = \sqrt{3}/2$ and $b = 1/2$.

```
a, b = quantum_state = (-math.sqrt(3)/2, complex(0, 1/2))
abs(a)**2 + abs(b)**2
```

```
0.9999999999999999
```

We had a **float** rounding error, so the sums of the squares of the absolute values did not equal 1.0.

```
circuit = QuantumCircuit(1)
circuit.initialize(quantum_state, 0)
circuit.measure_all()
```

```
circuit.draw(**draw_kwargs)
```

q |0> ⟶ |ψ⟩ [−0.866, 0.5j] ⟶ ⟶ measurement

$meas_0$ 0

When we run this circuit on the simulator, we expect to see 0 three-quarters of the time and 1 one-quarter of the time.

Exercise 9.43

Why do we expect these probabilities?

```
result = execute(circuit, simulator, shots=5000).result()
```

```
counts = result.get_counts(circuit)
counts
```

```
{'0': 3717, '1': 1283}
```

Given our discussion of pseudo-random numbers in section 5.7, these values are reasonable.

```
histogram = plot_histogram(counts, color=histogram_color,
                           title="IBM Quantum Simulator")
```

```
<Figure size 504x360 with 1 Axes>
```

```
histogram
```

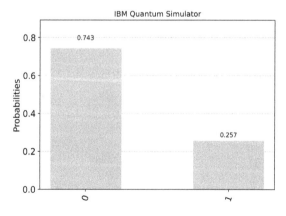

We see results balance in the same way when we run the circuit on quantum hardware.

```
quantum_hw = provider.get_backend('ibmq_belem')
result = execute(circuit, quantum_hw, shots=5000).result()
counts = result.get_counts(circuit)
```

```
counts
```

```
{'0': 3950, '1': 1050}
```

```
histogram = plot_histogram(counts, color=histogram_color,
                           title="ibmq_belem Hardware")
```

```
<Figure size 504x360 with 1 Axes>
```

```
histogram
```

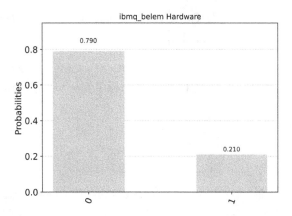

If you are curious, you can use the transpiler to see how Qiskit sets that quantum state for the qubit.

```
circuit_hw = transpile(circuit, quantum_hw, optimization_level=1)
```

```
circuit_hw.draw(**draw_kwargs)
```

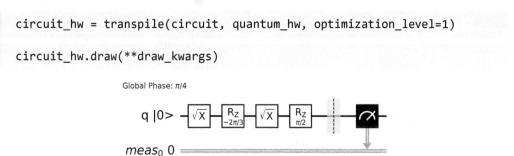

9.7.8 Entangling qubits

In this section, we work with two qubits. If qubit q_0 has the quantum state

$$|\psi_0\rangle = a_0\,|0\rangle + b_0\,|1\rangle$$

and qubit q_1 has the quantum state

$$|\psi_1\rangle = a_1\,|0\rangle + b_1\,|1\rangle,$$

then their 2-qubit quantum state is

$$|\psi_0\rangle \otimes |\psi_1\rangle = (a_0\,|0\rangle + b_0\,|1\rangle) \otimes (a_1\,|0\rangle + b_1\,|1\rangle) =$$
$$a_0 a_1\,|00\rangle + a_0 b_1\,|01\rangle + b_0 a_1\,|10\rangle + b_0 b_1\,|11\rangle.$$

Can we write all 2-qubit quantum states in this form? **No.**

If a 2-qubit quantum state can be written as $|\psi_0\rangle \otimes |\psi_1\rangle$ for single-qubit states $|\psi_0\rangle$ and $|\psi_1\rangle$, then the qubits are *separable*. Otherwise, the qubits are *entangled*. Entanglement is a fundamental property of quantum computing.

A simple separable state is $(\sqrt{2}/2)|00\rangle + (\sqrt{2}/2)|01\rangle$ because it is $|0\rangle \otimes ((\sqrt{2}/2)\,|0\rangle + (\sqrt{2}/2)\,|1\rangle)$. Here $|\psi_0\rangle = |0\rangle$ and $|\psi_1\rangle = (\sqrt{2}/2)\,|0\rangle + (\sqrt{2}/2)\,|1\rangle$. When I measure $|\psi_0\rangle$, it tells me nothing about $|\psi_1\rangle$.

In contrast, consider

$$(\sqrt{2}/2)\,|00\rangle + (\sqrt{2}/2)\,|11\rangle.$$

Let's measure the first qubit and leave the second alone. If we get 0, we must have gotten it from $|00\rangle$ because the first qubit in $|11\rangle$ would have delivered a 1. So the second qubit **must** measure to 0, the second entry in $|00\rangle$. Similarly, if we measure the second qubit and get a 1, the first qubit also must measure to 1 coming from $|11\rangle$. The qubits' behavior is entangled, and they have lost their independence.

The quantum state $(\sqrt{2}/2)\,|00\rangle + (\sqrt{2}/2)\,|11\rangle$ is one of the four entangled Bell states:

$$|\Phi^+|\ (\sqrt{2}/2)\,|00\rangle + (\sqrt{2}/2)\,|11\rangle \qquad |\Psi^+|\ (\sqrt{2}/2)\,|01\rangle + (\sqrt{2}/2)\,|10\rangle$$
$$|\Phi^-|\ (\sqrt{2}/2)\,|00\rangle - (\sqrt{2}/2)\,|11\rangle \qquad |\Psi^+|\ (\sqrt{2}/2)\,|01\rangle - (\sqrt{2}/2)\,|10\rangle$$

There are two things here I want you to remember:

1. When you see $\sqrt{2}/2$, think of the **H** gate.
2. When you need entanglement, think of the **CNOT** gate.

The circuit fragment to the right creates the Bell states. Think of it as a reusable quantum function. We can easily build it using Qiskit in Python.

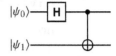

```
circuit = QuantumCircuit(2)
circuit.h(0)
circuit.cnot(0, 1)

circuit.draw(**draw_kwargs)
```

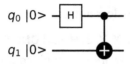

 Exercise 9.44

Which of $|11\rangle$, $|01\rangle$, $|00\rangle$, and $|10\rangle$ produce each of the four Bell states?

We use entanglement in many places in quantum computing, including algorithms where you might otherwise have made copies of data. Einstein did not initially believe in entanglement and called it "spooky action at a distance." Entangled quantum particles maintain their correlation over large, sometimes astronomical, distances.

Entanglement ties together different parts of a quantum system so that you cannot learn about all the information contained within it by looking at individual components. Professor John Preskill compares this to having to read a section of a book all at once to understand its content rather than reading it page by page. [QEF]

 I cover qubit entanglement in section 8.2 of *Dancing with Qubits*. [DWQ]

9.8 Universality for quantum gates

The existence of a universal set of gates for quantum computing is mostly academic for a coding book, but it is worth a short discussion to complement the classical situation.

Unlike the classical case, there is an infinite number of 1-qubit quantum gates. The $\mathbf{RX}_\varphi$, $\mathbf{RY}_\varphi$, and $\mathbf{RZ}_\varphi$ give us such a set, where $0 \leq \varphi < 2\pi$. If we can't generate these precisely, we need to approximate them as closely as we wish. We modify our definition to say that a set of quantum gates is *universal* if we can create a circuit with those gates and approximate any *n*-qubit quantum gate to arbitrary precision. [TCN]

We need qubit entanglement, which does not have a classical counterpart.

Finally, below the elegant gates like **X**, **Y**, **Z**, **H**, and **CNOT**, we need to understand which gates engineers implemented in the hardware in any given quantum computing technology. These vary across "superconducting transmon," "ion trap," "quantum dot," and "photonic" qubit approaches, for example, and may vary within each category. [DRQ] Can we approximately generate any gate we wish from the hardware gates?

Researchers have discovered several universal sets of quantum gates including

all 1-qubit gates plus **CNOT**,
Toffoli **CCX**, Hadamard **H**, and **S**

and

CNOT, Hadamard **H**, and **T**.

In practice, the set of gates available through Qiskit or that you can build yourself with circuits is large. We focus on them.

9.9 Summary

This chapter explored the gate and circuit foundation of classical computing. It showed how low-level coders could build bit addition and higher-level functions. From there, we moved to quantum gates and circuits. While we used the same words, the programming models are fundamentally different. In particular, superposition and entanglement give us behavior and tools that are not available using classical techniques alone.

In the next chapter, we look at techniques to test and optimize your code, be it classical or quantum. We saw some of this already when we used Qiskit's *transpile* function to optimize quantum gates for general and hardware-specific use.

10

Optimizing and Testing Your Code

Experience is the name everyone gives to their mistakes.

—Oscar Wilde, *Lady Windermere's Fan*

It would be wonderful if the code we wrote were perfect. It would run in the most efficient way possible, and it would perform exactly as we intended from the first time. There would not be one unused line or expression. It would execute correctly for the standard cases and every bizarre twist our software users can throw at us.

In this chapter, we face reality. Code usually has one or more bugs in it at the beginning. Your programming editor, such as Visual Studio Code, may help you with the syntax errors, but the semantic ones can be harder to find. We first learn about widely used techniques for testing your code, from simple sanity checks to organized and repeatable processes. We then look at timing how long your code takes to run and then suggestions for improving performance. Is all your code being used? We examine the idea of "coverage," where we see if all that code we wrote is being tested and executed. Finally, we investigate Python decorators and learn how they can add extra features to your functions and methods without modifying the original code.

Topics covered in this chapter

10.1 Testing your code

The code you write must work in all possible situations for all possible inputs. Therefore, you must devise a test system that assures you that every possibility is covered and the results are correct.

That's a powerful statement, and you might relax the rules if you are coding something "quick and dirty" to do once or for your own purposes. If you are sharing your code or it will be part of a production environment, testing is essential.

10.1.1 __debug__

By default, the system variable __debug__ is `True`. This setting allows you to write code like

```
from src.code.unipoly import UniPoly

x = UniPoly(1, "x", 1)

def square_poly(p):
    if __debug__:
        print(f"Argument {p = }\n")
    return p*p

square_poly(x**2 - 3*x + 7)

 Argument p = x**2 - 3*x + 7

 x**4 - 6*x**3 + 23*x**2 - 42*x + 49
```

Note that I imported **UniPoly** from the specific folder location "`src.code.unipoly`". You may need to change this to a different path for your use of this module. I introduced **UniPoly** in *Chapter 7, Organizing Objects into Classes*.

This feature has limited utility because if you try to change the value of __debug__, you get an error.

```
__debug__ = False

File "<stdin>", line 1
SyntaxError: cannot assign to __debug__
```

If you start Python from the operating system command line with either the -O or -OO option, Python sets __debug__ to False.

```
python -O
python -OO
python3 -O
python3 -OO
```

Depending on how someone installed Python on your system, you should use **python** or **python3**. See Appendix section A.2 for details. The "O" in -O and -OO is the capital letter and not zero "0".

In my opinion, if you want this functionality, use a global variable that you can change when you wish.

```
I_AM_DEBUGGING = True

def square_poly(p):
    if I_AM_DEBUGGING:
        print(f"Argument {p = }\n")
    return p*p

square_poly(x**2 - 3*x + 7)

Argument p = x**2 - 3*x + 7

x**4 - 6*x**3 + 23*x**2 - 42*x + 49
```

Now I can change the variable and stop seeing the debug output.

```
I_AM_DEBUGGING = False

square_poly(x**2 - 3*x + 7)

x**4 - 6*x**3 + 23*x**2 - 42*x + 49
```

10.1.2 assert

When you write a function or method, how can you be sure another developer will call it with the correct arguments? For example, if you expect the first argument to be an **int** and you get a **str**, what should you do?

One way is to validate the arguments and raise exceptions if you get unexpected object types or values. We did that in section 7.2 when we validated the arguments to __init__ for **UniPoly**. The exception types you use and the messages you give should be specific so that a user of your code knows what happened if anything goes wrong.

The names and arguments of the functions, classes, and methods you make available to other developers are part of the *API*, or "Application Programming Interface," of your code. You should always document and validate arguments within your API.

What about the functions and methods internal to your code and not part of its API? In this case, you might think, "I know what I'm doing, I know the type of every object, I know the values are acceptable, why should I validate the argument?"

If you are a good coder, you are likely correct, but not all the time. While you are debugging and testing your code, you should insert some tests ensuring conditions are as you expect. It might be an errant argument value, or it could be the result of a computation that you suspect might occasionally be problematic.

The assert keyword tests whether a condition is True. If the test fails, it raises an **AssertionError**.

In the original definition of *iterative_factorial* in section 6.12.3, I raised a **ValueError** if the argument was not a non-negative integer. Here is the version with an **assert**:

```
def iterative_factorial(n):
    assert isinstance(n, int)
    assert n >= 0

    product = 1
    for i in range(1, n + 1):
        product *= i
    return product

iterative_factorial(12)

479001600
```

```
iterative_factorial(1.3)

File "<stdin>", line 1, in "<module>"
File "<stdin>", line 2, in iterative_factorial
AssertionError
```

To be more specific, include a message after the condition:

```
def iterative_factorial(n):
    assert isinstance(n, int), "n not an int"
    assert n >= 0, "n < 0"

    product = 1
    for i in range(1, n + 1):
        product *= i
    return product

iterative_factorial(-1.3)

File "<stdin>", line 1, in "<module>"
File "<stdin>", line 2, in iterative_factorial
AssertionError: n not an int

iterative_factorial(-17)

File "<stdin>", line 1, in "<module>"
File "<stdin>", line 3, in iterative_factorial
AssertionError: n < 0
```

You can test for multiple conditions in a single `assert` statement, but I prefer to separate them to see which test failed.

There is overhead to using `assert` since Python must evaluate each condition. If you add assertions early in your code development, you can remove them once you are sure everything is working correctly. Using -O or -OO when you start Python turns off all `assert` condition checking.

10.1.3 pytest

Suppose you have written a function or a class. Typically, these do not exist by themselves but are part of a containing module, package, or project. Nevertheless, it makes sense for you to test each such component for correct behavior. Software that runs and stresses a piece of code is called a *unit test*.

Though Python includes **unittest** in its standard library, the **pytest** module has gained traction among developers in recent years. I introduce some of its many features in this section via the **UniPoly** class we developed in *Chapter 7, Organizing Objects into Classes*. [PYTST]

The source for **UniPoly** is in *Appendix C*. At the end of the listing are two functions:

```
1  def test_good_addition():
2      """Tests that an addition is correct."""
3      x = UniPoly(1, 'x', 1)
4      p = x + 1
5      q = x - 1
6      assert p + q == 2 * x
```

```
1  def test_bad_multiplication():
2      """Tests that a multiplication fails."""
3      x = UniPoly(1, 'x', 1)
4      p = x + 1
5      q = x - 1
6      assert p * q == x * x + 1
```

Note that they each begin with `test_`.

After some initial setup, they each test via an `assert` on line 6 whether an algebraic operation has the right answer. For *test_good_addition*, the addition is correct, but for *test_bad_multiplication*, I deliberately coded a wrong answer.

I keep the source code in the **unipoly.py** file. When I run `pytest unipoly.py` from the operating system command line, I get this output:

```
========================= test session starts =========================
platform win32 -- Python 3.9.5, pytest-6.2.3, py-1.10.0, pluggy-0.13.1
rootdir: D:\dropbox\github\writings\non-fiction\unify\src\code
plugins: pylama-7.7.1
collected 2 items

unipoly.py .F                                                    [100%]

=============================== FAILURES ===============================
_____ test_bad_multiplication _____

    def test_bad_multiplication():
```

```
          """Tests that a multiplication fails."""
          x = UniPoly(1, 'x', 1)
          p = x + 1
          q = x - 1
>         assert p * q == x * x + 1
E         assert x**2 - 1 == x**2 + 1
E             Use -v to get the full diff

unipoly.py:732: AssertionError
======================= short test summary info =========================
FAILED unipoly.py::test_bad_multiplication - assert x**2 - 1 == x**2 + 1
====================== 1 failed, 1 passed in 0.10s ======================
```

There's a lot of information there, including the operating system, the version of **pytest**, the installed testing plugins, the code in which the assertion failed, the original assertion, and the evaluated assertion. The output does not mention *test_good_addition* because the assertion condition was True.

pytest looked through the input file and processed all functions whose names started with "test_".

Normally, I would not define a function like *test_bad_multiplication*, but would instead code something like

```
def test_multiplication():
    """Tests polynomial multiplication."""
    x = UniPoly(1, 'x', 1)
    p = x + 1
    q = x - 1
    assert 0 * p == 0
    assert p * 0 == 0
    assert 1 * p == p
    assert -1 * p == -p
    assert p * 1 == p
    assert 2 * p == p * 2
    assert p * q == q * p
    assert p * q == x * x - 1
    # ... more multiplication tests
```

I would do this for every function and method.

Here's an example with a quantum circuit. If we initialize a qubit to |0⟩ and then apply a Hadamard **H**, we have an equal chance of getting |0⟩ or |1⟩ when we measure the qubit. However, since probability is involved, we will likely not get exactly as many of one quantum state as another. This checks if we do not get an equal number:

```python
1  from qiskit import QuantumCircuit, execute, Aer
2
3  def test_50_50():
4      circuit = QuantumCircuit(1)
5      circuit.h(0)
6      circuit.measure_all()
7      simulator = Aer.get_backend("aer_simulator")
8      result = execute(circuit, simulator, shots=1000).result()
9      counts = result.get_counts(circuit)
10     assert counts['0'] == counts['1']
```

```
pytest quantum-test.py

========================= test session starts =========================
platform win32 -- Python 3.9.5, pytest-6.2.3, py-1.10.0, pluggy-0.13.1
rootdir: D:\dropbox\github\writings\non-fiction\unify\src\code
plugins: pylama-7.7.1
collected 1 item

quantum-test.py F                                              [100%]

=============================== FAILURES ===============================
_____ test_50_50 _____

    def test_50_50():
        circuit = QuantumCircuit(1)
        circuit.h(0)
        circuit.measure_all()
        simulator = Aer.get_backend("aer_simulator")
        result = execute(circuit, simulator, shots=1000).result()
        counts = result.get_counts(circuit)
>       assert counts['0'] == counts['1']
E       assert 494 == 506

quantum-test.py:10: AssertionError
```

```
======================= short test summary info ========================
FAILED quantum-test.py::test_50_50 - assert 494 == 506
=========================== 1 failed in 1.54s ==========================
```

A better test could check whether the quantum states' counts differed by some percentage. That might indicate a problem with the simulator or unexpectedly poor hardware fidelity or measurement.

Developers usually put their tests in a separate `test` folder to keep them separate from the source. **pytest** will look for all files that start with "`test_`" and end with ".`py`" in the folder and run them. I encourage you to use many small test files that check individual aspects of your code versus one monolithic file that tests them all.

pytest has dozens of options, plugins, and ways to test your code across multiple directories. I recommend that you start small and then automate your testing on a set schedule. For example, **pytest-watch** can notify you when tests fail so you can take corrective action.

10.2 Timing how long your code takes to run

If I want to improve my code to run faster, I must measure how long it takes to run before and after my changes. In section 6.12.4, I covered computing execution time with the **time** module. Now let's look at *timeit*.

The *timeit* function takes a string containing a Python expression and evaluates it many times. It returns the total time for the entire execution.

```python
import timeit

timeit.timeit("2**1000")
```

```
0.8334714000000001
```

In this example, Python evaluated `2**1000` one million times and returned the number of seconds for the entire computation. You can use keyword arguments to specify the code and the number of iterations.

```python
timeit.timeit(stmt="2**1000", number=500000)
```

```
0.4071590999999999
```

Exercise 10.1

In the graph of *timeit* times for 2^n with $60 \le n \le 65$ in Figure 10.1, something strange happens when $n = 64$. Explain why we see the large increase in execution time. (Hint: see section 5.2.)

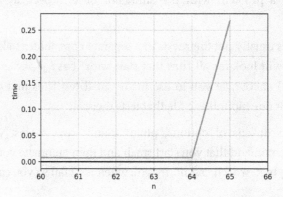

Figure 10.1: timeit times for computing 2^n

You can evaluate more code than can easily fit within the stmt string by defining a function.

```python
def f(n):
    assert n > 6

    the_list = []
    for x in range(n):
        the_list.append(x)
    for _ in range(10):
        the_list += the_list
        the_list = the_list[5:]
```

f is an inefficient nonsense function that creates a list, appends it to itself, slices it, and repeats ten times. Use the globals keyword argument to allow *timeit* to use your local definitions.

```python
count = 1000
timeit.timeit(stmt="f(100)", number=count, globals=globals())

0.9215112999999997
```

In many languages, including C and C++, the programmer is responsible for asking the system for memory, filling that memory with data, and then releasing the memory back to the system for reuse. If you do not manually "collect the garbage," your computing environment fills with unused and unreleased memory, and eventually, you run out. Managing memory yourself can be very efficient but also error-prone. Python implements garbage collection and handles the acquisition and release of memory for you.

Although modern software platforms do not stop and reclaim memory all at once, they may do it incrementally as they perform other tasks. Do you want garbage collection times included when you use *timeit*? By default, Python turns off garbage collection within *timeit*, but you can enable it.

```
import gc

timeit.timeit(stmt="f(110)", number=count,
              globals=globals(), setup="gc.enable()")
```

```
1.0470327
```

Depending on the code you are testing, turning garbage collection on or off may increase or decrease execution time. You might think garbage collection would always increase the time, but having readily accessible memory to reuse may improve performance.

10.3 Optimizing your code

Optimizing your software means to write, rewrite, or otherwise process your code so that it runs as efficiently as possible on your intended hardware.

Why did I say "as efficiently as possible" instead of "as fast as possible"? Optimization often involves trade-offs, such as running faster but using more memory or conserving memory while executing a little slower. This section primarily looks at methods to speed up your software.

Code optimization is an entire discipline within computer science and software engineering. The topics I cover here give you a jumping-off point to learn more. You'll have a better idea of how to write good code and when to make it better. That said, bear in mind this quote from Donald Knuth: [CPA]

"The real problem is that programmers have spent far too much time worrying about efficiency in the wrong places and at the wrong times; premature optimization is the root of all evil (or at least most of it) in programming."

Make your code correct, and then make your code faster when necessary.

10.3.1 Use existing code

Is there a Python library that provides functionality close to what you want or need? If so, research what people think about the library and its implementation. If you can use code written by experts, there's a good chance it will be better than what you create from scratch. [P.Y.L] [P.Y.P.I]

10.3.2 Choose a good algorithm

When we looked at implementations of the factorial and Fibonacci functions in *Chapter 6, Defining and Using Functions*, we saw that selecting the right algorithm can make a large difference in execution time or memory usage. Note that you might not see this until the problem gets big enough, so you may not need to over-analyze what to do in small cases.

For example, do you really need to call *sorted* for a list of two items?

Suppose you want to add together all positive integers less than or equal to a given integer. You might code:

```python
def adder_1(n):
    sum = 0
    for j in range(1, n + 1):
        sum += j
    return sum

adder_1(100)

5050

t = timeit.timeit(stmt="adder_1(10000000)",
                    number=3, globals=globals())
print(f"{t:.10f}")

2.0824539000
```

Or, you could note that this sum is $n(n + 1)/2$.

```python
def adder_2(n):
    return n * (n + 1) // 2
```

```
adder_2(100)
```

```
5050
```

```
t = timeit.timeit(stmt="adder_2(10000000)",
                  number=3, globals=globals())
print(f"{t:.10f}")
```

```
0.0000015000
```

Exercise 10.2

Why is $n(n-1)$ always even for **int** n?

10.3.3 Use the right collection type

We spent a lot of time in *Chapter 3, Collecting Things Together*, examining lists, sets, and dictionaries and when to use each. The choice of whether to use a dictionary or not is probably the easiest since you know if you have something that could be a unique key pointing to related data.

It's very slow to look up items in a list. If item order is not significant, you frequently need to test for membership, and your collection has unique immutable items, use a **set**. If your items are immutable, but you might have multiple copies of each, use a **Counter** object from **collections**.

```
from collections import Counter

my_guitars = Counter(["Fender", "Taylor", "Fender", "Gibson"])
my_guitars
```

```
Counter({'Fender': 2, 'Taylor': 1, 'Gibson': 1})
```

```
"Fender" in my_guitars
```

```
True
```

```
your_guitars = Counter(["Taylor", "Taylor", "Gibson", "Gibson"])
your_guitars
```

```
Counter({'Taylor': 2, 'Gibson': 2})
```

```
my_guitars + your_guitars

Counter({'Fender': 2, 'Taylor': 3, 'Gibson': 3})

my_guitars - your_guitars

Counter({'Fender': 2})
```

> **?**
>
> **Exercise 10.3**
>
> What does len(your_guitars) return, and why?

> **?**
>
> **Exercise 10.4**
>
> Discuss how you would define **Counter** as a child class of **dict**.

10.3.4 Favor iteration over recursion

Python's developers built it to do fast iteration. Recursion involves multiple function calls and significant resources to set up and execute those calls. Python limits the recursion depth, so your algorithms may fail to complete for large problems.

Python allows you to build new iterators and generators to construct your algorithms in very natural ways. These objects also enable you to produce objects on demand rather than create a large collection before processing.

10.3.5 Use comprehensions

A comprehension in Python is an elegant way to create a collection. It is often much faster than adding items one at a time.

```
def build_list_1(n):
    the_list = []
    for j in range(n):
        the_list.append(j)

t_1 = timeit.timeit(stmt="build_list_1(100000)",
                number=3, globals=globals())
print(f"Execution time = {t_1:.10f}")

Execution time = 0.0178861000
```

Now let's see how fast we can create the same list with a comprehension.

```python
def build_list_2(n):
    [j for j in range(n)]

t_2 = timeit.timeit(stmt="build_list_2(100000)",
                    number=3, globals=globals())
print(f"Execution time = {t_2:.10f}")

Execution time = 0.0098197000
```

If you are filtering items or computing with them, the execution time difference might be much less than what we just saw for building simple lists. A comprehension may not be your fastest solution in every case, as Python knows how to create its built-in collections from functions like *range*.

```python
def build_list_3(n):
    list(range(n))

t_3 = timeit.timeit(stmt="build_list_3(100000)",
                    number=3, globals=globals())
print(f"Execution time = {t_3:.10f}")

Execution time = 0.0062501000

def speed(n, m):
    ratio = eval(f"t_{n}") / eval(f"t_{m}")
    print(f"build_list_{m} is " +
          f"{ratio:.2f} times faster than build_list_{n}")

speed(1, 2)
speed(1, 3)
speed(2, 3)

build_list_2 is 1.82 times faster than build_list_1
build_list_3 is 2.86 times faster than build_list_1
build_list_3 is 1.57 times faster than build_list_2
```

Exercise 10.5

What does *eval* do in the definition of *speed*?

10.3.6 Don't do extra work in loops

Consider the function *looper_1*, which accepts a positive integer n. Inside *looper_1*, we execute a loop n times. Inside the loop, we compute 2^{100} and then add it to the loop variable j.

```
def looper_1(n):
    for j in range(n):
        a = 2**100
        b = a + j

t_1 = timeit.timeit(stmt="looper_1(100000)",
                    number=3, globals=globals())
print(f"Execution time = {t_1:.10f}")

Execution time = 0.0875948000
```

Python evaluates the assignment a = 2**100 within every iteration of the loop. Nothing happens inside the loop that could change the value! **Move statements and expression evaluations that are not affected by the loop out of the loop.** When we move the assignment before the loop, our function speeds up considerably.

```
def looper_2(n):
    a = 2**100
    for j in range(n):
        b = a + j

t_2 = timeit.timeit(stmt="looper_2(100000)",
                    number=3, globals=globals())
print(f"Execution time = {t_2:.10f}")

Execution time = 0.0128910000
```

We now add two numbers inside the loop, but we never use the sum! If we called a function or method that has a side effect, this might make sense, but not in this case. **Don't execute useless loops unless you are showing examples of useless loops!**

The time to do nothing at all is much less than executing the useless loop.

```
def looper_3(n):
    pass
```

```
t_3 = timeit.timeit(stmt="looper_3(100000)",
                    number=3, globals=globals())
print(f"Execution time = {t_3:.10f}")

Execution time = 0.0000008000
```

Python has a philosophy that coders should be able to do useless and inefficient things if they want to do so. It's your responsibility to do necessary manual optimizations. If you use a static code checker like **pylint** or **flake8**, you can see warnings and suggestions for code improvements. If I put the source for *looper_1* in the file **looper_1.py**, I will see the following when I run flake8 looper_1.py from the operating system command line:

```
looper_1.py:4:9: F841 local variable 'b' is assigned to but never used
```

pylint gives a similar message, together with disparaging comments about how I named my variables. These and other static checkers have many options and allow you to specify which diagnostics they do and the messages they display. Incidentally, the static checking process is called "linting."

10.3.7 Don't sort until you have to

In section 7.3, we looked at how to represent the terms within our **UniPoly** polynomial objects. We concluded that there was no need to keep them in a list in sorted order, and we could put them in the right sequence when we printed them.

The general lesson here is that there may be extra implementation work you do along the way as you write your code that is not necessary or optimal. We decided that a dictionary would be a much faster way to represent the polynomial data.

Yes, get your code working, but pause occasionally and ask yourself if you are approaching the problem the correct way. Don't be afraid to experiment and use functions like *timeit* to understand if performance is getting better. Is performance speed your primary consideration? Perhaps a better, more elegant, and slightly slower design is your best coding solution.

Exercise 10.6

What did Knuth mean in his quote at the start of this section?

10.4 Looking for orphan code

You have not thoroughly tested your code if you have not executed each line at least once. Checking for this is called *coverage testing*, and **coverage** is the tool to use. Install **coverage** from the operating system command line via

```
pip install coverage
```

Suppose I have the function *plane_location*, which accepts two arguments: the x coordinate of a point in the plane and the y coordinate. The function prints whether the point is

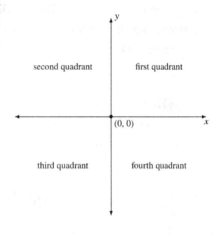

- at the origin $(0, 0)$
- on the x-axis (if $y = 0$)
- on the y-axis (if $x = 0$)
- in the first quadrant (if x and $y > 0$)
- in the second quadrant (if $x < 0$ and $y > 0$)
- in the third quadrant (if $x < 0$ and $y < 0$)
- in the fourth quadrant (if $x > 0$ and $y < 0$)

The implementation is straightforward, and I follow it with some examples:

```
1  def plane_location(x, y):
2      if x == 0:
3          if y == 0:
4              message = "at the origin"
5          else:
6              message = "on the y-axis"
7      elif y == 0:
8          message = "on the x-axis"
9      elif x > 0:
10          if y > 0:
11              message = "in the first quadrant"
12          else:
13              message = "in the fourth quadrant"
14      elif y > 0:
15          message = "in the second quadrant"
16      else:
```

```
17              message = "in the fourth quadrant"
18
19      print(f"({x}, {y}) is located {message}")
20
21
22  plane_location(2, 3)
23  plane_location(-42, 7)
24  plane_location(0, 9)
25  plane_location(0, -39)
26  plane_location(3, 0)
27  plane_location(-4, 4)
28  plane_location(-4, -4)
```

```
(2, 3) is located in the first quadrant
(-42, 7) is located in the second quadrant
(0, 9) is located on the y-axis
(0, -39) is located on the y-axis
(3, 0) is located on the x-axis
(-4, 4) is located in the second quadrant
(-4, -4) is located in the fourth quadrant
```

Do these examples cover all possible cases? *plane_location* is a small function, and I'm sure you can find the two cases I didn't test. What if my code had spanned several files, functions, and classes?

This code is in the file **coverage-example.py**. From the operating system command line (**not** from within the Python interpreter), I run two commands:

```
coverage run coverage-example.py
coverage report -m
```

and the tool displays the following information:

```
Name                    Stmts   Miss  Cover   Missing
---------------------------------------------------------
coverage-example.py        22      2    91%   4, 13
---------------------------------------------------------
TOTAL                      22      2    91%
```

I used the -m option so **coverage** would print which lines were missing. Here, "missing" means that the lines with those numbers did not run when I called *plane_location* with the points above. I did not test for the point at the origin nor one in the fourth quadrant.

We can get more information and have **coverage** place it in a JSON file. We covered reading and writing JSON-format files in section 8.8.2.

```
coverage json
```

Once you run this command, **coverage** produces the file **coverage.json**. Its contents will look something like the following:

```
{
    "meta": {
        "version": "5.4",
        "timestamp": "2021-02-12T17:31:54.234031",
        "branch_coverage": false,
        "show_contexts": false
    },
    "files": {
        "coverage-example.py": {
            "executed_lines": [
                1, 2, 3, 6, 7, 8, 9, 10, 11, 14, 15,
                17, 19, 22, 23, 24, 25, 26, 27, 28
            ],
            "summary": {
                "covered_lines": 20,
                "num_statements": 22,
                "percent_covered": 90.9090909090909,
                "missing_lines": 2,
                "excluded_lines": 0
            },
            "missing_lines": [
                4, 13
            ],
            "excluded_lines": []
        }
    },
    "totals": {
        "covered_lines": 20,
        "num_statements": 22,
        "percent_covered": 90.9090909090909,
        "missing_lines": 2,
        "excluded_lines": 0
    }
}
```

Developers usually run **coverage** on their **pytest** files to ensure they are thoroughly testing everything the code implements.

coverage has many more options and can produce interactive reports in HTML format. You can then view the information in your browser. Use the `--help` option to see the possibilities.

```
coverage --help
coverage report --help
coverage json --help
coverage html --help
```

10.5 Defining and using decorators

We've used Python decorators several times so far, but it's time we learned how to code them ourselves. We saw the decorators

- `@abstractmethod` in section 7.13.2 to indicate a method from an abstract base class that coders must override

- `@classmethod` and `@staticmethod` in section 7.12 to define class and static methods

- `@property` in section 7.5 to define properties from parameter-less class instance methods

Here's the core idea with decorators: given a function `my_function`, I want to wrap it with additional code that calls `my_function`. I'm *decorating* the outside of `my_function` with extra capability. For extra credit, I want to use `my_function` to access my original code with the new decorations.

I'll begin with a simple function that prints 100.

```
def one_hundred():
    return 100

one_hundred()

100

one_hundred

<function __main__.one_hundred()>
```

one_hundred is a function and a first-class Python object. I want to print the current time whenever *one_hundred* runs. I could just change the function:

```python
import datetime

def one_hundred():
    now = datetime.datetime.now()
    print(f"The time is {now.strftime('%H:%M:%S')}")
    return 100
```

```python
one_hundred()
```

```
The time is 12:48:28
100
```

A decorator simplifies this process considerably, and I do not have to alter the original function body. I reset *one_hundred* to its original definition and create *say_the_time*.

```python
def one_hundred():
    return 100

def say_the_time(func):
    def wrapper():
        now = datetime.datetime.now().strftime("%H:%M:%S")
        print(f"The time is {now}")
        return func()
    return wrapper
```

say_the_time takes a single argument `func`, which is a function with no arguments. It defines an inner function *wrapper*, which prints the time, evaluates `func()`, and returns the result. *say_the_time* then returns the function *wrapper*. Let's see it in action for *one_hundred*:

```python
h = say_the_time(one_hundred)
```

```python
h
```

```
<function __main__.say_the_time.<locals>.wrapper()>
```

```python
h()
```

```
The time is 12:48:28
100
```

I didn't need to change the definition of *one_hundred* at all. If I define

```
def one_thousand():
    return 1000
```

then I can decorate and call it directly.

```
say_the_time(one_thousand)()
```

```
The time is 12:48:28
1000
```

I think this is quite elegant. Even better, we can prefix *say_the_time* with an @ on the line before a function definition and get the time printed whenever we call a function by its original name.

```
@say_the_time
def one_million():
    return 1000000
```

```
one_million()
```

```
The time is 12:48:28
1000000
```

I can use more than one decorator on a function.

```
def say_the_date(func):
    def wrapper():
        now = datetime.datetime.now().strftime("%B %d, %Y")
        print(f"The day is {now}")
        return func()
    return wrapper
```

```
@say_the_date
@say_the_time
def one_million():
    return 1000000
```

```
one_million()
```

```
The day is August 11, 2021
The time is 12:48:28
1000000
```

Python printed the date and the time. When we reverse the decorators, the time and then the date are displayed.

```
@say_the_time
@say_the_date
def one_million():
    return 1000000


one_million()

The time is 12:48:28
The day is August 11, 2021
1000000
```

You can use a decorator with a function with regular and keyword arguments (section 6.5). In the next example, I use *args and **kwargs to collect positional and keyword arguments and pass them to the inner wrapper.

```
# first define without the decorator
def adder_with_carry(a, b, carry=0):
    return a + b + carry


adder_with_carry(10, 20, carry=1)

31

def say_the_date(func):
    def wrapper(*args, **kwargs):
        now = datetime.datetime.now().strftime("%B %d, %Y")
        print(f"The day is {now}")
        return func(*args, **kwargs)
    return wrapper


# now use the decorator

@say_the_date
def adder_with_carry(a, b, carry=0):
  return a + b + carry


adder_with_carry(10, 20, carry=1)

The day is August 11, 2021
31
```

It is as simple as adding or deleting the @ decorator line to get or remove the extra functionality.

Exercise 10.7

Write a decorator function *assert_first_arg_is_int* that inserts an `assert` statement that checks whether the first argument passed to a function is an `int`.

Although I coded `return func(*args, **kwargs)`, I could have saved the result in a variable, done more work in the wrapper function, and then returned the value.

```
def enter_and_exit(func):
    def wrapper(*args, **kwargs):
        print(f">>> Entering {func.__name__}")
        result = func(*args, **kwargs)
        print(f">>> Exiting {func.__name__} with {result}")
        return result
    return wrapper

@enter_and_exit
def adder_with_carry(a, b, carry=0):
  return a + b + carry

adder_with_carry(10, 20, carry=1)

>>> Entering adder_with_carry
>>> Exiting adder_with_carry with 31
31
```

Exercise 10.8

Extend the last definition of *enter_and_exit* to print the argument values in the `"Entering ..."` message.

Exercise 10.9

Use one or more functions from the **time** module to create a decorator that displays how long it takes to execute *func* within *wrapper*.

10.6 Summary

In this chapter, we looked at testing code and improving it. Developers have created many modules and tools like **timeit**, **coverage**, **pytest**, **pylint**, and **flake8** to make your job easier. It's your responsibility to incorporate them into the development process.

In the next chapter, we survey some classical and quantum algorithms for searching and learn how to implement them in Python.

11

Searching for the Quantum Improvement

The improvement of understanding is for two ends:
first, our own increase of knowledge; secondly, to
enable us to deliver that knowledge to others.

—John Locke

By considering new approaches and getting clever, we can develop classical algorithms that are faster than you might have expected. Using quantum techniques, we can go a step further: perform some operations faster than seems possible.

This chapter compares classical and quantum search techniques to see how extending our basic information unit from the bit to the qubit can show remarkable improvements. Note that I only discuss mainstream "universal" quantum computing and not limited-purpose systems that perform operations like simulated annealing.

Topics covered in this chapter

11.1 Classical searching

In coding, "searching" is attempting to find a specific object in a collection. In Python, the collection is often a list, though it could be a NumPy array, as we shall see in section 13.1.7. The object may or may not be in the collection.

Python sets and dictionaries are optimized for search, so we focus on collections with less secondary structure for locating objects. In particular, we look at lists.

Let numbers be a list of 16 unique integers between 1 and 50 in random order.

```
numbers = [4, 46, 40, 15, 50, 34, 32, 20, 24, 30, 22, 49, 36, 16, 5, 2]
numbers
```

```
[4, 46, 40, 15, 50, 34, 32, 20, 24, 30, 22, 49, 36, 16, 5, 2]
```

The first value in the list is 4, so it does not take long to find it if we start searching from the beginning. On the other hand, 2 is the last list item, so we need to check every member of the list to confirm its presence. This is a *linear search*.

The following simple search function returns a tuple. The first tuple object is the index at which the function finds the number, or None, if it is not present. The second tuple object is the number of "==" comparisons made.

```
def find_number(n, list_of_numbers):
    for j in range(len(list_of_numbers)):
        if n == list_of_numbers[j]:
            return (j, j + 1)
    return (None, len(list_of_numbers))

find_number(4, numbers)

(0, 1)

find_number(2, numbers)

(15, 16)
```

The number 17 is not in the list.

```
find_number(17, numbers)
```

```
(None, 16)
```

Can we do better if we sort the list?

```
numbers.sort()
numbers
```

```
[2, 4, 5, 15, 16, 20, 22, 24, 30, 32, 34, 36, 40, 46, 49, 50]
```

In the next version of *find_number*, we stop searching when the number in the list is greater than the search value. We use both "==" and "<".

```
def find_number(n, list_of_numbers):
    # list_of_numbers must be sorted in ascending order
    for j in range(len(list_of_numbers)):
        if n == list_of_numbers[j]:
            return (j, j + 1)
        if n < list_of_numbers[j]:
            break
    return (None, len(list_of_numbers))

for n in [2, 17, 30, 50]:
    print(f"n = {n:2}: {find_number(n, numbers)}")
```

```
n =  2: (0, 1)
n = 17: (None, 16)
n = 30: (8, 9)
n = 50: (15, 16)
```

This approach is frequently better, but searches for missing numbers or those at the end of the list are still inefficient. The worst case is still that we make `len(number)` "==" comparisons.

Exercise 11.1

How many "<" comparisons do we do in the worst case?

Exercise 11.2

For a sorted list of length *n*, show that each of the previous two versions of *find_number* is O(*n*) in the number of comparisons made in the worst case.

We are not taking full advantage of the list being sorted. Our next version of *find_number* takes a divide-and-conquer *binary search* approach.

1. Look at the value in the "middle of the list." If it is the number we seek, we are finished.

2. If the list has length one, the number is not in the list, so we return None.

3. If our number is less than the value, go to step 1 but only consider the first half of the list.

4. If our number is greater than the value, go to step 1 but only consider the second half of the list.

In this way, we repeatedly cut the size of the list we search in half, hence the name "binary search."

```python
def find_number(n, list_of_numbers):
    """Returns index of found number via binary search or None

    Parameters
    ----------
    n : `int` or `float` or `Fraction`
        The number we seek.
    list_of_numbers: `list`
        Sorted list of numbers.

    Returns
    -------
    `int` or None
        The index of n in list_of_numbers if found,
        None otherwise.
    """

    def print_num(j):
        # Utility function to display search progress
        if j is not None:
            print(f"-> {list_of_numbers[j]}", end=' ')
        else:
            print("-> None", end=' ')

    def binary_find(start, end, count):
        # Recursive binary search
        # start and end are list indices, count is the
        # number of times this function is called.

        count += 1

        # Handle the case when the list has one item.
        if start + 1 == end:
            if n == list_of_numbers[start]:
                print_num(start)
```

```
                return (start, count)

            print_num(None)
            return (None, count)

        # Find the index of the item in the middle of the list.
        mid_point = (start + end) // 2

        # Have we succeeded?
        if n == list_of_numbers[mid_point]:
            print_num(mid_point)
            return (mid_point, count)

        print_num(mid_point)

        # Should we check the second half of the list?
        if n > list_of_numbers[mid_point]:
            return binary_find(mid_point + 1, end, count)

        # Check the first half of the list.
        return binary_find(start, mid_point, count)

    # Begin the search on the entire list.
    return binary_find(0, len(list_of_numbers), 0)

numbers
```

```
[2, 4, 5, 15, 16, 20, 22, 24, 30, 32, 34, 36, 40, 46, 49, 50]
```

Let's see how the new *find_number* performs on several numbers:

```
for n in [2, 4, 17, 30, 34, 50]:
    print(f"n = {n:>2}:", end=' ')
    result = find_number(n, numbers)
    print('\n', 9*' ', result)
```

```
n =  2: -> 30 -> 16 -> 5 -> 4 -> 2
          (0, 5)
n =  4: -> 30 -> 16 -> 5 -> 4
          (1, 4)
n = 17: -> 30 -> 16 -> 22 -> None
          (None, 4)
n = 30: -> 30
          (8, 1)
n = 34: -> 30 -> 40 -> 34
          (10, 3)
n = 50: -> 30 -> 40 -> 49 -> 50
          (15, 4)
```

Exercise 11.3

Explain what is happening with *print* in the last for loop and in *find_number* to obtain the formatting you see.

The search sequence in *find_number* is in Figure 11.1.

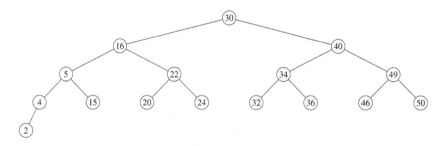

Figure 11.1: Binary search sequence for the list of numbers

Figure 11.1 shows a *graph*. The numbers in circles are *nodes*, and the lines connecting them are *edges*. When you travel from a node along a left edge, you get to a node of lesser value. When you take a right edge, you arrive at a node of greater value. Let's use the graph to locate **36**.

We begin at the node at the very top, which is the *root node*. It has the value 30. Since **36** > 30, we follow the right edge from 30 to get to the node with the value 40. Since **36** < 40, we follow the left edge from 40 to get to 34. We have **36** > 34, so we take the left edge to land on **36**. We are done!

Exercise 11.4

Devise a formula involving *math.log2* to predict the second value of the tuple returned by *find_number*.

Exercise 11.5

Though the definition of *find_number* uses the recursive *binary_find* function, that is not likely to be a problem because the recursion depth never gets too deep. Why? Nevertheless, rewrite *binary_find* as an iterative function. Do not use slices.

Sorting a collection does not necessarily involve physically moving the items. Imagine, for example, that you wanted to "sort" boxes of products in a warehouse. You can leave the boxes where they are, but create an *index* of them. The first entry in the index is the location of the box whose product name comes first alphabetically, and so on. In this way, you can create multiple indices by which you access the products by different characteristics.

While binary search is a significant improvement, the requirement that the list is sorted is substantial. It seems like we are stuck with being $O(n)$ for items in arbitrary order. Or are we?

11.2 Quantum searching via Grover

So how can we do better than looking at every one of n items in an unordered collection? If we had ten classical systems, we could divide the collection ten ways and search each chunk. This process is still $O(n)$ because the time is still proportional to the collection's size. We would also be adding overhead to parallelize the search over multiple systems.

The remainder of this chapter shows the basic ideas of how the Grover search algorithm finds the result in $O(\sqrt{n})$ time. [GSA] The detailed mathematics of the algorithm is well described elsewhere.

 I cover the Grover search algorithm in sections 9.5 through 9.7 of *Dancing with Qubits*. [DWQ]

Suppose I have a box of 8 dimmable light bulbs. They are all distinct in some way so that I can tell them apart. I number them 0 through 7, or, in binary, 000 through 111. Even better, I label them with the kets $|000\rangle$ to $|111\rangle$, as we saw in section 9.6.5. I start with them each in a half-dimmed state, as shown in Figure 11.2. This example is an analogy, so the bulbs do not really behave in a quantum manner.

Figure 11.2: Eight light bulbs, half-dimmed

If I use qubit terminology, I say that we have a balanced "dimming" superposition of the light bulbs. In this analogy, if a bulb is closer to being fully on, the likelihood of my "measuring" it to be fully on is greater than measuring it to be off. If it is closer to being fully off, I will more likely measure it to be that way.

Since they are half-dimmed, if I "measure" them using the laws of probability to turn them on or off, I expect roughly half of the lights to be on and half of them to be off. Figure 11.3 shows this situation.

Figure 11.3: Eight light bulbs, half off, half on

Before we continue with the dimming, I'm going to drop the light bulbs.

What?

If I drop the bulbs sequentially onto the floor one at a time, the process is $O(n)$, where n is 8 in this case. If I get eight people to drop one bulb each at the same time, distributing the bulbs is $O(n)$, and so the whole process is also $O(n)$. If I drop the entire box at once, I am doing something to every bulb simultaneously. This is $O(1)$.

I now have a new box of eight dimmable bulbs. If we can somehow search for a specific bulb in a way that does something to the bulbs simultaneously, perhaps I can do better than $O(n)$. This is the idea of Grover's search algorithm. I perform some action $O(\sqrt{n})$ times so that when I "measure" all the bulbs afterward, the brightest bulb is the one I seek.

Suppose I want to find the bulb labeled $|000\rangle$. After one iteration of the action, the bulbs look as in Figure 11.4. One bulb is a little brighter, and the other seven are dimmer.

Figure 11.4: Eight light bulbs, one getting brighter than the others

The "thing that I do to all bulbs simultaneously" is somehow affecting the |000⟩ bulb in a special way.

After $O(\sqrt{n})$, or approximately 2, iterations, the bulbs look like those in Figure 11.5. The |000⟩ bulb has a very high probability of being measured on, and all the rest have a high probability of being measured off.

|000⟩ |001⟩ |010⟩ |011⟩ |100⟩ |101⟩ |110⟩ |111⟩

Figure 11.5: Eight light bulbs, one brightest

This is the way I search and find the correct bulb. Is it always successful? Note that I said, "|000⟩ bulb has a very high probability of being measured on." To get a high confidence level that I have found the correct bulb, I should repeat the search several times. This is still better than the brute force sequential method.

Exercise 11.6

I am working on a jigsaw puzzle, and there are one hundred pieces spread out on the table in front of me. There is one particular piece I want among them. Explain how the Grover algorithm can help you find the piece in ten tries instead of one hundred.

11.3 Oracles

An *oracle* determines whether some condition is true or not. We have seen these many times, for example, in the code if x < y. When we use in to test list membership, we are using an oracle. We also use an oracle to answer the question, "Is this item in that collection?".

In practice, the code using the oracle doesn't necessarily have to return True or False. If I am searching for an integer in a list of positive numbers, the code might negate the integer's sign if it is a match. If it is not the value I am seeking, it leaves the sign alone.

In section 11.1, I focused on how many "==" comparisons we made in each version of the *find_number* function. Put another way, "how many times do I need to call the '==' oracle? ".

In the Grover search algorithm, the oracle is tuned to do something special for the item we seek, which is $|000\rangle$ in the light bulb example. What is extraordinary about quantum computing is that I can call the oracle **simultaneously** on all items in the collection. In this case, the oracle, together with a technique called "amplitude amplification," makes the sought-after bulb brighter and all the others dimmer. After a few calls, we see the much brighter bulb as the one we want.

11.3.1 1-qubit oracles

As we saw in section 9.6.1, we can represent a qubit as $a\,|0\rangle + b\,|1\rangle$, where a and b are complex number *amplitudes*, and $|a|^2 + |b|^2 = 1$.

In this situation, we have two items, $|0\rangle$ and $|1\rangle$, and we want to search for one of them. If we are looking for $|1\rangle$, what oracle circuit changes the sign of $|1\rangle$ but leaves the sign of $|0\rangle$ alone? It is the circuit with a single **Z** gate:

$$\mathbf{Z}: a\,|0\rangle + b\,|1\rangle \longmapsto a\,|0\rangle - b\,|1\rangle$$

What circuit changes the sign of $|0\rangle$ but leaves the sign of $|1\rangle$ alone? That is, we want to change a to $-a$ but keep b as it is.

When we apply **X**, we swap a and b. A **Z** gate then negates a, and another **X** gate swaps them again:

$$\mathbf{XZX}: a\,|0\rangle + b\,|1\rangle \longmapsto -a\,|0\rangle + b\,|1\rangle$$

More simply,

$$-\mathbf{Z}: a\,|0\rangle + b\,|1\rangle \longmapsto -a\,|0\rangle + b\,|1\rangle$$

This also follows from the first version because $\mathbf{XZ} = -\mathbf{ZX}$ and $\mathbf{XX} = \mathbf{id}$.

As we did in section 9.7.4, we use Qiskit to draw these circuits. The first is the oracle for $|1\rangle$:

```
from qiskit import QuantumCircuit
import math
import matplotlib

# Set up the options to draw the circuit
draw_kwargs = {
    "output": "mpl",          # use matplotlib
    "idle_wires": False,      # don't show unused wires
```

```
    "style": {
        "name": "bw",          # black-and-white for book
        "subfontsize": 9,      # font size of subscripts
        "dpi": 600             # image resolution
    }
}
```

Create and define a 1-qubit circuit:

```
circuit = QuantumCircuit(1)
circuit.z(0)
```

```
<qiskit.circuit.instructionset.InstructionSet at 0x286bb6dc220>
```

```
circuit.draw(**draw_kwargs)
```

$$q - \boxed{Z} -$$

The oracle for $|0\rangle$ is this circuit:

```
circuit = QuantumCircuit(1)
circuit.x(0)
circuit.z(0)
circuit.x(0)
```

```
<qiskit.circuit.instructionset.InstructionSet at 0x286bd40ac10>
```

```
circuit.draw(**draw_kwargs)
```

$$q - \boxed{X} - \boxed{Z} - \boxed{X} -$$

To summarize, **Z** acts as an oracle to distinguish $|1\rangle$ and $-$**Z** distinguishes $|0\rangle$. In practice, we would put $a = b = \sqrt{2}/2$. That is, we move the qubit into a balanced superposition.

Exercise 11.7

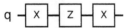

After reviewing section 9.6.7, show that the **S** = **RZ**$_{\pi/2}$ gate transforms $a\,|0\rangle + b\,|1\rangle$ to $a\,|0\rangle + i\,b\,|1\rangle$. What gate combination changes $a\,|0\rangle + b\,|1\rangle$ to $i\,a\,|0\rangle + b\,|1\rangle$?

11.3.2 2-qubit oracles

Next, suppose we have four items. We use two qubits and put the items in one-to-one correspondence with $|00\rangle$, $|01\rangle$, $|10\rangle$, and $|11\rangle$. We put the qubits in a balanced superposition state:

$$(1/2)\,|00\rangle + (1/2)\,|01\rangle + (1/2)\,|10\rangle + (1/2)\,|11\rangle$$

and ask, "What are the four oracle circuits that create each of the following from this quantum state?".

$$- (1/2)\,|00\rangle + (1/2)\,|01\rangle + (1/2)\,|10\rangle + (1/2)\,|11\rangle$$
$$(1/2)\,|00\rangle - (1/2)\,|01\rangle + (1/2)\,|10\rangle + (1/2)\,|11\rangle$$
$$(1/2)\,|00\rangle + (1/2)\,|01\rangle - (1/2)\,|10\rangle + (1/2)\,|11\rangle$$
$$(1/2)\,|00\rangle + (1/2)\,|01\rangle + (1/2)\,|10\rangle - (1/2)\,|11\rangle$$

Exercise 11.8

Confirm that the sum of the squares of the absolute values of the amplitudes of $|00\rangle$, $|01\rangle$, $|10\rangle$, and $|11\rangle$ in each of the previous five quantum states equals 1.

We start with the oracle for $|11\rangle$ because it is the simplest. The oracle appears to change the sign of the amplitude of $|11\rangle$ if and only if the first qubit is $|1\rangle$. This is a hint that we need a controlled gate. Flipping the sign of $|1\rangle$ and not $|0\rangle$ for the second qubit means we need a **Z** gate. Thus the oracle to distinguish $|11\rangle$ is **CZ**.

```
circuit = QuantumCircuit(2)
circuit.cz(0, 1)
```

```
<qiskit.circuit.instructionset.InstructionSet at 0x286bcfbdc70>
```

```
circuit.draw(**draw_kwargs)
```

Note how Qiskit draws **CZ**. We discussed controlled gates and their symbols at the end of section 9.6.4.

Exercise 11.9

Since

$$|00\rangle = |0\rangle \otimes |0\rangle$$
$$|01\rangle = |0\rangle \otimes |1\rangle$$
$$|10\rangle = |1\rangle \otimes |0\rangle$$
$$|11\rangle = |1\rangle \otimes |1\rangle$$

verify that the circuit has the desired effect by setting each of q_0 and q_1 equal to all combinations of $|0\rangle$ and $|1\rangle$.

I have represented multi-qubit states the way authors usually write them in the literature. This means that if q_0 is $|0\rangle$ and q_1 is $|1\rangle$, I write their tensor product as $|01\rangle$. Qiskit reverses this ordering and writes it as $|10\rangle$. Since I mostly display circuits, I have tried to minimize any confusion.

Exercise 11.10

Show that the oracle for $|11\rangle$ is the same as this circuit:

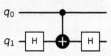

Let's think about the oracle for $|10\rangle$. By the logic for $|11\rangle$ and the oracle for $|0\rangle$, we should expect three controlled gates to work.

```
circuit = QuantumCircuit(2)
circuit.cx(0, 1)
circuit.cz(0, 1)
circuit.cx(0, 1)
```

```
<qiskit.circuit.instructionset.InstructionSet at 0x286bd1cb760>
```

```
circuit.draw(**draw_kwargs)
```

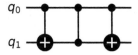

This circuit leaves $|00\rangle$ and $|01\rangle$ alone because q_0 is $|0\rangle$. For $|11\rangle$ and $|10\rangle$, the three controlled gates activate the gates for q_1 because q_0 is $|1\rangle$. $|11\rangle$ goes to $|10\rangle$, then $|10\rangle$, and then to $|11\rangle$. $|10\rangle$ goes to $|11\rangle$, then to $-|11\rangle$, and finally to $-|10\rangle$. So this works.

Exercise 11.11

Show that the oracle for $|10\rangle$ is the same as this circuit:

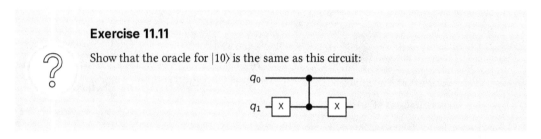

We should expect that the remaining two oracles are similar to the last two, except that we swap $|0\rangle$ and $|1\rangle$ for q_0. We do this with an **X** gate, but we must apply another one at the end to change the state back to what it was.

Exercise 11.12

Show that this circuit is the oracle for $|01\rangle$:

```
circuit = QuantumCircuit(2)
circuit.x(0)
circuit.cz(0, 1)
circuit.x(0)
```

```
<qiskit.circuit.instructionset.InstructionSet at
0x286bd103fa0>
```

```
circuit.draw(**draw_kwargs)
```

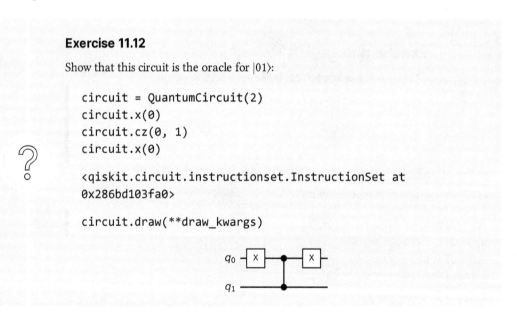

Finally, the oracle for $|00\rangle$ is:

```
circuit = QuantumCircuit(2)
circuit.x(0)
circuit.x(1)
circuit.cz(0, 1)
```

```
circuit.x(0)
circuit.x(1)
```

```
<qiskit.circuit.instructionset.InstructionSet at 0x286bea29d30>
```

```
circuit.draw(**draw_kwargs)
```

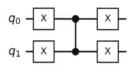

11.3.3 3-qubit oracles

The controlled-controlled-**Z** gate (**CCZ**) is the oracle for $|111\rangle$ when we have three qubits. The double-control means that both q_0 and q_1 must be $|1\rangle$ for **Z** to be applied to q_2.

We can use a **CX** gate with a Hadamard **H** on each side of it on q_1 to create the **CZ** oracle for $|11\rangle$. Similarly, we can use a **CCX** Toffoli gate to make a **CCZ**. We discussed the Toffoli gate in section 9.6.5.

Here is the Qiskit oracle circuit for $|111\rangle$:

```
circuit = QuantumCircuit(3)
circuit.h(2)
circuit.ccx(0, 1, 2)
circuit.h(2)
```

```
<qiskit.circuit.instructionset.InstructionSet at 0x286be92efd0>
```

```
circuit.draw(**draw_kwargs)
```

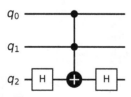

This is the oracle for $|000\rangle$:

```
circuit = QuantumCircuit(3)
for q in range(3): circuit.x(q)
circuit.h(2)
```

```
circuit.ccx(0, 1, 2)
circuit.h(2)
for q in range(3): circuit.x(q)

circuit.draw(**draw_kwargs)
```

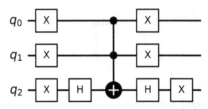

Exercise 11.13

Verify that the last two circuits are oracles for $|111\rangle$ and $|000\rangle$. What is the oracle for $|010\rangle$? $|101\rangle$?

11.4 Inversion about the mean

Suppose I have four data points and their mean:

```
data = [3, 1, 2, 5]
mean = float(sum(data)) / len(data)
mean
```

```
2.75
```

I show these values in the bar chart on the left-hand side of Figure 11.6.

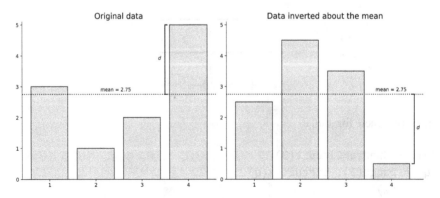

Figure 11.6: Inversion about the mean for four data points

On the right-hand side, I've *inverted the data about the mean.* Geometrically, if a value x is above the mean with difference d from the mean, the inverted value is the mean minus $|d|$. Look at the fourth bar. The value is well above the mean. The inverted value on the right is below the mean by the same amount.

 I cover inversion about the mean in section 9.6.2 of *Dancing with Qubits.* [D.W.Q]

For a value below the mean, again look at the difference d from the mean. The inverted value is the mean plus $|d|$. The second bar is a good example of this. Given a non-empty list of numbers, I can define the function *invert_about_the_mean* like this:

```
1  def invert_about_the_mean(data):
2      assert data
3      mean = sum(data) / float(len(data))
4      return [2.0 * mean - x for x in data]
5
6  invert_about_the_mean(data)
```

```
[2.5, 4.5, 3.5, 0.5]
```

You can see that the results match what I've shown on the right of Figure 11.6.

 Exercise 11.14

Replace `assert data` on line 2 with `raise` and an exception. What kind of exception do you use?

 Exercise 11.15

Why do I use *float* on line 3?

 Exercise 11.16

Explain the inversion formula on line 4. Why is the mean the same before and after inversion?

Let's repeat this process for four new data points.

```
data = [0.5, -0.5, 0.5, 0.5]
invert_about_the_mean(data)
```

```
[0.0, 1.0, 0.0, 0.0]
```

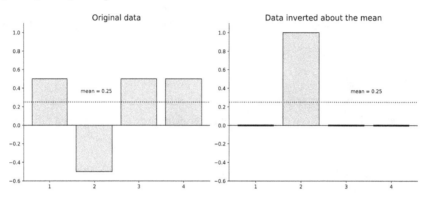

Figure 11.7: Inversion about the mean for [0.5, -0.5, 0.5, 0.5]

I plot the original data and their inversions in Figure 11.7. The points initially have the same absolute value, but only one has a negative sign. When I invert about the mean, the new values are all 0 except for the one corresponding to the original negative value. That new value is 1.

Exercise 11.17

What do you know about that negates the sign of one and only one value in a list of numbers? What is the sum of the squares of the absolute values of the original data values? The inverted values?

While inversion may seem like an arbitrary calculation, it is a crucial component in our implementation of Grover search, as we shall see next.

11.5 Amplitude amplification

I now stop talking about "data values" and call them what they are for Grover search: *amplitudes*. In this section, we work with three qubits and begin by placing them in a balanced superposition:

$$\frac{1}{\sqrt{8}}|000\rangle + \frac{1}{\sqrt{8}}|001\rangle + \frac{1}{\sqrt{8}}|010\rangle + \frac{1}{\sqrt{8}}|011\rangle + \frac{1}{\sqrt{8}}|100\rangle + \frac{1}{\sqrt{8}}|101\rangle + \frac{1}{\sqrt{8}}|110\rangle + \frac{1}{\sqrt{8}}|111\rangle$$

These are the same quantum states we saw with the light bulbs in section 11.2. The qubits being in balanced superposition translates to the light bulbs all being in the half-dimmed state.

 I cover amplitude amplification in section 9.6 of *Dancing with Qubits*. [DWQ]

When I apply the oracle for $|000\rangle$, the amplitude for that ket becomes negative:

$$\boxed{-\frac{1}{\sqrt{8}}|000\rangle} + \frac{1}{\sqrt{8}}|001\rangle + \frac{1}{\sqrt{8}}|010\rangle + \frac{1}{\sqrt{8}}|011\rangle + \frac{1}{\sqrt{8}}|100\rangle + \frac{1}{\sqrt{8}}|101\rangle + \frac{1}{\sqrt{8}}|110\rangle + \frac{1}{\sqrt{8}}|111\rangle$$

Figure 11.8 shows these two sets of amplitudes. The mean on the left-hand side is $1/\sqrt{8}$, which is approximately 0.354.

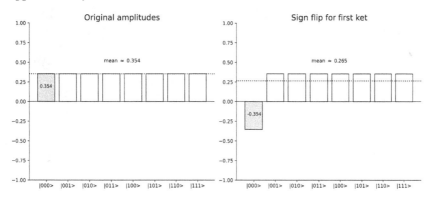

Figure 11.8: Amplitudes for three qubits: Original and sign flipped for $|000\rangle$

If we measure the quantum states on either side of Figure 11.8, the probabilities of getting any of the eight kets are equal. Recall from section 9.6.2 that the probability of getting a ket is the square of the absolute value of the ket's amplitude.

```
import math

amplitudes = [1.0/math.sqrt(8) for _ in range(8)]
amplitudes[0] = -amplitudes[0]

# Amplitude for |000>
amplitudes[0]
```

```
-0.35355339059327373
```

```
abs(amplitudes[0])**2
```

```
0.12499999999999997
```

We see a round-off error, but the probability is $1/8 = 0.125$.

```
# Amplitude for |001>
amplitudes[1]
```

```
0.35355339059327373
```

```
abs(amplitudes[1])**2
```

```
0.12499999999999997
```

First inversion

We now invert the amplitudes on the right-hand side about their mean. This action yields what we see on the left in Figure 11.9.

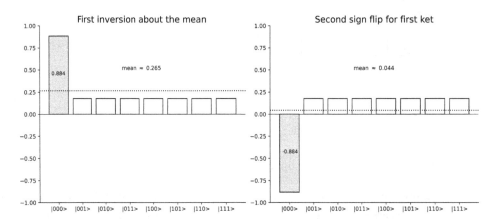

Figure 11.9: First inversion about the mean, followed by a sign flip

See how we increased or *amplified* the amplitude for $|000\rangle$? If we measure now, the probability of getting that ket is approximately $|0.884|^2 \approx 0.781$.

```
inverted_amplitudes = invert_about_the_mean(amplitudes)
inverted_amplitudes[0]
```

```
0.8838834764831843
```

```
abs(inverted_amplitudes[0])**2
```

```
0.7812499999999999
```

```
inverted_amplitudes[1]
```

```
0.17677669529663687
```

```
abs(inverted_amplitudes[1])**2
```

```
0.031249999999999993
```

This result is good, but as you can see in the last expression, the probability of getting any other particular ket is still more than 3%.

Second inversion

Let's repeat the sign flip and inversion, giving us the left side of Figure 11.10.

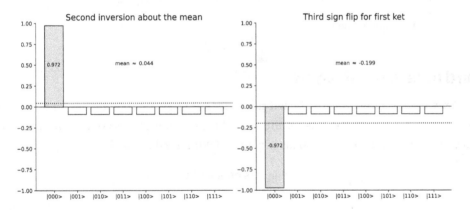

Figure 11.10: Second inversion about the mean, followed by a sign flip

The chance of getting |000⟩ is now almost 95%, and the likelihood and getting any other specific ket is less than 1%.

```
amplitudes = inverted_amplitudes
amplitudes[0] = -amplitudes[0]
inverted_amplitudes = invert_about_the_mean(amplitudes)
inverted_amplitudes[0]
```

```
0.9722718241315027
```

```
abs(inverted_amplitudes[0])**2
```

```
0.9453124999999998
```

```
inverted_amplitudes[1]
```

```
-0.08838834764831843
```

```
abs(inverted_amplitudes[1])**2
```

```
0.007812499999999998
```

Exercise 11.18

Relate what we did here to the process of finding a designated light bulb in section 11.2. In particular, explain the sentence

> The "thing that I do to all bulbs simultaneously"
> is somehow affecting the |000⟩ bulb in a special way.

from that section.

Third inversion attempt

Can we keep going? Look at the left side of Figure 11.10. The amplitudes for all kets other than |000⟩ are negative. If we repeat the sign-flip-inversion process one more time, we get what you see in Figure 11.11. The odds of getting |000⟩ have decreased!

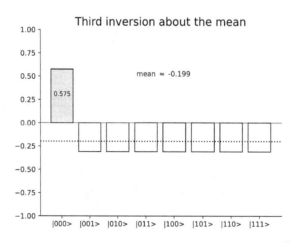

Figure 11.11: Third inversion about the mean

√n inversions

This is the Grover search algorithm in practice. We are searching for some object X within a collection of size n. We put the items in the collection in correspondence with the standard basis kets for m qubits, where $n \leq 2^m$. In the example in this section, $n = 8$ and $m = 3$.

Now assume someone gives us an oracle that acts in this way: when given X, the oracle flips the sign of the amplitude of the ket corresponding to X. When given any object other than X, it does not change the amplitude.

We now do the sign-flip-inversion process approximately $\sqrt{n}$ times. At the end of this iteration, we get the correct ket corresponding to X with very high probability. The probability of the answer being wrong is $O(1/n)$. We repeat the search until we are confident we have found the correct result.

I have skipped over the parts "put the items in the collection in correspondence with the standard basis kets for m qubits" and "assume someone gives us an oracle." These are particular to each specific problem. In our example, the collection is the set of standard basis kets for three qubits, and we code the oracle as we saw in section 11.3.3.

Exercise 11.19

What should we do if the size of our collection is not a power of two?

11.6 Searching over two qubits

Go back and look at Figure 11.7. What does this tell you about how many iterations you need for finding one item among four and the probability of your success?

```python
# Set the drawing options
draw_kwargs = {
    "output": "mpl",           # use matplotlib
    "initial_state": True,     # show |0> and 0
    "idle_wires": False,       # don't show unused wires

    "style": {
        "name": "bw",          # black-and-white for book
        "subfontsize": 9,      # font size of subscripts
        "dpi": 600             # image resolution
    }
}
```

Let's build the complete circuit to locate $|11\rangle$. We use Qiskit to create a 2-qubit circuit and place the qubits in superposition with two Hadamard **H** gates.

```
circuit = QuantumCircuit(2)
circuit.h(0)
circuit.h(1)
circuit.barrier()
```

```
<qiskit.circuit.instructionset.InstructionSet at 0x286bfc11e50>
```

```
circuit.draw(**draw_kwargs)
```

$q_0\ |0>$ —[H]—

$q_1\ |0>$ —[H]—

The vertical dashed line on the right is a *barrier*. I'm using it here to visually separate the parts of the circuit, but its real purpose is to limit how Qiskit optimizes adjacent gates. We need not worry about that here.

Now we insert a **CZ** gate. As we saw in section 11.3, this is the oracle for $|11\rangle$. It negates the sign of the amplitude for $|11\rangle$ and leaves the other amplitudes alone.

```
circuit.cz(0, 1)
circuit.barrier()
```

```
<qiskit.circuit.instructionset.InstructionSet at 0x286bfc02e50>
```

```
circuit.draw(**draw_kwargs)
```

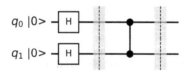

Next, we extend the circuit to invert about the mean. I'll explain how these gates do the inversion in a few moments. I use the **id** gates to format and align the gates. At the end of the circuit, I measure all qubits.

```
circuit.h(0)
circuit.h(1)
```

```
circuit.x(0)
circuit.x(1)
circuit.id(0)
circuit.h(1)
circuit.cx(0, 1)
circuit.id(0)
circuit.h(1)
circuit.x(0)
circuit.x(1)
circuit.h(0)
circuit.h(1)
```

```
<qiskit.circuit.instructionset.InstructionSet at 0x286be8fe8e0>
```

```
circuit.measure_all()
```

```
circuit.draw(**draw_kwargs)
```

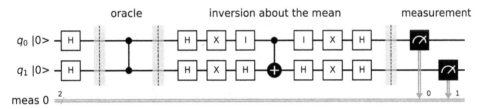

Aside: the result from *draw* is a **matplotlib** figure. I used the *text* function to label each portion of the circuit. See *Chapter 13, Creating Plots and Charts*, for details.

Before I explain how the inversion about the mean part of the circuit works, let's verify the result by running the circuit on a simulator.

```
from qiskit import execute, Aer
from qiskit.visualization import import plot_histogram


simulator = Aer.get_backend("aer_simulator")

result = execute(circuit, simulator, shots=1000).result()
counts = result.get_counts(circuit)
counts
```

```
{'11': 1000}
```

For each of the 1,000 shots (executions) of the circuit, the simulator returned $|11\rangle$ every time. The histogram reflects this as well.

```
histogram_color = "#82caaf"

histogram = plot_histogram(counts, color=histogram_color,
                         title="Searching for |11>")

<Figure size 504x360 with 1 Axes>

histogram
```

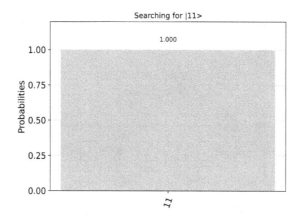

Now let's look at how the inversion about the mean portion of the circuit does its work. I warn you that this is the most complicated "qubit and quantum state" math in the book, so work through each step carefully.

 I cover the full Grover search algorithm with all its math in section 9.7 of *Dancing with Qubits*. [D.W.Q]

The initial state is

$$|00\rangle$$

and this becomes

$$(1/2)\,|00\rangle + (1/2)\,|01\rangle + (1/2)\,|10\rangle + (1/2)\,|11\rangle$$

after the two **H** gates.

After the **CZ** oracle which does a sign flip on $|11\rangle$, the 2-qubit quantum state is

$$(1/2)\,|00\rangle + (1/2)\,|01\rangle + (1/2)\,|10\rangle - (1/2)\,|11\rangle$$

We next apply **H** gates to each qubit. This is the 2-qubit $\mathbf{H}^{\otimes 2}$ gate we saw at the end of section 9.6.4.

$$(1/2)\,(\mathbf{H}^{\otimes 2})\,|00\rangle + (1/2)\,(\mathbf{H}^{\otimes 2})\,|01\rangle + (1/2)\,(\mathbf{H}^{\otimes 2})\,|10\rangle - (1/2)\,(\mathbf{H}^{\otimes 2})\,|11\rangle$$

This is a very messy calculation with 16 terms. The applications of $\mathbf{H}^{\otimes 2}$ on the standard basis kets are

$$\mathbf{H}^{\otimes 2}\,|00\rangle = (1/2)\,(|00\rangle + |01\rangle + |10\rangle + |11\rangle)$$
$$\mathbf{H}^{\otimes 2}\,|01\rangle = (1/2)\,(|00\rangle - |01\rangle + |10\rangle - |11\rangle)$$
$$\mathbf{H}^{\otimes 2}\,|10\rangle = (1/2)\,(|00\rangle + |01\rangle - |10\rangle - |11\rangle)$$
$$\mathbf{H}^{\otimes 2}\,|11\rangle = (1/2)\,(|00\rangle - |01\rangle - |10\rangle + |11\rangle)$$

Inserting these values and collecting terms, the result is

$$(1/2)\,|00\rangle + (1/2)\,|01\rangle + (1/2)\,|10\rangle - (1/2)\,|11\rangle$$

It may surprise you that this is the original value, but it is nevertheless the result in this case.

Next, we look at the effect of $\mathbf{X}^{\otimes 2}$ on the standard basis kets:

$$\mathbf{X}^{\otimes 2}\,|00\rangle = |11\rangle$$
$$\mathbf{X}^{\otimes 2}\,|01\rangle = |10\rangle$$
$$\mathbf{X}^{\otimes 2}\,|10\rangle = |01\rangle$$
$$\mathbf{X}^{\otimes 2}\,|11\rangle = |00\rangle$$

Applying this to the previous result, we get

$$- (1/2)\,|00\rangle + (1/2)\,|01\rangle + (1/2)\,|10\rangle + (1/2)\,|11\rangle$$

Now we examine **id** ⊗ **H**. Since

$$\mathbf{id} \otimes \mathbf{H}\,|00\rangle = (\sqrt{2}/2)\,|00\rangle + (\sqrt{2}/2)\,|01\rangle$$
$$\mathbf{id} \otimes \mathbf{H}\,|01\rangle = (\sqrt{2}/2)\,|00\rangle - (\sqrt{2}/2)\,|01\rangle$$
$$\mathbf{id} \otimes \mathbf{H}\,|10\rangle = (\sqrt{2}/2)\,|10\rangle + (\sqrt{2}/2)\,|11\rangle$$
$$\mathbf{id} \otimes \mathbf{H}\,|11\rangle = (\sqrt{2}/2)\,|10\rangle - (\sqrt{2}/2)\,|11\rangle$$

the result of the next step is

$$- (1/2)\,((\sqrt{2}/2)\,|00\rangle + (\sqrt{2}/2)\,|01\rangle) + (1/2)\,((\sqrt{2}/2)\,|00\rangle - (\sqrt{2}/2)\,|01\rangle)$$
$$+ (1/2)\,((\sqrt{2}/2)\,|10\rangle + (\sqrt{2}/2)\,|11\rangle) + (1/2)\,((\sqrt{2}/2)\,|10\rangle - (\sqrt{2}/2)\,|11\rangle)$$

or

$$-(\sqrt{2}/2)\,|01\rangle + (\sqrt{2}/2)\,|10\rangle$$

Applying the 2-qubit **CX** gate yields

$$-(\sqrt{2}/2)\,|01\rangle + (\sqrt{2}/2)\,|11\rangle$$

followed by the second **id ⊗ H**

$$-(1/2)\,|00\rangle + (1/2)\,|01\rangle + (1/2)\,|10\rangle - (1/2)\,|11\rangle$$

Almost done! Apply the second **X ⊗ X**

$$-(1/2)\,|00\rangle + (1/2)\,|01\rangle + (1/2)\,|10\rangle - (1/2)\,|11\rangle$$

And finish with the **H**$^{\otimes 2}$, yielding

$$-\,|11\rangle$$

When we measure this, the result must be $|11\rangle$, subject to any hardware noise. So with just one call to the oracle and one inversion about the mean, we find the search value with 100% accuracy.

Exercise 11.20

Verify that the Grover search algorithm works when we use the oracles for $|00\rangle$, $|01\rangle$, and $|10\rangle$.

While I just showed you that a series of gates on two qubits computes the inversion about the mean, who discovered the circuit? The simple answer is a "quantum algorithm developer." [GSA] From qubit idioms to small circuits to algorithmic components, we learn and extend our set of techniques for attacking problems.

11.7 Summary

In this chapter, we looked at classical linear and binary search. In the first case, the items in the collection were not ordered initially, and so we were forced to look at each item in turn to locate the one we sought. When we could assume the items were sorted, the binary search method was much more efficient as we could iteratively divide the problem size in half until we got our result. We then turned to one of the most accessible quantum algorithms, Grover search. Using Grover, the search time became proportional to the square root of the number of items.

An important caveat with the Grover algorithm is that quantum computers today cannot process large enough collections of data so that it is practically more efficient than classical methods. Nevertheless, it is an excellent example of where and how we might get a quantum advantage.

PART III

Advanced Features and Libraries

12

Searching and Changing Text

*If I cease searching, then, woe is me, I am lost. That is
how I look at it – keep going, keep going come what
may.*

—Vincent van Gogh

We represent much of the world's information as text. Think of all the words in all the digital
newspapers, e-books, PDF files, blogs, emails, texts, and social media services such as Twitter
and Facebook. Given a block of text, how do we search it to see if some desired information is
present? How can we change the text to add formatting or corrections or extract information?

Chapter 4, Stringing You Along, covered Python's functions and methods. This chapter
begins with regular expressions and then proceeds to natural language processing (NLP)
basics: how to go from a string of text to some of the meaning contained therein.

Topics covered in this chapter

12.1 Core string search and replace methods

This section lists the primary **str** methods for finding and replacing substrings. Though we discussed several of these in *Chapter 4, Stringing You Along*, it is helpful to have their descriptions together before discussing more advanced regular expression functions.

In each of the following methods that includes optional arguments `start=0` and `end=len(string)`, the search is within the substring given by the slice coordinates `start:end`.

substring in string

Returns `True` if `substring` is in `string`, and `False` otherwise. See section 4.2.

```
"z Fe" in "Franz Ferdinand"
```
```
True
```

substring not in string

Returns `True` if `substring` is not in `string`, and `False` otherwise. See section 4.2.

```
"Ferd" not in "Franz Ferdinand"
```
```
False
```

string.count(substring, start=0, end=len(string))

Returns the number of times that `count` occurs in `string`. *count* does not include overlapping instances of `substring`, so `"ZZZ".count("ZZ")` is 1 and not 2.

```
"a aa aaa aaaa aaaaa".count("aa")
```
```
6
```
```
"a aa aaa aaaa aaaaa".count("aa", 8, 13)
```
```
2
```

string.find(substring, start=0, end=len(string))

Returns -1 if `substring` is not contained in `string`. Otherwise, it returns the index of the first occurrence of `substring` in `string`. See section 4.3.

```
len("a aa aaa aaaa aaaaa")
```
```
19
```
```
"a aa aaa aaaa aaaaa".find("B")
```
```
-1
```

```
"a aa aaa aaaa aaaaa".find("aa")

2

"a aa aaa aaaa aaaaa".find("aa", 8, 13)

9
```

string.rfind(substring, start=0, end=len(string))

Returns -1 if substring is not contained in string. Otherwise, it returns the index of the last occurrence of substring in string.

```
"a aa aaa aaaa aaaaa".rfind("B")

-1

"a aa aaa aaaa aaaaa".rfind("aa")

17

"a aa aaa aaaa aaaaa".rfind("aa", 8, 13)

11
```

string.index(substring, start=0, end=len(string))

Raises **ValueError** if substring is not contained in string. Otherwise, it returns the index of the first occurrence of substring in string.

```
"a aa aaa aaaa aaaaa".index("B")

ValueError: substring not found

"a aa aaa aaaa aaaaa".index("aa")

2
```

string.rindex(substring, start=0, end=len(string))

Raises **ValueError** if substring is not contained in string. Otherwise, it returns the index of the last occurrence of substring in string.

```
"a aa aaa aaaa aaaaa".rindex("B")

ValueError: substring not found

"a aa aaa aaaa aaaaa".rindex("aa")

17
```

string.startswith(substring_or_tuple, start=0, end=len(string))

If `substring_or_tuple` is a **str**, return `True` if string begins with `substring`. If `substring_or_tuple` is a tuple of strings, return `True` if string begins with any of them. Return `False` otherwise. See section 4.7.

```
"# This is a Python comment".startswith('#')
```

```
True
```

```
"abcdef".startswith(('b', 'B'), 1)
```

```
True
```

string.endswith(substring_or_tuple, start=0, end=len(string))

If `substring_or_tuple` is a **str**, return `True` if string terminates with `substring`. If `substring_or_tuple` is a tuple of strings, return `True` if string terminates with any of them. Return `False` otherwise. See section 4.7.

```
"Is this a question?".endswith('?')
```

```
True
```

```
"This could be a sentence!".endswith(('.', '!', '?'))
```

```
True
```

You might want to call one of *strip*, *lstrip*, or *rstrip*, before using *startswith* or *endswith*.

Exercise 12.1

Use *startswith* and *endswith* to test whether a string is a palindrome. Show the code you would use to make the test case-insensitive.

string.replace(old_substring, new_substring, count=-1)

Without the count argument, *replace* substitutes all instances of `old_substring` with `new_substring` in string. When count is given and is a non-negative **int**, *replace* does, at most, count substitutions, beginning at the start of string. See section 4.4.2.

```
"-A-A-A-A-A-A-A-".replace("-A-", "=B=")
```

```
'=B=A=B=A=B=A=B='
```

```
"-A-A-A-A-A-A-A-".replace("-A-", "=B=", 2)
```

```
'=B=A=B=A-A-A-A-'
```

Exercise 12.2

Define what a *rreplace* function should do and implement it.

12.2 Regular expressions

Suppose you post a question on an online software developer channel or website asking about some complex text search problem. Someone is likely to give a regular expression answer written as a combination of most of the letters and punctuation marks you can imagine.

Python's engine to process the expression and find your text via regular expressions is in the **re** module. [PYR2] We also use regular expressions to perform sophisticated text substitutions.

Let's begin by thinking about how we would find the substring "ecos" in

```
string = "Economic, ecological ecosystem"
```

Our process should return None if substring is not present in string. If it is, the function returns a tuple where the first item is the index of the first position at which it occurs, and the second is the length of the substring. We call this tuple the *match information*.

We code this in a function, *my_search*. Though it is similar to *find* and *index*, we do not use them in our implementation. That would be cheating!

```
1   def my_search(string, substring):
2       if not string or not substring:
3           return None
4
5       def have_a_match(n):
6           for m in range(0, len(substring)):
7               if n >= len(string):
8                   return False
9               if string[n] != substring[m]:
10                  return False
11
12              n += 1
13          return True
14
```

```
15        for j in range(len(string)):
16            if have_a_match(j):
17                return (j, len(substring))
18
19        return None
```

I show the characters of string and their indices in Figure 12.1.

E	c	o	n	o	m	i	c	,		e	c	o	l	o
0	1	2	3	4	5	6	7	8	9	10	11	12	13	14

g	i	c	a	l		e	c	o	s	y	s	t	e	m
15	16	17	18	19	20	21	22	23	24	25	26	27	28	29

Figure 12.1: The string "Economic, ecological ecosystem" and its indices

Now we try some searches. Check the "edge cases" of the start and end of the string:

```
print(my_search(string, "Eco"))
```

```
(0, 3)
```

```
print(my_search(string, "stem"))
```

```
(26, 4)
```

Test when none of the characters of the substring are in the string:

```
print(my_search(string, "ABC"))
```

```
None
```

Test when we have a partial match that is ultimately not in the string:

```
print(my_search(string, "ecom"))
```

```
None
```

Test searching for our specific substring:

```
print(my_search(string, "ecos"))
```

```
(21, 4)
```

Exercise 12.3

What line would you change if you wanted a case-insensitive search? Rewrite *my_search* to take a keyword parameter, `case_insensitive=False`, and use this option to respect letter case or ignore it.

I now change terminology: instead of "finding a substring," we are "matching a pattern." In the previous example, the pattern is `"ecos"`. This pattern is an initial `'e'`, followed by a `'c'`, followed by an `'o'`, followed by a terminating `'s'`.

If I change "followed by a terminating `'s'`" to "followed by a terminating lowercase character," it is more apparent that I am not matching a literal substring. If we modify *my_search* to support this form of pattern matching, either `"ecom"` or `"ecos"` as the second argument will return the same result.

We write regular expressions in a definition language that allows us to specify both simple and sophisticated patterns with options and conditions. The following section begins our discussion of the core features of regular expressions in Python.

my_search shows the general idea of finding a match to a pattern: move to the right until you are at the beginning of a candidate, keep trying to match characters, and then either succeed or backtrack to where you last made a choice. Repeat as necessary.

Implementations of regular expressions vary across programming languages and applications. Code editors such as Visual Studio Code, vi/vim, Emacs, Sublime Text, TextMate, Notepad++, Bluefish, and PyCharm support varying forms of regular expression search and replace.

12.2.1 Searching and basic character classes

A regular expression often contains special characters and backslashes. Here, I adopt the convention of writing a pattern as a raw Python string, even if it is not strictly necessary. For example, I now write r`"ecos"` instead of `"ecos"`.

Given a pattern, we first transform it into a form the **re** matching engine can use by calling *compile*.

```
import re

string = "Economic, ecological ecosystem"
```

```
pattern = re.compile(r"ecos")
pattern
```

```
re.compile(r'ecos', re.UNICODE)
```

We can now use *search* to locate a substring that matches the pattern.

```
result = pattern.search(string)
print(result)
```

```
<re.Match object; span=(21, 25), match='ecos'>
```

The result is either None or an instance of the **re.Match** class. For convenience, here is a function that displays the key information in a match object:

```
def print_match(result, group_number=0):
    if result is not None:
        # Display the matched substring
        print(f"group({group_number}) = "
            f'{result.group(group_number)}"   ', end="")

        # Display the starting index and index after end of match
        print(f"start({group_number}) = {result.start(group_number)}   "
            f"end({group_number}) = {result.end(group_number)}")
    else:
        print(None)
```

```
print_match(result)
```

```
group(0) = ecos"   start(0) = 21   end(0) = 25
```

I will later describe the role of the group_number argument. For now, note that when group_number is 0, we get information about the entire match.

The function's comments explain what is displayed, but do note that result.end() shows the index *after* the matched substring.

Exercise 12.4

How would you use a slice with result.start() and result.end() to extract the matched substring from the original string?

The pattern r"ecom" does not match a substring:

```
pattern = re.compile(r"ecom")
print_match(pattern.search(string))
```

```
None
```

Two substrings begin with "eco". To find them, we insert a *character class* into our pattern using square brackets:

```
pattern = re.compile(r"eco[ls]")
pattern
```

```
re.compile(r'eco[ls]', re.UNICODE)
```

[ ]

This pattern indicates that we should match the explicit characters 'e', 'c', and 'o', followed by either 'l' or 's'.

```
result = pattern.search(string)
print_match(result)
```

```
group(0) = ecol"   start(0) = 10   end(0) = 14
```

We matched the first occurrence, "ecol". Use *findall* to get a list of all matching substrings:

```
pattern.findall(string)
```

```
['ecol', 'ecos']
```

Exercise 12.5

Why does the following return the same result as the last example?

```
pattern = re.compile("[e][c][o][ls]")
pattern.findall(string)
```

```
['ecol', 'ecos']
```

If we pass *compile* the re.IGNORECASE argument, the search is case-insensitive. You may also use re.I.

Exercise 12.6

Explain what is returned by *compile*, *search*, *findall*, and *finditer* in the following code:

```
pattern = re.compile(r"eco[lns]", re.IGNORECASE)
pattern

re.compile(r'eco[lns]', re.IGNORECASE|re.UNICODE)

print_match(pattern.search(string))

group(0) = Econ"  start(0) = 0  end(0) = 4

pattern.findall(string)

['Econ', 'ecol', 'ecos']

for match in pattern.finditer(string):
    print(match)

<re.Match object; span=(0, 4), match='Econ'>
<re.Match object; span=(10, 14), match='ecol'>
<re.Match object; span=(21, 25), match='ecos'>
```

What character class are we using?

Exercise 12.7

Let

```
string = "[1] Sutor, Robert S. (2019). Dancing with
Qubits."
```

What does each of the following return?

```
re.compile(r"[1]").search(string)
re.compile(r"\[1\]").search(string)
```

What happens if you omit the r before the pattern?

To simplify the code ahead, I now define this helper function:

```
def print_matches(string, pattern):
    for match in re.compile(pattern).finditer(string):
        print(match)
```

12.2.2 More general patterns

When we have a character class containing a single character, we may omit the surrounding square brackets. This is why the pattern r"ABC" is the same as "[A][B][C]".

Alphanumeric characters

Use a hyphen "-" to specify a range of characters within a character class in brackets. Instead of using this pattern for matching a lowercase letter:

r"[abcdefghijklmnopqrstuvwxyz]"

the simpler

r"[a-z]"

suffices. To match an upper or lowercase letter, use the pattern

r"[a-zA-Z]"

To match a decimal digit, use r"[0-9]" or r"\d". To match any character that is **not** a decimal digit, use either r"[^0-9]" or r"\D". When you use "^" inside square brackets, it negates what immediately follows.

\d
\D

To match an alphanumeric character, which in Python is a letter or a decimal digit or an underscore, use \w. Use \W to match a non-alphanumeric character.

\w
\W

To match any character, use a period, ".".

Beginnings and endings

When "^" starts a pattern like r"^Eco", it means "match "Eco" only at the start of the string." A dollar sign "$" at the end of a pattern means to match at the end of the string.

^
$

```
string = "economic ecosystem center"

# match 'e' followed by any lowercase letter

re.compile(r"e[a-z]").findall(string)

['ec', 'ec', 'em', 'en', 'er']
```

```
# match 'e' at the beginning of the string
# followed by any lowercase letter

re.compile(r"^e[a-z]").findall(string)

['ec']

# match any character that is not 'e' that is
# followed by any lowercase letter

re.compile(r"[^e][a-z]").findall(string)

['co', 'no', 'mi', ' e', 'co', 'sy', 'st', ' c', 'nt']

# match 'E' at the beginning of the string
# followed by any lowercase letter

re.compile(r"^E[a-z]").findall(string)

[]
```

Whitespace characters

A whitespace character is a space " ", tab \t, newline \n, carriage return \r, form feed \f, or vertical tab \v. The last two are historic control characters for printers.

To match a specific whitespace character, use a space or one of these escape characters in a pattern. To match any whitespace character, use \s. For anything that is not whitespace, use \S.

```
\s
\S
```

```
string = "economic ecosystem center"

pattern = re.compile(r"c e")
print_match(pattern.search(string))

group(0) = c e"   start(0) = 7   end(0) = 10

pattern = re.compile(r"c\se")
print_match(pattern.search(string))

group(0) = c e"   start(0) = 7   end(0) = 10
```

Optionality and repetition

What if we want to specify that part of a pattern is optional? Use a question mark "?" after the optional part to say that either zero or one occurrence is acceptable.

```
print_matches(string, r"c\s?e")
```

```
<re.Match object; span=(7, 10), match='c e'>
<re.Match object; span=(19, 21), match='ce'>
```

Using the question mark in a pattern is shorthand for {0, 1}. {x, y} means, "match at least x instance but no more than y." If you want to match precisely x instances, use {x}.

```
?
{ }
```

```
string = "economic ecosystem bookkeeping center"
```

```
print_matches(string, r"[eko]{2}")
```

```
<re.Match object; span=(20, 22), match='oo'>
<re.Match object; span=(22, 24), match='kk'>
<re.Match object; span=(24, 26), match='ee'>
```

Exercise 12.8

Why didn't we match `"ok"` or `"ke"`?

There are two other repetition modifiers we frequently use. The first is "*", which specifies that the **re** engine should match zero or more instances. Use a "+" to match one or more.

```
*
+
```

```
re.compile(r"[cb]o*").findall(string)
```

```
['co', 'c', 'co', 'boo', 'c']
```

```
re.compile(r"[cb]o+").findall(string)
```

```
['co', 'co', 'boo']
```

You can use regular expressions for data format verification. This example shows how to validate a 10-digit phone number using hyphens and the North American Numbering Plan:

```
string = "914-555-1234"

pattern = re.compile(r"[2-9]\d{2}-[2-9]\d{2}-\d{4}")
print_match(pattern.search(string))

group(0) = 914-555-1234"  start(0) = 0  end(0) = 12
```

The first and fourth digits cannot be 0 or 1.

Word boundaries

The following string has three matches for "red":

```
string = "too much credit reduction can put you in the red"

print_matches(string, r"red")

<re.Match object; span=(10, 13), match='red'>
<re.Match object; span=(16, 19), match='red'>
<re.Match object; span=(45, 48), match='red'>
```

The first is in the middle of the word "credit", the second is at the start of the word "reduction", and the third is "red" itself at the end of the sentence. What if we want the free-standing third case? Use \b at the beginning or end of a pattern to signify that any match should be at a word boundary.

```
\b
\B
```

```
print_matches(string, r"\bred\b")

<re.Match object; span=(45, 48), match='red'>
```

The match engine only returned the third case, "red", at the end of the sentence. If we put \b at the start but not the end, we get both "reduction" and "red".

```
print_matches(string, r"\bred")

<re.Match object; span=(16, 19), match='red'>
<re.Match object; span=(45, 48), match='red'>
```

Use \b to specify that the match should **not** be at a word boundary.

```
print_matches(string, r"\Bred\B")

<re.Match object; span=(10, 13), match='red'>

print_matches(string, r"red\B")

<re.Match object; span=(10, 13), match='red'>
<re.Match object; span=(16, 19), match='red'>
```

Escaping special characters

Characters such as "?", "*", "+", ".", "|", "{", "}", "(", ")", "[", and "]" have special meanings inside patterns, and you must escape them with a backslash if you wish to match them. You must also escape a backslash to match a backslash. This is why we use Python raw strings for regular expression patterns!

Exercise 12.9

Let

```
fstring = 'f"{x} {y} {A_1}"'
```

be a Python f-string. Use *findall* to get a list of all Python identifiers within the braces.

12.2.3 Groups

Use parentheses when you need to group a part of the pattern. In this example, I'm using grouping and the "or" special character "|" to match "red" or "green" or "blue".

( )

```
string = "Do you have the red piece, the green piece, and the blue
piece?"

print_matches(string, r"(red)|(green)|(blue)")

<re.Match object; span=(16, 19), match='red'>
<re.Match object; span=(31, 36), match='green'>
<re.Match object; span=(52, 56), match='blue'>
```

What if we want "red piece" or "green piece" or "blue piece"?

```
print_matches(string, r"(red)|(green)|(blue) piece")

<re.Match object; span=(16, 19), match='red'>
<re.Match object; span=(31, 36), match='green'>
<re.Match object; span=(52, 62), match='blue piece'>
```

That didn't work as you might have expected because the **re** engine tried to match "red" or "green" or "blue piece". It took everything after the second "|" to be one of the "or" possibilities.

Use an extra pair of parentheses to get what we intended:

```
print_matches(string, r"((red)|(green)|(blue)) piece")

<re.Match object; span=(16, 25), match='red piece'>
<re.Match object; span=(31, 42), match='green piece'>
<re.Match object; span=(52, 62), match='blue piece'>
```

When specifying alternatives with "|", put those most likely to match furthest to the left. In this case, if we doubt that "red" will appear much, put it at the end instead of the beginning. [FIT]

The **re** engine gives us access to what it matched in each group in a pattern. Suppose I define a simple pattern for a name where we try to match a capitalized first name, a capital middle initial, and a capitalized last name:

```
pattern = re.compile(r"\b([A-Z][a-z]*)\s+([A-Z])\.\s+([A-Z][a-z]*)\b")
result = pattern.search("My name is Robert S. Sutor")
print_match(result)

group(0) = Robert S. Sutor"  start(0) = 11  end(0) = 26
```

I know that this overly simple example does not cover all possible names using all possible alphabets.

Exercise 12.10

Explain in words the meaning of each part of the `pattern` regular expression.

A successful match always has at least one group, and it is group number 0, the entire match. That's why my earlier definition of *print_match* takes a default value for the group_number argument:

```python
def print_match(result, group_number=0):
    if result is not None:
        # Display the matched substring
        print(f"group({group_number}) = "
            f'{result.group(group_number)}"   ', end="")

        # Display the starting index and index after end of match
        print(f"start({group_number}) = {result.start(group_number)}   "
            f"end({group_number}) = {result.end(group_number)}")
    else:
        print(None)
```

Since I put explicit groups in the pattern with parentheses, I can access them as well, starting with number 1:

```python
for g in range(1, 4):
    print_match(result, g)
```

```
group(1) = Robert"   start(1) = 11   end(1) = 17
group(2) = S"   start(2) = 18   end(2) = 19
group(3) = Sutor"   start(3) = 21   end(3) = 26
```

Use *len* on the value returned from the *groups* method to find out how many groups were matched.

```python
len(result.groups())
```

```
3
```

Exercise 12.11

Define a function *number_of_groups* that returns the obvious value whether a search is successful or not.

If we define a group within a group, the numbering descends before continuing to the right.

```python
pattern = re.compile(r"\b([A-Z]([a-z]*))\s+([A-Z])\.\s+([A-Z][a-z]*)\b")
result = pattern.search("My name is Robert S. Sutor")

for g in range(1 + len(result.groups())):
    print_match(result, g)
```

```
group(0) = Robert S. Sutor"  start(0) = 11  end(0) = 26
group(1) = Robert"  start(1) = 11  end(1) = 17
group(2) = obert"  start(2) = 12  end(2) = 17
group(3) = S"  start(3) = 18  end(3) = 19
group(4) = Sutor"  start(4) = 21  end(4) = 26
```

Here, group 2 is the part of the first name after the first letter. Group 3 is the middle initial.

We also use group numbers for *back-references*. In the following pattern, "\1" means "what you matched for group 1 must also appear here verbatim":

```python
pattern= re.compile(r"\b(.+)\s+City\s+is\s+in\s+\1\b")
```

\1

```python
print_match(pattern.search("New York City is in New York"))
```

```
group(0) = New York City is in New York"  start(0) = 0  end(0) = 28
```

```python
print_match(pattern.search("Iowa City is in Iowa"))
```

```
group(0) = Iowa City is in Iowa"  start(0) = 0  end(0) = 20
```

```python
print_match(pattern.search("Jefferson City is in Missouri"))
```

```
None
```

If you want to use a group but have no intention of referencing it later, insert "?:" immediately after the opening parenthesis.

```python
pattern = re.compile(r"\b(?:[A-Z][a-z]*)\s+([A-Z])\.\s+(?:[A-Z][a-z]*)\b")
result = pattern.search("My name is Robert S. Sutor")
for g in range(1 + len(result.groups())):
    print_match(result, g)
```

```
group(0) = Robert S. Sutor"  start(0) = 11  end(0) = 26
group(1) = S"  start(1) = 18  end(1) = 19
```

In this example, I only want the middle initial when it appears in the context of a name.

Exercise 12.12

If you understand that last regular expression, congratulate yourself!

The Python **re** module has an extension that allows you to name groups. You can still reference them with numbers, but using names makes your regular expressions (slightly) more self-documenting.

?P

```
pattern= re.compile(r"\b(?P<city>.+)\s+City\s+is\s+in\s+(?P=city)\b")

result = pattern.search("New York City is in New York")

result.group("city")
'New York'
```

In this example, I named and defined the group with (?P<city>.+). I made a back-reference to it with (?P=city). The ?P indicates that this syntax is a Python-specific extension and may not work with other languages or applications.

re makes it easy to get a dictionary of all named groups in a match:

```
result.groupdict()
{'city': 'New York'}
```

12.2.4 Substitution

Once we have a match with a regular expression, what do we do with it? One application is substituting new text into the string where the matching engine located the pattern. This does not change the original string, but returns a new string with the changes. Think of the **re** *sub* function as a super-charged version of *replace*.

Since we now know about word boundaries and groups, we can improve our regular expression pattern for matching phone numbers using the North American Numbering Plan:

```
pattern = re.compile(r"\b([2-9]\d{2}-[2-9]\d{2}-\d{4})\b")
```

Now we match against a string. Since I used parentheses, group 1 holds the phone number.

```
string = "My phone number is 914-555-1234."
result = pattern.search(string)
print_match(result, 1)

group(1) = 914-555-1234"  start(1) = 19  end(1) = 31
```

With replace, I would have needed to match against a specific phone number. With the regular expression, I can match against anything that *looks* like a phone number.

For privacy reasons, I might want to mask the number with asterisks. This substitution will do that:

```
pattern.sub("***-***-****", string)

'My phone number is ***-***-****.'
```

I can use a back-reference to a group to modify the general text that the engine matched. Here, I am adding a US country code:

```
pattern.sub(r"+1-\1", string)

'My phone number is +1-914-555-1234.'
```

By default, *sub* works on all non-overlapping matches. If you give it a third positive **int** argument, it will perform at most that many substitutions. The default value for this argument is 0, which means "make all substitutions."

```
string = "212-555-0123   585-555-9876   320-555-3467"
pattern.sub("***-***-****", string)

'***-***-****   ***-***-****   ***-***-****'

pattern.sub(r"+1-\1", string, 2)

'+1-212-555-0123   +1-585-555-9876   320-555-3467'
```

The *subn* function returns the new string in a tuple with the number of substitutions performed.

```
string = "212-555-0123   185-555-9876   320-555-3467"
pattern.subn("***-***-****", string)

('***-***-****   185-555-9876   ***-***-****', 2)
```

Exercise 12.13

Why were there two substitutions and not three?

Our final regular expression function is *split,* an advanced version of the core **str** method we saw in section 4.8.

I want to split a string that consists of several of my cats' names and the ages they would have been now. The result I expect is a list of the cats' names.

```
string = "Hilbert 40 Charlie 35 Gideon 28"
```

I will split at any match that is optional whitespace followed by two digits, followed by optional whitespace.

```
pattern = re.compile(r"\s*\d{2}\s*")
pattern.split(string)
```

```
['Hilbert', 'Charlie', 'Gideon', '']
```

To include the matches in the list, put parentheses in the pattern.

```
pattern = re.compile(r"(\s*\d{2}\s*)")
pattern.split(string)
```

```
['Hilbert', ' 40 ', 'Charlie', ' 35 ', 'Gideon', ' 28', '']
```

Exercise 12.14

What is different about the patterns and the results between the last example and the following one?

```
pattern = re.compile(r"\s*(\d{2})\s*")
pattern.split(string)
```

```
['Hilbert', '40', 'Charlie', '35', 'Gideon', '28', '']
```

Exercise 12.15

Why do we see the empty string in the result? Can you remove it by modifying the pattern or by some other method?

12.2.5 Advice for working with regular expressions

Though I have covered much of the Python implementation of regular expressions, there are several additional features and subtleties that you can read about in the official documentation. [PYR1] [PYR2] Here are a few points you should remember as you start using this powerful and sometimes complicated tool:

- If **str** methods, such as *find* and *replace*, accomplish your task, use those instead of bringing in the extra **re** machinery.

- Remember that *search* returns None when it does not find a match.

- Keep in mind the size of the string you are searching. You might choose a different approach if the text is 50 characters versus 5 megabytes. For example, you might use *finditer* instead of *findall* in the latter case.

- Be aware of diminishing returns when you are creating the regular expression pattern. For example, you can use *strip* on the strings in the list returned by *split*.

- Search the web for regular expressions that may come close to doing what you want.

- Regular expression matching can be very slow with badly written patterns. Be as specific as you can. Try not to use ".*" instead of \w* if you know you are looking for alphanumeric characters, for example. [FIT]

- It's easy to forget the regular expression syntax and all the special characters. Don't be scared to look them up when you need them.

- When creating a pattern, ask yourself, "should I be using '\b' or '\w' somewhere?".

12.3 Introduction to Natural Language Processing

Natural Language Processing (NLP) takes text and its contained characters to meaning. From meaning, we get knowledge and insights and possibly take action. This section surveys some of the tools available to you in Python to do NLP. Please follow the references to find more information and examples about the specific tools.

Suppose I stand in my kitchen and say,

"I'm making more fried green tomatoes."

How does that translate into my cooking the food?

The first step is knowing if anyone or anything is listening to me, but that's not a coding issue. Perhaps, my daughter is with me in the kitchen, or a virtual personal assistant (VPA) like Apple's Siri®, Amazon's Echo® with Alexa®, Google Home™, or Microsoft Cortana®.

12.3.1 Speech-to-text

Computer speech-to-text takes spoken words and produces text that we can then analyze. The speech recognition engines are often on the cloud and use sophisticated AI techniques. The result's accuracy can vary significantly by spoken language, rate of speaking, accent, pronunciation, and voice volume. Python libraries are frontends to these cloud services: you import a library and call functions or methods, and they connect to the cloud and invoke the speech recognition APIs, and you get the result as a string.

Examples of cloud-based services with Python libraries are IBM Watson™ Speech to Text **ibm-watson** [IST], Amazon Transcribe **amazon-transcribe** [AST], and Google Speech-to-Text API **google-cloud-speech** [GST].

12.3.2 Data cleaning

Now, let's assume we have text in the form of a Python string. It may have come from speech-to-text, from reading a file or website, from a database, or from the user typing on a mobile device or keyboard. While you might need to analyze a single sentence or utterance, it's also possible that you could need to look through megabytes of documents. Performance may or may not be a significant issue or bottleneck for you.

Suppose the request for the recipe came to me as:

```
the_text ="I'm making mor freid green tomatos."
the_text

"I'm making mor freid green tomatos."
```

This text has three spelling mistakes. As part of my data cleaning process, I will fix common misspellings.

```
corrections = {
  "accross":    "across",
  "definately": "definitely",
  "freid":      "fried",
  "mor":        "more",
  "neeed":      "need",
  "tomatos":    "tomatoes",
  "wierd":      "weird"
}
```

Maybe we can use string replacement:

```
def clean_text(text):
    for wrong, right in corrections.items():
        text = text.replace(wrong, right)
    return text

result = clean_text(the_text)
result
```

```
"I'm making more fried green tomatoes."
```

Not bad! Let's see what it does if I call it again:

```
clean_text(result)
```

```
"I'm making moree fried green tomatoes."
```

Whoops. We forgot about word boundaries. We should try regular expressions.

```
import re

def clean_text(text):
    for wrong, right in corrections.items():
        pattern = re.compile(fr"\b({wrong})\b")
        text = pattern.sub(f"{right}", text)
    return text

result = clean_text(the_text)
result
```

```
"I'm making more fried green tomatoes."
```

```
clean_text(result)
```

```
"I'm making more fried green tomatoes."
```

 Exercise 12.16

Why did I use `fr` in the regular expression pattern string?

The regular expression approach worked. In practice, we would use a much larger list of spelling corrections. We might also convert city, state, or country abbreviations to their full names. We could also expand contractions so that "I'll" becomes "I will", for example. Another common task is removing extraneous whitespace.

Exercise 12.17

Extend *clean_text* to reduce any occurrence of two or more consecutive spaces to a single space. Do the same for successive newlines.

We use *sub* in *clean_text* to scan the entire string for every wrong word-right word pair. The **flashtext** module can be much more efficient for hundreds of substitutions on large amounts of text. **flashtext** is not part of the Python Standard Library, and you must install it from the operating system command line.

```
pip install flashtext
```

To process the replacements, import **KeywordProcessor** from **flashtext**, load the corrections, and call *replace_keywords*:

```
from flashtext import KeywordProcessor

keyword_processor = KeywordProcessor(case_sensitive=True)

for wrong, right in corrections.items():
    keyword_processor.add_keyword(wrong, right)

result = keyword_processor.replace_keywords(the_text)
result

"I'm making more fried green tomatoes."

keyword_processor.replace_keywords(result)

"I'm making more fried green tomatoes."
```

For more sophisticated matching beyond exact text within word boundaries, use regular expressions.

12.3.3 NLP with spaCy

To go beyond characters, words, numbers, whitespace, and patterns, we need more sophisticated tools trained on many documents to recognize parts of speech, word roots, phrases, structure, names of people and locations, and other language features. One such Python tool is spaCy, which you install from the command line as the **spacy** package.

```
pip install spacy
python -m spacy download en_core_web_sm
```

This section introduces the basic functionality of **spacy**, but I encourage you to read the tutorial and documentation to see more examples and the breadth of functionality. [SCY]

Tokenization

We begin by importing **spacy** and loading the "small English database" into the `nlp` engine.

```
import spacy

nlp = spacy.load("en_core_web_sm")
```

We now process our document, which is the string we have been using for the previous examples. In practice, the document might be a paragraph, an article, a book, a web page, a collection of social media entries, a customer's purchase history, chatbot questions, or any *corpus* of text we wish to analyze. The result is a sequence of *tokens*:

```
doc = nlp("I'm making more fried green tomatoes.")

print(" | ".join([token.text for token in doc]))

I | 'm | making | more | fried | green | tomatoes | .
```

A token is a word or word fragment, and *tokenization* is the process of creating tokens.

Lemmas and parts of speech

In addition to its contained characters, each `token` in the document includes language properties. Here is a formatted table of the document's tokens and some of their properties:

```
for token in doc:
    print(f"{token.text:9} {token.lemma_:8} "
          f"{token.pos_:8} {token.dep_:8}")

I         I        PRON     nsubj
'm        be       AUX      aux
```

making	make	VERB	ROOT
more	more	ADV	advmod
fried	fried	ADJ	amod
green	green	ADJ	amod
tomatoes	tomato	NOUN	dobj
.	.	PUNCT	punct

Column 1 contains the tokens themselves. The only surprising entries are the first and second, where **spacy** split "I'm" into "I" and "'m". If we had first expanded the contraction to "I am", we would have seen two tokens as well.

Column 2 lists the word roots, also known as *lemmas*. In *lemmatization*, we choose a unique word among all the forms it can take on. Here, "'m" has the lemma "be", as would "am," "are," "is," "be," "was," and so forth. The lemma for "making" is "make," and the lemma for "tomatoes" is "tomato."

Column 3 shows the parts of speech. [SCA]

- ADJ: adjective
- ADV: adverb
- AUX: auxiliary verb
- NOUN: noun
- PRON: pronoun
- PUNCT: punctuation
- VERB: verb

There are a dozen other parts, including ADP for adposition (often a preposition), PROPN for proper noun, CONJ for conjunction ("and," "but," "or"), and DET for determiner ("a," "an," "the").

Dependencies

Column 4 contains the dependency tags.

- advmod: adverbial modifier
- amod: adjectival modifier
- aux: auxiliary
- dobj: the direct object
- nsubj: the nominal subject
- punct: punctuation
- root: the root verb of the sentence

These and the forty other dependency labels go beyond what most people learn in school. When you are trying to understand a sentence and its structure, you may need this level of detail.

This breakdown of our sentence allows us to analyze it:

- Start at the root and ask, "What is being done?": **making**
- "Is it extended with an auxiliary verb?" [aux]: yes – **am**
- "Who is making?" [nsubj]: **I**
- "What is being made?" [dobj]: **tomatoes**
- "Do you have more information on **tomatoes**?" [amod]: yes – **green tomatoes**
- "Do you have more information on **green tomatoes**?" [amod]: yes – **fried green tomatoes**
- "Do you have more information on **fried green tomatoes**?" [advmod]: yes – **more fried green tomatoes**
- "Are we done?" [punct]: yes!

It's easier to see this in the dependency visualization in Figure 12.2 generated by **spacy**:

```
from spacy import displacy


nlp = spacy.load("en_core_web_sm")
doc = nlp("I'm making more fried green tomatoes.")
format_options = {"distance": 125, "add_lemma": True}


displacy.render(doc, style="dep", options=format_options)
```

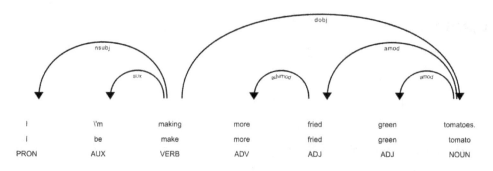

Figure 12.2: Sentence visualization generated by spaCy

render produces a Scalable Vector Graphics (SVG) file, which you can either use directly or convert to a jpg or png image. The `options` keyword argument is a dictionary that controls several aspects of the diagram's formatting. `distance` determines how far apart the words should be in pixels. The default is 175, but I chose 125 to have a more compact display. Via `add_lemma`, I told *render* to display the lemmas under the words in the sentence.

Named entities

Now, let's look at named entities. **spacy** has an extensive database it uses to guess when a word or a phrase is a specific person, organization, place, quantity, date, amount of money, and so on. I say "guess" because choosing the correct entity depends on context. If I have a sentence discussing Fiona Apple's songs and then start the next sentence with "Apple," am I still talking about the singer, or am I now referring to the fruit or the computer company? A better and more accurate phrase than "guess" is "statistically predicts."

spacy automatically does named entity recognition while digesting text.

```
doc = nlp("William Shakespeare ate crumpets in England in 1597.")

for ent in doc.ents:
    print(f'{ent.label_}: "{ent.text}"')
PERSON: "William Shakespeare"
GPE: "England"
DATE: "1597"
```

The three named entities in the sentence are:

- William Shakespeare: PERSON
- England: GPE (geopolitical entity)
- 1597: DATE

spacy can help you visualize entities by generating annotated HTML, as shown in Figure 12.3.

```
displacy.render(doc, style="ent")
```

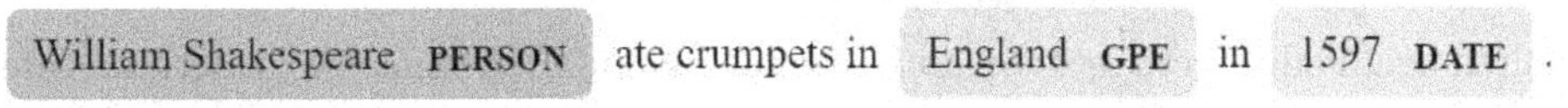

William Shakespeare **PERSON** ate crumpets in England **GPE** in 1597 **DATE** .

Figure 12.3: Entity visualization generated by spaCy

Exercise 12.18

What entities does **spacy** recognize in the sentence: [CNW]

"American Bill Gates was estimated to be worth $60 billion in 2000."

Matching

Now that we know some of what **spacy** can recognize, let's do some matches based on literal text, lemmas, and parts of speech. We begin by importing the **Matcher** class.

```
from spacy.matcher import Matcher
```

My goal is to create a pattern that can determine whether someone in my family is cooking, did cook, or will cook some food. A **spacy** pattern is a list of Python dictionaries. From top to bottom, I am looking for parts of the document that match these components:

```
1  pattern = [
2      {"TEXT": {"IN": ["I", "brother", "father", "mother", "sister"]}},
3      {"OP": "*"},
4      {"POS": "VERB", "LEMMA": {"IN": ["bake", "cook", "make"]}},
5      {"OP": "*"},
6      {"LEMMA": {"IN": ["cake", "peach", "bean", "pea", "tomato"]}}
7  ]
```

The pattern begins on line 2 with explicit TEXT that can be IN a list of possibilities. If I only wish to match "brother", I would have coded this as {"TEXT": "brother"}. You can use ORTH instead of TEXT if you wish to be more linguistic-technical.

On line 3, I allow zero or more instances of anything via {"OP": "*"}. This form is similar to the regular expression r".*".

Line 4 is the most interesting so far. It says, "match a verb with the root-form "bake", cook, or make."

I then again allow zero or more arbitrary tokens, and I terminate the pattern on line 6 by trying to match one of a list of foods. I use the lemma forms so I can match the singular or plural forms of the foods.

We define and use a *callback function* when we want to be notified when something happens in another function. For example, we might pass the callback to a function containing a conditional test. That function invokes the callback when the test evaluates to True. The keyword argument for the callback usually has a None default value.

When **spacy** finds a match, it creates a tuple of an internal identifier for the rule that matched, the starting token index, and the index after the ending token. We code the callback function *on_ match* to print the document part that matches:

```
def on_match(matcher, doc, id, matches):
    # Callback function for matcher
    # Show the span of text that we matched
    for id, start, end in matches:
        print(doc[start:end])
```

In your work, you may want to display more information, such as the parts of speech. Now we build a matcher object and add the pattern to it. The first argument to *add* is a name for a list of patterns, and the second is that list. I also pass in *on_ match*.

```
matcher = Matcher(nlp.vocab)
matcher.add("Cooking", [pattern], on_match=on_match)
```

Now we see what **spacy** yields for three sentences about family cooking:

```
doc = nlp("I'm making more fried green tomatoes.")
matches = matcher(doc)

I'm making more fried green tomatoes

matches = matcher(nlp("My sister cooked peas."))

sister cooked peas

matches = matcher(nlp("My father will bake three cakes."))

father will bake three cakes
```

Since I no longer need the "Cooking" list of patterns, I remove it from the matcher.

```
matcher.remove("Cooking")
```

I could just as quickly have created a new matcher with no rules since there was only one rule in the set.

To conclude this section, we look at phone numbers again. **spacy** has a basic facility for showing and specifying the "shape" of a token. For example, the shape of "I" is "X", the shape of "can" is "xxx", and the shape of "123" is "ddd". In the last case, 'd' represents a decimal digit. We can match against shapes, but only up to four characters.

```
pattern = [
    {"SHAPE": "ddd"}, {"TEXT": "-"},
    {"SHAPE": "ddd"}, {"TEXT": "-"},
    {"SHAPE": "dddd"}
]

matcher.add("PhoneNumber", [pattern], on_match=on_match)
matches = matcher(nlp("My phone number is 914-555-1234."))
```

```
914-555-1234
```

We found the phone number in the sentence, but unlike our match using regular expressions, we cannot use a shape to say that a digit must be in a given range. **spacy** does allow you to incorporate regular expressions into more advanced patterns.

```
pattern = [
    {"TEXT": {"REGEX": r"[2-9]\d{2}"}},
    {"TEXT": "-"},
    {"TEXT": {"REGEX": r"[2-9]\d{2}"}},
    {"TEXT": "-"},
    {"TEXT": {"REGEX": r"\d{4}"}}
]

matcher = Matcher(nlp.vocab)
matcher.add("PhoneNumber", [pattern], on_match=on_match)
matches = matcher(nlp("My phone number is 914-555-1234."))
```

```
914-555-1234
```

Exercise 12.19

Write an interactive Python chatbot that carries on a conversation with a user about their favorite foods and when they first ate them. Use the function *input* to prompt for keyboard input.

12.3.4 Other NLP tools to consider

I used spaCy in this chapter because it is relatively new and quite fast. You have other NLP options in Python, and even more if you use tools written in other languages or for specialized purposes. You should choose the package that gives you the ease of use, speed, and multiple language support that you need for your work. I think the two other top candidates for your consideration are

- TextBlob – easy to use and to get started, perhaps not as fast as spaCy. [TBL]
- Natural Language Toolkit (NLTK) – extensive and has almost everything you can imagine for linguistic analysis. [NLTK]

I consider spaCy as roughly between TextBlob and NLTK, though their authors might debate that. Since the three packages are constantly evolving, experiment with each to see which performs best and most easily for your tasks.

Another option is a cloud-based NLP service. The major providers, such as IBM that I mentioned for speech-to-text in section 12.3.1, have NLP APIs and can do all computation and data storage on the cloud. [INL]

12.4 Summary

In this chapter, we looked at text. We started with strings and the characters in them and moved on to regular expressions to perform sophisticated matching and substitution. While character patterns are interesting, they do not tell us much about linguistic content. For that, we saw many examples of natural language processing (NLP) using the **spacy** Python package.

Understanding text is complicated, even for people sometimes. Use these tools carefully, and do not assume your results represent the absolute truth. For example, compare

"I love your new apartment!"

and

"I love that my boss bought a €150,000 sports car
and I bought a 20-year old rusty van."

There is "love" in each, but it is easy to miss the sarcasm in the second.

13

Creating Plots and Charts

My last word is that it all depends on what you
visualize.

—Ansel Adams

Among mathematicians and computer scientists, it's said that a picture is worth 2^{10} words. Okay, that's a bad joke, but it's one thing to manipulate and compute with data, but quite another to create stunning visualizations that convey useful information.

While there are many ways of building images and charts, Matplotlib is the most widely used Python library for doing so. [MAT] Matplotlib is very flexible and can produce high-quality output for print or digital media. It also has great support for a wide variety of backends that give you powerful mouse-driven interactivity. Generally speaking, if you have a coding project and you need to visualize numeric information, see if Matplotlib already does what you want. This chapter covers the core functionality of this essential library.

Topics covered in this chapter

13.1 Function plots

This section introduces **matplotlib** and develops the techniques to create the plots in Figure 13.1.

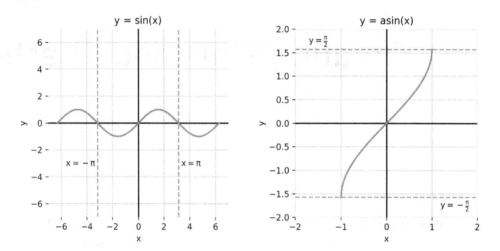

Figure 13.1: Plots of the sine and inverse sine functions

We use the **matplotlib.pyplot** module, which we always import as `plt`.

13.1.1 Plotting a point

In its most basic form, the *plot* function takes a point's *x* coordinate, *y* coordinate, and some keyword arguments that state how to display the point.

```
import matplotlib.pyplot as plt
```

```
the_plot = plt.plot(1, 0.5, color="black", marker='o')
```

```
<Figure size 432x288 with 1 Axes>
```

The *show* function displays the plot.

```
plt.show()
```

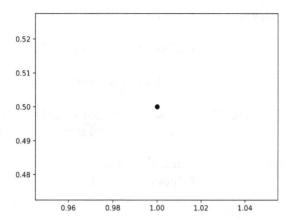

The `color` keyword argument sets the color of the point, and you can use `"black"`, `"blue"`, `"cyan"`, `"green"`, `"magenta"`, `"red"`, `"white"`, and `"yellow"`. You can also use HTML color names like `"ForestGreen"` or `"MediumSlateBlue"`, or hexadecimal RGB color specifications like `"#4169E1"`.

Exercise 13.1

What is the HTML color name corresponding to `"#4169E1"`?

The `marker` keyword argument sets the shape of the point. In this example, I use `"o"`, which is a filled circle. The **matplotlib.pyplot** module documentation lists the 22 marker shapes, including `"s"` for a square, `"*"` for a star, and `"x"` for an "×".

We call all the data and the style information the *figure*. **matplotlib** uses many assumptions and default settings when it creates this very bare plot. Let's see how we can begin to customize how the plot looks.

13.1.2 Adding the grid, axes, and title

The plot has no drawn axes, nor does it have a grid in the background to better see our point's position. Since we will have many plots, we should insert a title.

```python
import matplotlib.pyplot as plt

plt.clf()                      # clear the previous figure

plt.grid()                     # include a background grid

plt.axhline(color="black")  # draw a horizontal axis in black
plt.xlabel('x')                # label the horizontal axis as 'x'

plt.axvline(color="black")  # draw a vertical axis in black
plt.ylabel('y')                # label the vertical axis as 'y'

plt.title("Plot with grid, axes, and title")  # set the title

the_plot = plt.plot(1, 0.5, color="black", marker='o')

<Figure size 432x288 with 1 Axes>

plt.show()
```

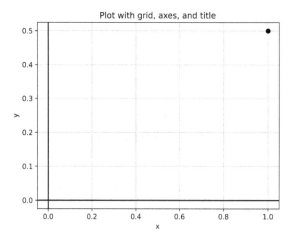

The flow of processing in this example is:

- Clear the previous data and style information with the *clf* function. *clf* is short for "clear figure."

- State that we want to display a *grid* by using the function with that name. **matplotlib** is sensitive to when you call *grid* and some other functions. If you do not see a visual feature displayed, move the function call higher or lower in the code sequence.

- Call *axhline* to get a **horizontal line**. Label it with 'x' via *xlabel*.

- Call *axvline* to get a **vertical line**. Label it with 'y' via *ylabel*.

- Use the *title* function with a string to display a title caption over the plot.

- Draw the point using a black "•" marker.

13.1.3 Adjusting the aspect ratio

In the previous sections, we thought of a **matplotlib** figure containing one plot, and the module makes it easy to add lines and points. A figure can include multiple plots, and each corresponds to an `axes` object. Although we will be working with a single plot for several more sections, we will now be more precise and use `axes` to specify the plot elements.

So that I don't need to repeat the same code multiple times, I now define the first version of a helper function to do my usual plot initializations.

```
 1  def initialize_plot(x_label, y_label, title_label=''):
 2      plt.clf()                       # clear the figure
 3
 4      figure, axes = plt.subplots()   # set the plot layout
 5
 6      axes.grid()                     # include a background grid
 7
 8      axes.axhline(color="black")     # draw a horizontal axis in black
 9      if x_label:
10          axes.set_xlabel(x_label)    # label the horizontal axis
11
12      axes.axvline(color="black")     # draw a vertical axis in black
13      if y_label:
14          axes.set_ylabel(y_label)    # label the vertical axis
15
16      if title_label:
17          axes.set_title(title_label) # set the title
18
19      return figure, axes
```

The assignment

$$\text{figure, axes = plt.subplots()}$$

returns the `figure` and an `axes` for our plot. By default, the *subplot* function returns a single `axes` object, but we'll use it later with arguments to get several.

initialize_plot calls methods like *grid, axhline,* and *axvline* on `axes` rather than using module-level functions. It also uses the *set_xlabel, set_ylabel,* and *set_title* methods to set text for the *x*-axis, *y*-axis, and plot title, respectively.

I have placed *initialize_plot* in the "`mpl_helper_1.py`" file so that I can import the **mpl_helper_1** module whenever I need it.

So far, **matplotlib** has been deciding the numeric ranges shown on the horizontal and vertical axes. It scaled those axes so that the length of a unit on one axis was not the same on the other axis. While this behavior is often convenient when the ranges are very different, I want our plot to have equal axes scaling. To do this, I call the *set_aspect* `axes` method.

```
import matplotlib.pyplot as plt
from mpl_helper_1 import initialize_plot

figure, axes = initialize_plot('x', 'y',
    "Setting the aspect ratio for the axes")


# -----------------------------------------------------------
# set the aspect ratio of vertical to horizontal
axes.set_aspect('equal')
# -----------------------------------------------------------


the_plot = axes.plot(1, 0.5, color="black", marker='o')

<Figure size 432x288 with 0 Axes>
<Figure size 432x288 with 1 Axes>

plt.show()
```

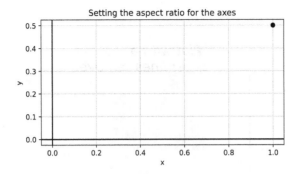

13.1.4 Removing the outer lines

Want to change the size and shape of a point? The style, width, and color of a line? The font for some text? **matplotlib** has a keyword argument for some method or function to do what you want stylistically. Its extensive documentation describes your many options. [MPL]

In this next example, I remove the lines surrounding the plot since I don't like how they look. These lines are *spines*.

```
import matplotlib.pyplot as plt
from mpl_helper_1 import initialize_plot

figure, axes = initialize_plot('x', 'y')

# set the aspect ratio of vertical to horizontal
axes.set_aspect('equal')

# -------------------------------------------------------------
for position in ("top", "right", "bottom", "left"):
    # turn off the extra lines around the plot
    axes.spines[position].set_visible(False)

axes.set_title("Plot with annoying\nouter lines removed",
               fontsize=20, fontstyle="italic")
# -------------------------------------------------------------

the_plot = axes.plot(1, 0.5, color="black", marker='o')

<Figure size 432x288 with 0 Axes>
<Figure size 432x288 with 1 Axes>
```

```
plt.show()
```

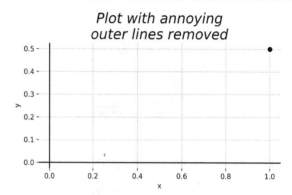

Notice that the title is larger and in italics. I set the title's size and style via the fontsize and fontstyle keyword arguments for *set_title*.

Let's update our initialization to remove spines:

```
1  def initialize_plot(x_label, y_label, title_label):
2      plt.clf()                       # clear the figure
3
4      figure, axes = plt.subplots()   # set the plot layout
5
6      axes.grid()                     # include a background grid
7
8      axes.axhline(color="black")     # draw a horizontal axis in black
9      if x_label:
10         axes.set_xlabel(x_label)    # label the horizontal axis
11
12     axes.axvline(color="black")     # draw a vertical axis in black
13     if y_label:
14         axes.set_ylabel(y_label)    # label the vertical axis
15
16     if title_label:
17         axes.set_title(title_label) # set the title
18
19     for position in ("top", "right", "bottom", "left"):
20         # turn off the extra lines around the plot
21         axes.spines[position].set_visible(False)
22
23     return figure, axes
```

13.1.5 Plotting two points

If I *plot* another point, **matplotlib** shows it and the original point but does not connect them.

```
import matplotlib.pyplot as plt
from mpl_helper_2 import initialize_plot

figure, axes = initialize_plot('x', 'y',
    "Plot with an extra point added")

# set the aspect ratio of vertical to horizontal
axes.set_aspect('equal')

axes.plot(1.0,  0.5,  color="black", marker='o')
axes.plot(0.25, 0.25, color="black", marker='o')

[<matplotlib.lines.Line2D at 0x23176f90af0>]
<Figure size 432x288 with 0 Axes>
<Figure size 432x288 with 1 Axes>

plt.show()
```

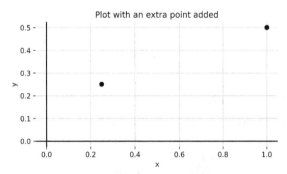

To connect points, give *plot* a list of *x* coordinates and a list of the corresponding *y* coordinates.

```
import matplotlib.pyplot as plt
from mpl_helper_2 import initialize_plot

figure, axes = initialize_plot('x', 'y',
    "Plot with a line between two points")

# set the aspect ratio of vertical to horizontal
```

```
axes.set_aspect('equal')

# -----------------------------------------------------------
x_values = [1.0, 0.25]
y_values = [0.5, 0.25]
# -----------------------------------------------------------

the_plot = axes.plot(x_values, y_values, color="black", marker='o')

<Figure size 432x288 with 0 Axes>
<Figure size 432x288 with 1 Axes>
```

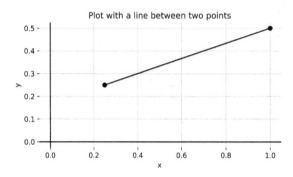

If I had omitted the keyword arguments color and marker, **matplotlib** would have shown no markers and drawn the line in blue. When you have many points, you likely will not want markers.

Every time you call *plot*, you get additional points, lines, and, as we shall see, curves. **matplotlib** chooses colors for these elements if you do not specify them.

```
import matplotlib.pyplot as plt
from mpl_helper_2 import initialize_plot

figure, axes = initialize_plot('x', 'y',
    "Plot with two lines and a labeled point")

# set the aspect ratio of vertical to horizontal
axes.set_aspect('equal')

x_values = [1.0, 0.25]
y_values = [0.5, 0.25]
axes.plot(x_values, y_values, color="black", marker='o')
```

```
# draw a point with a square shape
axes.plot(0.5, 0.4, marker='s')

# label the point with the text centered
axes.text(0.5, 0.43, "(0.5, 0.4)", horizontalalignment="center")

# draw another line
the_plot = axes.plot([0.1, 0.3], [0.4, 0.1])

<Figure size 432x288 with 0 Axes>
<Figure size 432x288 with 1 Axes>

plt.show()
```

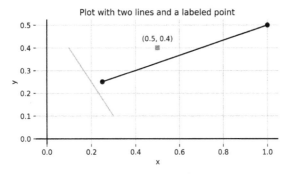

As you might guess, the *text* method displays textual markup at a specified location. The horizontalalignment keyword argument tells **matplotlib** to center the text at the point I specified.

13.1.6 Plotting a parabola

I now want to draw the graph of the parabola $y = x^2 - 3$. I create the list x_values with the nine integral values from -4 to 4, inclusive. For each of these, I use a comprehension to calculate the y_values.

```
import matplotlib.pyplot as plt
from mpl_helper_2 import initialize_plot

figure, axes = initialize_plot('x', 'y',
    r"Plot of the parabola $\mathregular{y=x^2-3}$")

# ------------------------------------------------------------------
x_values = list(range(-4, 5))
```

```
y_values = [x**2 - 3 for x in x_values]
# ------------------------------------------------------------

the_plot = axes.plot(x_values, y_values)

<Figure size 432x288 with 0 Axes>
<Figure size 432x288 with 1 Axes>
```

Figure 13.2 shows the plot **matplotlib** generates when we call `plt.show()`.

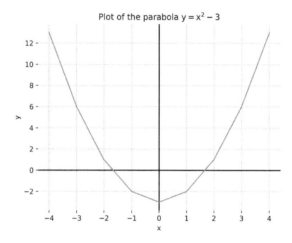

Figure 13.2: Plot of the parabola y = x^2 − 3

To get better formatting in the title, I used LaTeX markup between the dollar signs for the polynomial equation. The LaTeX \mathregular expression told **matplotlib** to use the same font as the surrounding text.

To smooth out this graph, we need to calculate many more points. We can use lists, but it is much more efficient to use 1-dimensional NumPy arrays.

13.1.7 NumPy arrays

How big is a Python object if we measure it in bytes? Let's look at the size of an empty string.

```
from sys import getsizeof

getsizeof("")
```

```
51
```

getsizeof is a function in the **sys** module that returns the number of bytes of memory an object uses. In the case of a string, it includes the length, the characters themselves, plus system overhead for processes like garbage collection. If we add some text, the size gets larger, as we expect.

```
a = "Better a witty fool, than a foolish wit."
len(a)
```

```
40
```

```
getsizeof(a)
```

```
89
```

The size of a string object with 100 times the number of characters is roughly 100 times bigger, allowing for the overhead.

```
b = 100 * a
len(b)
```

```
4000
```

```
getsizeof(b)
```

```
4049
```

What about a list containing the two strings?

```
getsizeof([a, b])
```

```
72
```

That's odd, and it is much less than the sum of the sizes of the strings. The size is the same if we have a list of two copies of the longer string.

```
getsizeof([b, b])
```

```
72
```

If we include a third copy, the number of bytes increases modestly.

```
getsizeof([b, b, b])
```

```
80
```

getsizeof measures an object's size, **but not the size of the objects to which it refers**. In this example, *getsizeof* is not including the sizes of the strings in the lists. We can get a better approximation of the true total amount of memory used with a function like the following:

```
def getsizeof_list(the_list):
    size = getsizeof(list)
    for item in the_list:
        size += getsizeof(item)
    return size

x_values = list(range(1000))
getsizeof_list(x_values)
```

28404

Exercise 13.2

Is the result from *getsizeof_list* correct if the list contains a set, dictionary, or another list? Write a new function that recursively measures the size of an object and everything it contains.

There are more efficient ways of storing numeric data by packing it closely together consecutively in memory. Such a structure is called an *array*, and NumPy is a widely used Python package that includes powerful multi-dimensional array support. NumPy calls itself "The fundamental package for scientific computing with Python," and I agree. [APN] [NPY] [NUM] By tradition, we import the NumPy package as **np.**

```
import numpy as np
```

The *arange* function is similar to *range*, but it builds an array more efficiently than using an iterator, generator, or list.

```
x_values_array = np.arange(1000)

getsizeof(x_values_array)
```

4104

See how much less memory the array of the integers from 0 to 999 uses compared to the list above? The *size* function returns the number of items in an array.

```
x_values_array.size
```

```
1000
```

Python can compute with arbitrarily large integers, as we saw in section 5.2. When we know the integers are bounded in value, we can use a fixed number of bytes to contain the numbers. The *itemsize* property gets the size used by each item.

```
x_values_array.itemsize
```

```
4
```

This result says each integer value uses 4 bytes. Multiply this by the number of items and you get the value of the *nbytes* property, which is the total number of bytes for all the array data.

```
x_values_array.nbytes
```

```
4000
```

Exercise 13.3

Why is x_values_array.nbytes different from getsizeof(x_values_array)?

If you have a list of a tuple of numbers, use *array* to create an array. NumPy forces all numbers in the array to have the same type, which it calls the *dtype*.

```
np.array([2, 3.4])
```

```
array([2. , 3.4])
```

```
np.array((-1, 3.1415926j))
```

```
array([-1.+0.j       ,  0.+3.1415926j])
```

Suppose you want the eleven floating-point numbers between 0 and 1, including the endpoints. `linspace(n, m, count)` gives you an array of `count` numbers starting with n and ending with m.

```
np.linspace(0, 1, 11)

array([0. , 0.1, 0.2, 0.3, 0.4, 0.5, 0.6, 0.7, 0.8, 0.9, 1. ])
```

We can create an array with the *x* coordinates we used in plotting the parabola in Figure 13.2 via:

```
x_values = np.linspace(-4, 4, 9)
```

One of NumPy's most powerful features is that you can use algebraic operations directly on an array. When we define *p* to evaluate the parabola's formula, we can apply it directly to `x_values`.

```
def p(x): return x*x - 3

p(x_values)

array([13.,   6.,   1.,  -2.,  -3.,  -2.,   1.,   6.,  13.])
```

We're now ready to draw our parabola with 1,000 points.

```
import matplotlib.pyplot as plt
from mpl_helper_2 import initialize_plot
import numpy as np

figure, axes = initialize_plot('x', 'y',
    r"Smooth plot of the parabola $\mathregular{y=x^2-3}$")

# Generate 1000 equally spaced points between -4 and 4
x_values = np.linspace(-4, 4, 1000)

def p(x): return x*x - 3

the_plot = plt.plot(x_values, p(x_values))

<Figure size 432x288 with 0 Axes>
<Figure size 432x288 with 1 Axes>
```

```
plt.show()
```

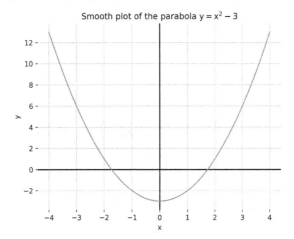

Smooth plot of the parabola $y = x^2 - 3$

13.1.8 Plotting the sine function

The technique of applying a function to each element of an array works for many mathematical operations, though you should use them from the **numpy** module and not the **math** module. Here I draw the sine function from $x = -2\pi$ to 2π:

```
import matplotlib.pyplot as plt
from mpl_helper_2 import initialize_plot
import numpy as np

figure, axes = initialize_plot('x', 'y', "y = sin(x)")

# set the aspect ratio of vertical to horizontal
axes.set_aspect('equal')

# --------------------------------------------------------------
# Generate 1000 equally spaced points between -2 pi and 2 pi
x_values = np.linspace(-2*np.pi, 2*np.pi, 1000)
# --------------------------------------------------------------

the_plot = plt.plot(x_values, np.sin(x_values))

<Figure size 432x288 with 0 Axes>
<Figure size 432x288 with 1 Axes>
```

```
plt.show()
```

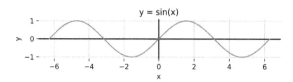

This graph is wide but short in height. I now use the *set_xlim* and *set_ylim* methods to set the range of displayed values for the x and y coordinates.

```
import matplotlib.pyplot as plt
from mpl_helper_2 import initialize_plot
import numpy as np


figure, axes = initialize_plot('x', 'y', "y = sin(x)")


# set the aspect ratio of vertical to horizontal
axes.set_aspect('equal')


# ------------------------------------------------------------------
# show x-coordinates from -7 to 7
axes.set_xlim([-7, 7])


# show y-coordinates from -7 to 7
axes.set_ylim([-7, 7])
# ------------------------------------------------------------------


# Generate 1000 equally spaced points between -2 pi and 2 pi
x_values = np.linspace(-2 * np.pi, 2 * np.pi, 1000)


the_plot = plt.plot(x_values, np.sin(x_values))

<Figure size 432x288 with 0 Axes>
<Figure size 432x288 with 1 Axes>

plt.show()
```

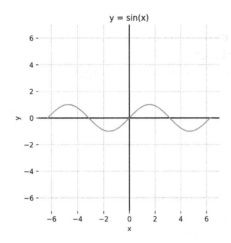

Finally, I use the *axvline* and *text* methods to draw and label gray dashed vertical lines at $x = -\pi$ and $x = \pi$.

```python
import matplotlib.pyplot as plt
from mpl_helper_2 import initialize_plot
import numpy as np

figure, axes = initialize_plot('x', 'y',
                               "y = sin(x)")

# set the aspect ratio of vertical to horizontal
axes.set_aspect('equal')

# show x-coordinates from -7 to 7
axes.set_xlim([-7, 7])

# show y-coordinates from -7 to 7
axes.set_ylim([-7, 7])

# ----------------------------------------------------------------
# draw and label a dashed vertical line at x = -pi
axes.axvline(x=-np.pi, color="#808080",
             linestyle="dashed", linewidth=1.5)

axes.text(-4.5, -3.2,
          r"$\mathregular{x = -\pi}$",
          horizontalalignment="center")
```

```
# draw and label a dashed vertical line at x = -pi
axes.axvline(x=np.pi, color="#808080",
            linestyle="dashed", linewidth=1.5)

axes.text(4.1, -3.2,
          r"$\mathregular{x = \pi}$",
          horizontalalignment="center")
# -------------------------------------------------------------

# Generate 1000 equally spaced points between -2 pi and 2 pi
x_values = np.linspace(-2 * np.pi, 2 * np.pi, 1000)

the_plot = axes.plot(x_values, np.sin(x_values))

<Figure size 432x288 with 0 Axes>
<Figure size 432x288 with 1 Axes>

plt.show()
```

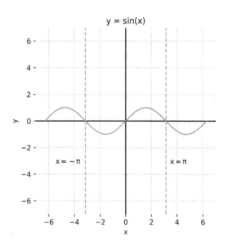

Exercise 13.4

Use the techniques in this section to draw the graph of the inverse sine shown on the right-hand side of Figure 13.1. Use the function *np.arcsin*.

13.1.9 Multiple plots

In section 13.1.3 we saw that

$$\text{figure, axes = plt.subplots()}$$

creates a figure with one subplot, and we could use axes to reference it. More generally,

$$\text{figure, axes = plt.subplots(number\_of\_rows, number\_of\_columns)}$$

returns a figure with multiple subplots and axes objects. To refer to a specific axes, use an index. The first subplot has index 0.

```python
import matplotlib.pyplot as plt

figure, axes = plt.subplots(1, 3)

for x in range(3):
    axes[x].set_title(f"Plot at position {x}")
    axes[x].plot(x, x, marker='o')

figure.tight_layout()

<Figure size 432x288 with 3 Axes>

plt.show()
```

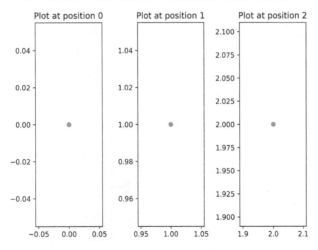

The call to *tight_layout* forced **matplotlib** to avoid overlapping the subplots.

Exercise 13.5

Change this example so each subplot uses equal aspect ratios and has the same range of x and y coordinates.

We can stack the subplots vertically as well as spreading them out horizontally. In the next example, we have two rows, each with three subplots. We now need two indices to access an `axes`, with the first being the row.

```python
import matplotlib.pyplot as plt

figure, axes = plt.subplots(2, 3, constrained_layout=True)

for row in range(2):
    for col in range(3):
        axes[row, col].set_title(f"Plot at position {row}, {col}")
        axes[row, col].plot(row, col, marker='o')

figure.suptitle("This is a title above the figure",
            y=1.05, fontweight="bold",
            family="serif", fontsize=16)
```

```
Text(0.5, 1.05, 'This is a title above the figure')
<Figure size 432x288 with 6 Axes>
```

```python
plt.show()
```

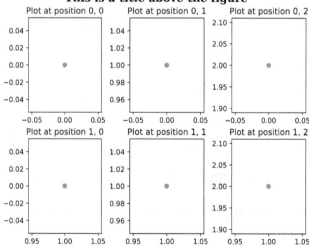

Instead of calling *tight_layout*, I used `constrained_layout=True` in the call to *subplots*. Experiment with each to see which gives you the better spacing. To manually control the horizontal and vertical padding, use *subplots_adjust*.

I set the title for the entire figure with *suptitle*. I specified a large, bold, serifed font and adjusted it higher with the y keyword argument to not overlap with the subplot titles.

Exercise 13.6

Use the techniques from this section and the sine and inverse sine plots from section 13.1.8 to draw Figure 13.1.

13.1.10 Semi-log plots

Every plot we have seen in this chapter has each axis with a linear scale. That means we add some constant value to get from one labeled location on the axis to the next. A semi-log plot has one axis with a linear scale and the other with a logarithmic scale. This means that instead of showing y, the vertical axis shows 10^z where $z = \log(y)$. Frequently, the horizontal axis is linear, and the vertical axis is logarithmic.

We saw the plot of three exponential functions in Figure 5.7, which I reproduce here as Figure 13.3.

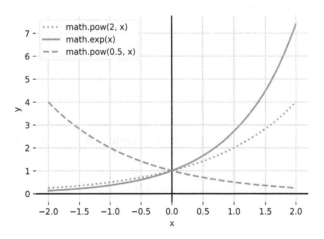

Figure 13.3: Plots of three exponential functions

The code I use to generate the chart uses three calls to *plot*:

```
figure, axes = plt.subplots(figsize=(6, 4))
# axes.set_aspect("equal")

for side in ("top", "right", "bottom", "left"):
    axes.spines[side].set_visible(False)

plt.grid()
plt.axvline(color="black")
plt.axhline(color="black")

xs = np.linspace(-2, 2.0, 400)
ys = np.power(2, xs)

plt.plot(xs, ys, linewidth=2.0, linestyle="dotted", label="math.pow(2,
x)", color="#1f77b4")

ys = np.exp(xs)
plt.plot(xs, ys, linewidth=2.0, label="math.exp(x)", color="#1f77b4")

ys = np.power(0.5, xs)
plt.plot(xs, ys, linewidth=2.0, linestyle="dashed", label="math.pow(0.
5, x)", color="#1f77b4")

plt.xlabel('x')
plt.ylabel('y')

axes.legend()
```

`axes.legend()` creates the box in the upper-left corner that tells the viewer which curve belongs to which function.

When I add the command `axes.set_yscale("log")`, we get the plot in Figure 13.4.

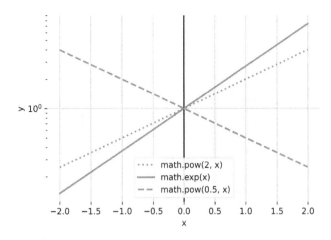

Figure 13.4: Semi-log plots of three exponential functions

Mathematically, the logarithm of an exponential function `c * a**x` produces the equation for a line `log(c) + x * log(a)`.

13.2 Bar charts

Bar charts are popular in the media to show count and measurement data. I suspect people create many of these in Microsoft Excel, but you can also use **matplotlib** to make bar charts for display or publication. We saw bar charts when we generated histograms for the results of quantum circuits in section 9.7.2.

Most of the methods, functions, and stylistic control we saw in the last section for plots also apply to bar charts. Experiment until you like what you see.

13.2.1 Basic bar charts

The most straightforward bar chart corresponds to a list of numeric data values. These values are the heights of the bars.

```
import matplotlib.pyplot as plt

data = [21, 14, 10, 9, 19]

# draw the bar chart
bar_chart = plt.bar(range(1, len(data) + 1), data)

<Figure size 432x288 with 1 Axes>
```

```
plt.show()
```

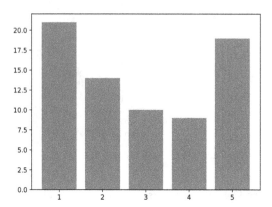

It's more informative for your bar chart viewers if you label the chart and the horizontal axis values.

```
import matplotlib.pyplot as plt

# give descriptive names to the values on the horizontal
# and vertical axes

days = ["Monday", "Tuesday", "Wednesday", "Thursday", "Friday"]
support_calls = [21, 14, 10, 9, 19]

# add a title to the top of the bar chart
plt.title("Support Call Data for Last Week")

# draw the bar chart
bar_chart = plt.bar(days, support_calls)

<Figure size 432x288 with 1 Axes>

plt.show()
```

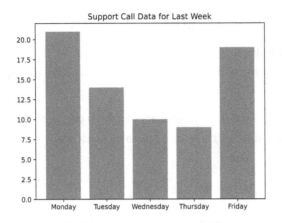

 Exercise 13.7

Extend this bar chart by adding 6 support calls for Saturday and 4 for Sunday.

Exercise 13.8

Label the horizontal axis with "Days" and the vertical axis with "Number of Support Calls."

13.2.2 Horizontal bar charts

If you use the *barh* function instead of *bar*, you get the chart in Figure 13.5.

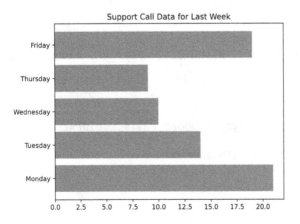

Figure 13.5: A horizontal bar chart

13.2.3 Adding colors and grid lines

By default, **matplotlib** displays every bar in blue. If you use the `color` keyword argument with a single color, every bar will use that. You can also give a list of colors, one for each bar on the horizontal axis.

In the next example, I use three gray shades to show bars representing three daily support call count ranges. I also draw a grid with horizontal black dashed lines. The `alpha` argument determines the lines' opaqueness.

```python
import matplotlib.pyplot as plt

days = ["Monday", "Tuesday", "Wednesday", "Thursday", "Friday"]
support_calls = [21, 14, 10, 9, 19]

# create a list of colors for ranges of daily call values
colors = []
for calls in support_calls:
    if calls >= 15:
        colors.append("#505050")
    elif calls >= 10:
        colors.append("#808080")
    else:
        colors.append("#B0B0B0")

# add the bar chart title
plt.title("Number of Support Calls Per Day")

# draw horizontal lines
plt.grid(color="black", alpha=0.6, axis='y', linewidth=1, linestyle='--
')

# draw the bar chart
bar_chart = plt.bar(days, support_calls, color=colors)

<Figure size 432x288 with 1 Axes>

plt.show()
```

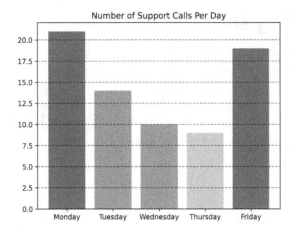

Exercise 13.9

Does `alpha` = `1.0` mean completely opaque or completely transparent?

Exercise 13.10

I ran the 3-qubit circuit in Figure 13.6 one thousand times and got the following results:

```
counts = {"000": 129, "001": 128, "010": 93, "011": 115,
          "100": 138, "160": 128, "113": 93, "111": 124}
```

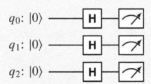

Figure 13.6: A quantum circuit with three Hadamard gates

Draw a bar chart with the count keys on the horizontal axis and the values on the vertical. Use *axhline* to draw a green horizontal line at $y = 125$. Color any bar whose value is less than 120 or greater than 130 in red.

13.2.4 Stacking and labeling bars

Stacked bar charts are a good way of visualizing individual and total results. The key to stacking is to use the bottom keyword argument when drawing the bars above the lowest levels. The bottom of the upper bar is the height of the bar below it.

You can use the label keyword argument for each plot or bar chart and then use the *legend* function. **matplotlib** displays a box in the corner that tells the viewer which label corresponds to which bars.

```
import matplotlib.pyplot as plt
import numpy as np

data = [[21, 14, 10, 9, 19],
        [17, 16, 4, 12, 14]]

# draw the bar chart for Team A
plt.bar(np.arange(1, len(data[0]) + 1), data[0],
        color="#505050", label="Team A")

# draw the bar chart for Team B above that of Team A
bar_chart = plt.bar(np.arange(1, len(data[1]) + 1), data[1],
                    color="#B0B0B0", label="Team B",
                    bottom=data[0])

# draw a legend identifying the bars for each team
plt.legend()

<matplotlib.legend.Legend at 0x23177034b20>
<Figure size 432x288 with 1 Axes>

plt.show()
```

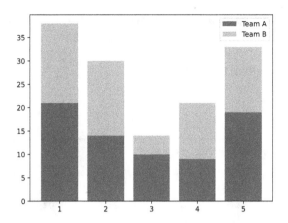

Exercise 13.11

Extend this example by stacking a third list of values $[12, 21, 7, 9, 16]$.

13.2.5 Side-by-side bars

A side-by-side bar chart makes it easy to see how corresponding results from different times vary. The trick to drawing a side-by-side bar chart is to specify the width of each bar and then shift the position of bars to the right by that width.

```
import matplotlib.pyplot as plt
import numpy as np

plt.clf()

data = [[21, 14, 10, 9, 19],
        [17, 16, 4, 12, 14]]

# set the width of a bar
width = 0.4

# draw the first bar chart with each bar the specified width
bar_chart = plt.bar(np.arange(1, len(data[0]) + 1), data[0],
                    width=width, color="#505050")

# draw the second bar chart with the bars to the right of the first
```

```
bar_chart = plt.bar(np.arange(1, len(data[1]) + 1) + width, data[1],
                    width=width, color="#B0B0B0")

<Figure size 432x288 with 1 Axes>

plt.show()
```

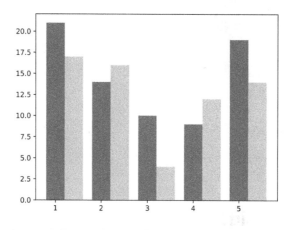

Exercise 13.12

Extend this example by placing new bars to the right of those in the last example for the data [12, 21, 7, 9, 16].

13.3 Histograms

A histogram is a form of bar chart for displaying statistical data. In Figure 5.12 in section 5.7.2, I showed a plot of one million random numbers in a normal distribution with mean μ = 1.0 and standard deviation σ = 0.2. This code creates a histogram with the same characteristics:

```
import matplotlib.pyplot as plt
import numpy as np

plt.clf()

np.random.seed(23)

# generate the random numbers in a normal distribution
```

```
mu = 1.0
sigma = 0.2
random_numbers = np.random.normal(mu, sigma, 1000000)

# set the number of bins in which to divide up the random numbers
bin_size = 100

# draw the histogram
the_plot = plt.hist(random_numbers, bins=bin_size)

<Figure size 432x288 with 1 Axes>

plt.show()
```

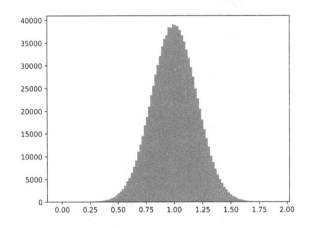

The bins argument determines how many bars are in the histogram.

Exercise 13.13

Does decreasing bin_size make the histogram smoother or blockier?

Exercise 13.14

Investigate the definitions of the keyword arguments density and orientation to *hist*. Create one histogram per option for each argument.

13.4 Pie charts

Another common chart type is the pie chart, where we visualize data as wedges. The size of each wedge is related to its data value compared to the sum of the values. We use the *pie* function to create a pie chart.

```
import matplotlib.pyplot as plt

days = ["Monday", "Tuesday", "Wednesday", "Thursday", "Friday"]
support_calls = [21, 14, 10, 9, 19]

# draw the pie chart
pie_chart = plt.pie(support_calls, labels=days)

<Figure size 432x288 with 1 Axes>

plt.show()
```

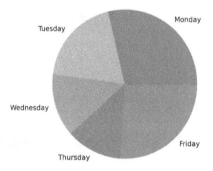

matplotlib automatically assigns colors to each wedge. The last chart had wedges in blue, purple, red, green, and orange, starting with Monday, though they may appear monochromatically to you depending on the medium in which you are reading this book.

We provide many of the stylistic controls for the display of the chart as keyword arguments to *pie*. If I add the argument `autopct="%1.1f%%"` to the call to *pie*, **matplotlib** displays the relative percentage sizes of each wedge. Figure 13.7 shows the result of using `autopct`.

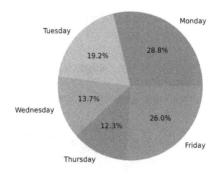

Figure 13.7: The pie chart with wedge percentages

The format specifier `"%1.1f%%"` comes from the time before Python supported f-strings. The first "**%**" says, "this is a format string." "**1.**" says, "show at least one digit followed by a decimal point." "**1f**" means, "show one digit after the decimal point of this floating-point number." Finally, "**%%**" will insert a percentage sign.

Use the `colors` keyword argument to give a sequence of color names or hex RGB values. Typically, the sequence is a list or tuple. If you give fewer colors than there are data values, **matplotlib** cycles through them again starting at the beginning of the sequence. I defined

```
gray_shades = ["#A0A0A0", "#B0B0B0", "#C0C0C0", "#D0D0D0", "#E0E0E0"]
```

and then passed them to *pie* via `colors=gray_shades`. The updated chart is in Figure 13.8.

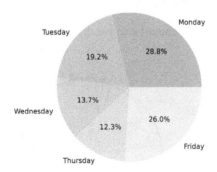

Figure 13.8: The pie chart drawn with gray shades

Notice that there are no lines around the wedges. The `wedgeprops` argument takes a dictionary of stylistic properties. First, I define

```
wedgeproperties = {'linewidth': 2, 'edgecolor': 'black'}
```

and then pass it to *pie* as `wedgeprops=wedgeproperties`.

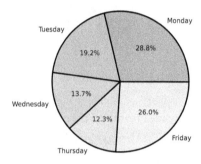

Figure 13.9: The pie chart with black wedge outlines

Exercise 13.15

In Figure 13.9, the lines are solid. What property would you add to `wedgeproperties` to make them dashed?

If you want to highlight one or more wedges, you can "explode" them from the pie. If I set:

```
explode_amounts = [0.1, 0, 0, 0, 0]
```

and then pass `explode=explode_amounts` to *pie*, the "Monday" wedge is pushed out slightly. You can see this in Figure 13.10.

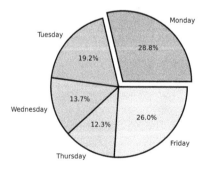

Figure 13.10: The pie chart with the "Monday" wedge exploded

Exercise 13.16

Push out any wedge corresponding to a data value less than or equal to 10.

Finally, you can rotate the chart to display the pie for the best visual effect.

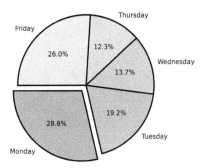

Figure 13.11: The rotated pie chart

In Figure 13.11, I rotated the pie 180° via the `startangle=180` argument.

Exercise 13.17

Explain what the `counterclock` keyword argument to *pie* does when you assign it `True` or `False`.

13.5 Scatter plots

We use a scatter plot to display many points. After doing that, we make it fancy with colors, markers, labels, titles, and fonts and adjust the `axes` so the plot conveys your data effectively and pleasantly.

Like *pie*, we control most of the look of a scatter plot via keyword arguments to *scatter*. The data for the plots is the high temperatures for three cities—A, B, and C—over the first ten days of March 2021. The temperatures are in Fahrenheit, and I use **numpy** arrays to hold the numeric days of the month and the data.

If I merge the data into single arrays for the days and temperatures via the **numpy** *concatenate* function, *scatter* shows all points in the same color with the same marker. The default color is blue.

```python
import matplotlib.pyplot as plt
import numpy as np

days = np.arange(1, 11)

city_high_temperatures = {
    'A': np.array([63, 57, 59, 67, 54, 53, 59, 62, 62, 58]),
    'B': np.array([45, 45, 39, 31, 34, 41, 66, 63, 60, 56]),
    'C': np.array([80, 82, 82, 79, 78, 76, 76, 82, 84, 83])
}

figure, axes = plt.subplots()

# draw the scatter charts

axes.set_xlabel("Day in March, 2021")
axes.set_ylabel("Temperature in F")
axes.set_title("Daily high temperatures for three cities")

d = np.concatenate([days, days, days])
t = np.concatenate([
    city_high_temperatures['A'],
    city_high_temperatures['B'],
    city_high_temperatures['C']])

the_plot = axes.scatter(d, t)


<Figure size 432x288 with 1 Axes>

plt.show()
```

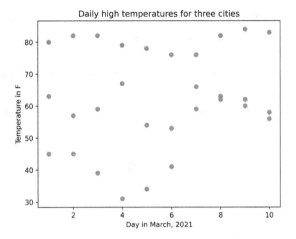

If we make three calls to *scatter,* we can display each city's points in different ways. Use the s argument to change the size of the marker as in this code:

```
axes.scatter(days, city_high_temperatures['A'], s=10)
axes.scatter(days, city_high_temperatures['B'], s=100)
axes.scatter(days, city_high_temperatures['C'], s=50)
```

s can either be a single **int** or **float** or an array of sizes, one per point. The result is in Figure 13.12.

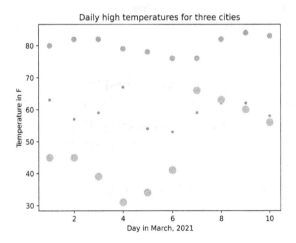

Figure 13.12: A scatter plot with different size circles per city

matplotlib uses a different color for each scatter plot, but we can choose the ones we wish to see. I used this code to set the gray colors for the plot in Figure 13.13:

```
axes.scatter(days, city_high_temperatures['A'], color="black")
axes.scatter(days, city_high_temperatures['B'], color="#909090")
axes.scatter(days, city_high_temperatures['C'], color="#D0D0D0")
```

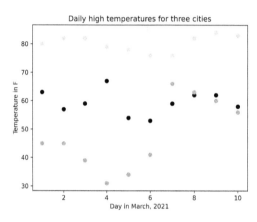

Figure 13.13: A scatter plot with different chosen colors per city

As with *plot*, we can change the point marker shape. By adding label keyword arguments and calling *legend*, I provide a guide to the user to distinguish among the datasets. Figure 13.14 shows the result.

```
axes.scatter(days, city_high_temperatures['A'],
            marker="o", color="black", label="City A")
axes.scatter(days, city_high_temperatures['B'],
            marker="*", color="black", label="City B")
axes.scatter(days, city_high_temperatures['C'],
            marker="x", color="black", label="City C")

axes.legend()
```

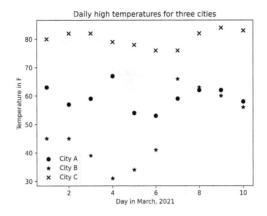

Figure 13.14: A scatter plot with different markers per city

By mixing and matching the stylistic keyword arguments to *scatter*, you can get the plot to look the way you wish.

Look at Figure 13.14 and notice how only the even-numbered days are labeled. We call the short vertical lines above the numbers "tick marks" or "ticks," and **matplotlib** includes the **ticker** module to control the display of labels and ticks. It's not nearly as straightforward as the plot features we have used in this chapter, so I will only show you the code to adjust the numbering. To change the horizontal ticks to be one per day and the vertical ticks to be one every five days, issue:

```
import matplotlib.ticker as ticker

axes.xaxis.set_major_locator(ticker.MultipleLocator(1))
axes.yaxis.set_major_locator(ticker.MultipleLocator(5))
```

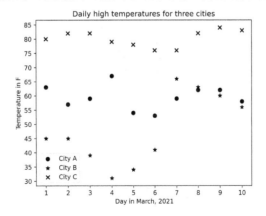

Figure 13.15: A scatter plot with adjusted tick marks

The adjusted plot is in Figure 13.15. This technique works with any plot that has **axes**. Horizontal grid lines also make it easier to determine data values from the chart.

```
plt.grid(color="black", alpha=0.6, axis='y',
        linewidth=0.5, linestyle='-')
```

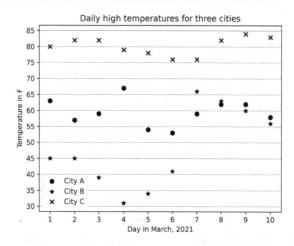

Figure 13.16: A scatter plot with horizontal grid lines

13.6 Moving to three dimensions

matplotlib can plot in three dimensions if we import the tools we need from **mpl_toolkits.mplot3d**. To help you see the main components of a 3D plot, here is an example where I have drawn the point at the origin (0, 0, 0) and specified the **axes** ranges and labels.

```
import matplotlib.pyplot as plt
from mpl_toolkits.mplot3d import Axes3D


# ----------------------------------------------------------------
# create the plot figure
figure = plt.figure()

# create a 3D axes
axes = plt.axes(projection='3d')
# ----------------------------------------------------------------


# label the x, y, and z axes in a large font
```

```
axes.set_xlabel('x', fontsize="16")
axes.set_ylabel('y', fontsize="16")
axes.set_zlabel('z', fontsize="16")

# set the limits of the values shown on the axes
axes.set_xlim(0, 1)
axes.set_ylim(0, 1)
axes.set_zlim(0, 1)

# label the origin
axes.text(0.1, 0, 0.03, "(0, 0, 0)")

# draw the axes
axes.plot([0, 0], [0, 0], [0, 1], linewidth=2, color="black")
axes.plot([0, 0], [0, 1], [0, 0], linewidth=2, color="black")
axes.plot([0, 1], [0, 0], [0, 0], linewidth=2, color="black")

# draw the point at the origin
the_plot = axes.plot([0, 0], [0, 0], [0, 0], marker='o',
                     markersize=6, color="black")
```

```
<Figure size 432x288 with 1 Axes>
```

```
plt.show()
```

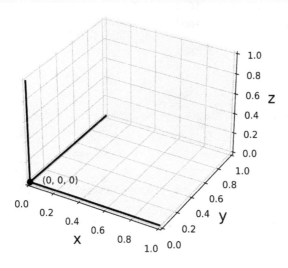

The newest feature is how we create the `axes`. First, we build the figure with

```
figure = plt.figure()
```

and then make 3-dimensional `axes` within the figure via

```
axes = plt.axes(projection='3d')
```

I use *set_zlabel* to label the *z*-axis, which is consistent with my earlier use of *set_xlabel* and *set_ylabel* in section 13.1.2.

13.6.1 Drawing a curve

Our first approach to 3D plotting is to take what we did with points, lines, and curves in two dimensions and give them a vertical component. Essentially, they are 0- and 1-dimensional objects in three dimensions. Let's draw the cosine:

```
import matplotlib.pyplot as plt
from mpl_toolkits.mplot3d import Axes3D
import numpy as np

# create the plot figure and 3D axes
figure = plt.figure()
axes = plt.axes(projection='3d')

# x will range from -10 to 10 for 1000 values
xs = np.linspace(-10, 10, 1000)

# y will always be 0
ys = np.full(1000, 0)

# z is the cosine of x
zs = np.cos(xs)

# draw the curve
the_plot = axes.plot(xs, ys, zs, color="black")

<Figure size 432x288 with 1 Axes>

plt.show()
```

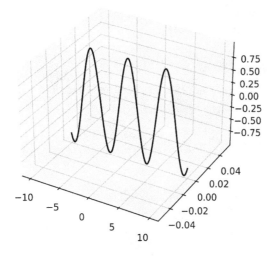

In this plot, we use *linspace* to create an array of one thousand floating-point numbers between −10 and 10. These are the *x* coordinates. For *y*, we fill an array with one thousand zeros via the *full* function. Finally, the *z* values are the cosines of the *x* values.

Exercise 13.18

Draw two other cosine curves in this plot with *y* equal to 1 and *y* equal to −1. Draw a line on the plot connecting the three curves where *x* is 0.

13.6.2 Drawing a scatter plot

A 3D scatter plot displays many points, like its 2D counterpart. Though **matplotlib** uses some color shading to give you an idea of depth, it can be difficult to determine a point's coordinates.

```
import matplotlib.pyplot as plt
from mpl_toolkits.mplot3d import Axes3D


# set the data coordinates
xs = [9, 10, 8, 4, 8, 5, 7, 2, 5, 3]
ys = [8, 1, 6, 2, 1, 3, 3, 10, 4, 3]
zs = [6, 10, 8, 1, 3, 20, 24, 26, 23, 25]


# create the plot figure and 3D axes
```

```
figure = plt.figure()
axes = plt.axes(projection='3d')

# draw the scatter plot
the_plot = axes.scatter(xs, ys, zs, color="black")


<Figure size 432x288 with 1 Axes>

plt.show()
```

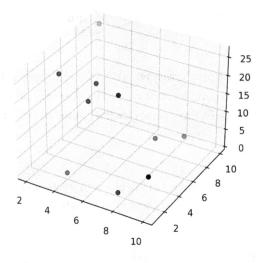

Consider using a grid or different markers or colors to make the plot easier to interpret.

Exercise 13.19

By calling *scatter* twice, draw this scatter plot so that points with z values less than 20 use a `'+'` marker and otherwise use an `'*'` marker.

Exercise 13.20

Draw this scatter plot so that points with x + y > 10 are shown in red and otherwise in green.

13.6.3 Drawing a square

Let's think about drawing a 10-by-10 square sitting flat at $z = 10$. There are four corners, but if we give a 1-dimensional array of their x coordinates and a 1-dimensional array of their y coordinates, *plot* will draw four points and lines connecting them.

```
import matplotlib.pyplot as plt
from mpl_toolkits.mplot3d import Axes3D
import numpy as np

# set the data coordinates
xs = np.array([0, 0, 10, 10])
ys = np.array([0, 10, 10, 0])
zs = np.array([10, 10, 10, 10])

# create the plot figure and 3D axes
figure = plt.figure()
axes = plt.axes(projection='3d')

axes.set_title("plot", fontstyle="italic")

# draw the plot for the corners
the_plot = axes.plot(xs, ys, zs, color="black")

<Figure size 432x288 with 1 Axes>

plt.show()
```

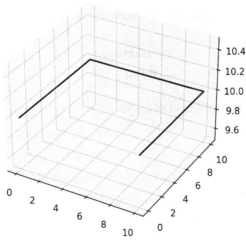

plot

Exercise 13.21

Why are there only three lines? How would you draw the fourth line?

13.6.4 Drawing wireframe and surface plots

A square is 2-dimensional, so we need to change our data representation and plotting functions. Suppose we have nine points with the following *x* and *y* coordinates.

```
xs = [4, 5, 6]
ys = [1, 2, 3]
```

The points correspond to all combinations of xs and ys values and are shown in Figure 13.17.

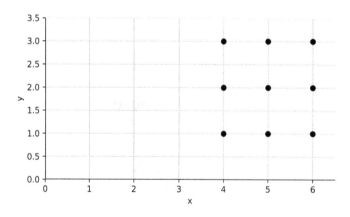

Figure 13.17: Nine points in a grid

To create a surface, we need 2-dimensional arrays. The *meshgrid* function takes lists or arrays of *x* and *y* values and produces 2D **numpy arrays**.

```
x_mesh, y_mesh = np.meshgrid(xs, ys)
x_mesh

array([[4, 5, 6],
       [4, 5, 6],
       [4, 5, 6]])
```

```
y_mesh
```

```
array([[1, 1, 1],
       [2, 2, 2],
       [3, 3, 3]])
```

The coordinates of marked points in Figure 13.17 are corresponding numbers in the arrays x_mesh and y_mesh.

Now I can draw our square properly. After establishing the grids, I show the square's outline with *plot_wireframe* and the filled surface with *plot_surface*.

```
import matplotlib.pyplot as plt
from mpl_toolkits.mplot3d import Axes3D
import numpy as np

# set the data coordinates
xs = [0, 1]
ys = [0, 1]
zs = np.array([[10, 10], [10, 10]])

# compute the grids
x_mesh, y_mesh = np.meshgrid(xs, ys)

# create the plot figure and two 3D axes
figure, axes = plt.subplots(1, 2,
                            figsize=(8, 4),
                            subplot_kw={'projection': "3d"},
                            constrained_layout=True)

# draw the wireframe plot
axes[0].set_title("plot_wireframe", fontstyle="italic")
axes[0].plot_wireframe(x_mesh, y_mesh, zs, color="black")

# draw the filled surface plot
axes[1].set_title("plot_surface", fontstyle="italic")
axes[1].plot_surface(x_mesh, y_mesh, zs, color="lightgray")
```

```
<mpl_toolkits.mplot3d.art3d.Poly3DCollection at 0x23178d009d0>
<Figure size 576x288 with 2 Axes>
```

```
plt.show()
```

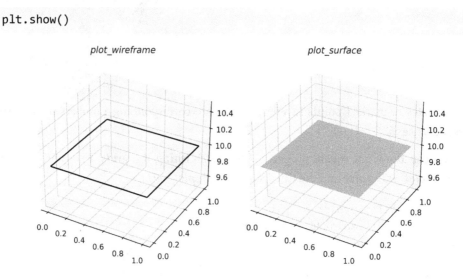

This last example demonstrates another way of stating that you want a 3D plot using the subplot_kw={'projection': "3d"} argument to *subplots*. To help **matplotlib** better lay out this wide but short figure, I specified figsize=(8, 4). This directive sizes the figure to be 8 inches wide by 4 inches high.

13.6.5 Drawing the cosine two ways

Our last example for 3-dimensional plotting is two surface plots involving cosine. On the left of Figure 13.18, I used *plot_surface* to display a surface where the *z* value does not depend on *y*. On the right, the plot is symmetrical around the origin in the *xy*-plane.

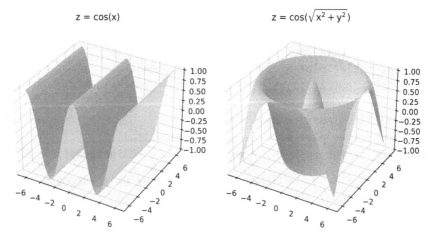

Figure 13.18: Drawing surface plots for cosine

Exercise 13.22

Draw the surface plots in Figure 13.18 in a single figure. Use the following lines of code in some way and in some order:

```
xs = np.linspace(-2 * np.pi, 2 * np.pi, 125)
ys = np.linspace(-2 * np.pi, 2 * np.pi, 125)
zs = np.cos(x_mesh)
zs = np.cos(np.sqrt(x_mesh**2 + y_mesh**2))
```

Don't forget to label the plots.

13.7 Summary

In this chapter, we looked at **matplotlib** and how we can draw points, lines, curves, surfaces, pie charts, bar charts, and scatter plots. While this is a powerful module, it violates the Zen of Python, especially regarding one and only one way of doing something. [ZEN] There are many things you can do at either the `plt` or `axes` level. I gave you examples of each, but I prefer the latter.

We also learned how to create and use 1- and 2-dimensional **numpy** arrays. These have many applications beyond plotting in scientific computing, AI, and data science.

14

Analyzing Data

It is a capital mistake to theorize before one has data.

—Arthur Conan Doyle, *Sherlock Holmes*

While we can use fancy names like "data science," "analytics," and "artificial intelligence" to talk about working with data, sometimes you just want to read, write, and process files containing many rows and columns of information. People have been doing this interactively for years, typically using applications like Microsoft Excel® and online apps like Google Sheets™.

To "programmatically" manipulate data, I mean that we use Python functions and methods. This chapter uses the popular **pandas** library to create and manipulate these collections of rows and columns, called *DataFrames*. [PAN] [PCB] We will later introduce other methods in *Chapter 15, Learning, Briefly*. Before we discuss DataFrames, let's review some core ideas from statistics.

Topics covered in this chapter

14.1 Statistics

This section introduces several core statistical functions for analyzing data. The data we use is the ages in 2021 of one hundred attendees at a fictional concert of 1990s music cover bands.

```python
import numpy as np

ages = np.array(
    [40, 41, 39, 35, 42, 37, 45, 43, 42, 38, 39, 45, 37, 36,
     45, 41, 41, 31, 42, 40, 39, 38, 40, 39, 41, 46, 42, 44,
     46, 48, 45, 39, 46, 43, 35, 38, 43, 41, 36, 40, 34, 44,
     42, 44, 40, 49, 47, 51, 52, 45, 44, 47, 39, 38, 43, 39,
     45, 40, 36, 43, 38, 43, 32, 35, 36, 42, 40, 38, 37, 36,
     41, 41, 31, 39, 51, 38, 42, 36, 35, 36, 40, 40, 37, 43,
     39, 42, 44, 50, 39, 38, 37, 33, 52, 35, 44, 29, 42, 39,
     40, 42]
)
```

14.1.1 Means and medians

The *mean* is the average of the data values. We often write the mean as the Greek letter mu = μ, though you may also see "*s*."

```python
mean = np.mean(ages)
mean
```

```
40.62
```

You can also compute the mean directly by dividing the sum of the values by the number of values.

```python
np.sum(ages)/ages.size
```

```
40.62
```

We use the **numpy** functions *max* and *min* to compute the maximum and minimum data values.

```
np.max(ages), np.min(ages)
```

```
(52, 29)
```

In section 13.3, we saw how to draw histograms using **matplotlib**. Here is a histogram of the age data, with the mean labeled:

```
import matplotlib.pyplot as plt
plt.hist(ages, bins=33, color="lightgray", edgecolor="black")

plt.xlabel("Age")
plt.ylabel("Count")
plt.title("Histogram of concert attendee ages")

mean = np.mean(ages)
plt.axvline(mean, color="black", linestyle="dotted")
plt.text(mean + 0.1, 11, "mean")
```

```
Text(40.72, 11, 'mean')
<Figure size 432x288 with 1 Axes>
```

```
plt.draw()
```

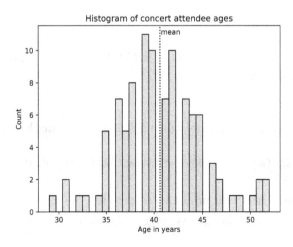

The value where we split the data in the middle, with half above and half below, is called the *median.*

```
np.median(ages)
```

```
40.0
```

 Exercise 14.1

How do you calculate the median if the number of data values is even? Odd?

The first quartile, commonly called *Q1*, is the median of the data values less than the median. Approximately 25% of the smallest data values are in the first quartile. We compute it using the *quantile* function.

```
Q1 = np.quantile(ages, 0.25)
```

The third quartile, commonly called *Q3*, is the median of the data values greater than the median. Approximately 25% of the largest data values are in the third quartile.

```
Q3 = np.quantile(ages, 0.75)
Q1, Q3
```

```
(38.0, 43.0)
```

 Exercise 14.2

What is np.quantile(ages, 0.1)? What does it represent? What about np.quantile(ages, 2.0/3.0)?

The interquartile range, commonly called *IQR*, is the difference between *Q3* and *Q1*.

```
IQR = np.quantile(ages, 0.75) - np.quantile(ages, 0.25)
```

Data is often associated with units like meters, feet, hours, nanoseconds, gallons, and liters. The unit for each value in ages is "years." Therefore, the unit for mean, median, and standard deviation is also years.

14.1.2 Box and whisker plots

A *box plot*, also called a *whisker plot*, shows all these descriptive statistics. We can draw one with **matplotlib**:

```
plt.boxplot(ages)
plt.draw()
```

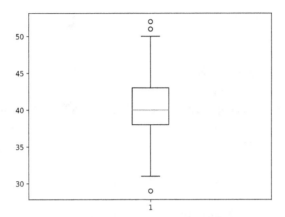

The box in the middle ranges vertically from *Q1* to *Q3*, with the median marked by the horizontal line within. **matplotlib** shows the minimum and maximum points at the bottom and top.

Below the maximum is something that looks like a "T," and above the minimum is an inverted "T." These are the "whiskers." The top whisker is at Q3 + 1.5 * IQR, rounded down if necessary, and the lower is at Q1 - 1.5 * IQR, rounded up if necessary. Beyond the whiskers are *outliers*, data values that you should inspect for correctness.

This plot is very plain, so let's embellish it to see what the plot is describing. The following code adds lines and labels to the box plot to make it easier to understand:

```
import matplotlib.pyplot as plt
import numpy as np
import math

ages = np.array(
    [40, 41, 39, 35, 42, 37, 45, 43, 42, 38, 39, 45, 37, 36,
     45, 41, 41, 31, 42, 40, 39, 38, 40, 39, 41, 46, 42, 44,
```

```
        46, 48, 45, 39, 46, 43, 35, 38, 43, 41, 36, 40, 34, 44,
        42, 44, 40, 49, 47, 51, 52, 45, 44, 47, 39, 38, 43, 39,
        45, 40, 36, 43, 38, 43, 32, 35, 36, 42, 40, 38, 37, 36,
        41, 41, 31, 39, 51, 38, 42, 36, 35, 36, 40, 40, 37, 43,
        39, 42, 44, 50, 39, 38, 37, 33, 52, 35, 44, 29, 42, 39,
        40, 42]
)

plt.title("Box plot of concert attendee ages")
plt.ylabel("Age in years")

# medianprops is a dict with the
# drawing options for the median line
plt.boxplot(ages, medianprops={"color": "black", "linewidth": 2})

quantiles = np.quantile(ages, [0.0, 0.25, 0.5, 0.75, 1.0])
labels = ["minimum", "Q1 = first quartile",
          "mean", "Q3 = third quartile", "maximum"]

for (q, label) in zip(quantiles, labels):
    plt.axhline(q, color="gray", linewidth=0.5)
    plt.text(0.55, q + 0.2, f"{label} = {q}", fontsize=9)

IQR = quantiles[3] - quantiles[1]

# Draw the lower whisker extreme. Below this are outliers.
q = math.ceil(quantiles[1] - 1.5 * IQR)
plt.axhline(q, color="gray", linewidth=0.5)
plt.text(0.55, q + 0.2, f"Q1 - 1.5 * IQR = {q}", fontsize=9)

# Draw the upper whisker extreme. Above this are outliers.
q = math.floor(quantiles[3] + 1.5 * IQR)
plt.axhline(q, color="gray", linewidth=0.5)
plt.text(0.55, q + 0.2, f"Q3 + 1.5 * IQR = {q}", fontsize=9)

Text(0.55, 50.2, 'Q3 + 1.5 * IQR = 50')
<Figure size 432x288 with 1 Axes>
```

```
plt.draw()
```

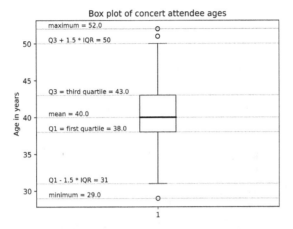

14.1.3 Standard deviation and variance

Let's look at two datasets of 99 numbers. The first contains each integer from 1 to 99. The second contains 99 copies of the integer 50. In each case, the mean is 50.

```
data_1 = np.arange(1, 100)
data_2 = np.repeat(50, 99)
(np.mean(data_1), np.mean(data_2))
```

```
(50.0, 50.0)
```

In the first case, the values are well spread out from the mean. In the second, every value is centered at the mean. We use the *standard deviation*, written as the Greek letter sigma = σ, to measure this spread. The **numpy** function *std* computes the standard deviation of a 1-dimensional array of values.

```
(np.std(data_1), np.std(data_2))
```

```
(28.577380332470412, 0.0)
```

There is no deviation from the mean in the second case. The box plots in Figure 14.1 show the different distributions in the two datasets.

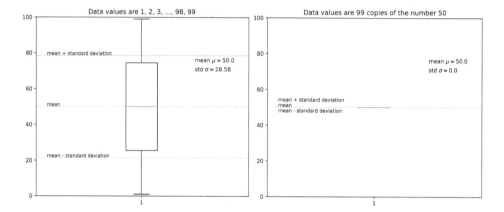

Figure 14.1: Box plots of two distributions of the numbers 1 to 99

Informally, we want to get an idea of how much each data value is different from the mean, on average. This would give us an "average deviation," and we could imagine computing it as the average of the absolute values of the values minus the mean. If the data values are numbers d_1 to d_n, then this would be the second line in

$$\text{mean} = \mu = \frac{d_1 + \cdots + d_n}{n} = \frac{1}{n}\sum_{j=1}^{n} d_j$$

$$\text{average deviation} = \frac{|\mu - d_1| + \cdots + |\mu - d_n|}{n} = \frac{1}{n}\sum_{j=1}^{n} |\mu - d_j|$$

$$\text{standard deviation} = \sigma = \sqrt{\frac{(\mu - d_1)^2 + \cdots + (\mu - d_n)^2}{n}} = \sqrt{\frac{1}{n}\sum_{j=1}^{n}(\mu - d_j)^2}$$

The average deviation is not easy to manipulate mathematically, and so we use the standard deviation defined in the third line. The standard deviation also has the advantage of giving more weight to values further from the mean.

The right-most expressions in the formulas for mean, average deviation, and standard deviation use the Σ sum notation if you have seen that in school. The definition of each Σ expression is the formula to its left.

Exercise 14.3

Define a function to compute the standard deviation of a 1-dimensional **numpy** array without using *std*. Can you do it in one line?

Exercise 14.4

Define a function to compute the average deviation of a 1-dimensional **numpy** array. Can you do it in one line? What is the average deviation of `data_1`?

The *variance* is the square of the standard deviation, and so is σ^2. Like standard deviation, it measures the spread of data values about their mean.

$$\text{variance} \; = \; \sigma^2 \; = \; \frac{(\mu - d_1)^2 + \cdots + (\mu - d_n)^2}{n} \; = \; \frac{1}{n} \sum_{j=1}^{n} (\mu - d_j)^2$$

In the dataset `ages` at the beginning of section 14.1 with the ages of one hundred concert-goers, the unit for variance is years^2 (years-squared).

We compute variance with the *var* function.

```
np.var(ages)
```

```
20.555599999999995
```

14.1.4 Covariance and correlation

We have some additional (fictional) data about our 1990s cover band concert attendees. In addition to their ages, we know how many concerts they attended in 1995.

```
concerts_attended_in_1995 = np.array(
    [0, 0, 0, 0, 2, 0, 4, 4, 4, 0, 0, 4, 0, 0, 6, 0, 2, 0, 4, 0,
     0, 0, 0, 0, 0, 5, 4, 4, 5, 5, 4, 0, 5, 4, 0, 0, 4, 0, 0, 0,
     0, 4, 3, 4, 0, 7, 5, 6, 6, 4, 4, 5, 0, 0, 4, 0, 4, 0, 0, 4,
     0, 4, 0, 0, 0, 4, 0, 0, 0, 0, 2, 0, 0, 0, 6, 0, 4, 0, 0, 0,
     1, 0, 0, 4, 0, 4, 3, 6, 0, 0, 0, 0, 6, 0, 4, 0, 6, 0, 0, 4] )
```

Figure 14.2 is a scatter plot of age versus 1995 concerts attended.

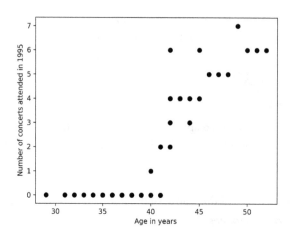

Figure 14.2: Scatter plot of age versus concerts attended in 1995

We expect to see few or no 1995 concerts attended by people who were very young then. Of those older than 16, we expect to see some non-zero counts. Also, if the people were older and working, they might have gone to more concerts. The scatter plot appears to support these hypotheses.

The *covariance* of two datasets of equal length gives us an idea of whether values in one increase or decrease linearly with values in the other. If the covariance is close to 0, we do not expect a linear relationship between the features measured in the datasets. A positive covariance means they increase or decrease together, while a negative one means that one increases as the other decreases. Figure 14.2 says the covariance is positive.

numpy can calculate a general covariance matrix, but all we need from it is the number in the second position in the first row. That is, the value at index [0, 1].

```
covariance_matrix = np.cov(ages, concerts_attended_in_1995)
covariance_matrix[0, 1]
```

```
8.792323232323243
```

I explain in section 14.5 why we need the [0, 1].

The *Pearson correlation coefficient,* often just called the correlation coefficient *r*, is the covariance divided by the standard deviation of each dataset. This division normalizes the covariance to be between `-1.0` and `1.0`. Values close to either extreme are "strongly correlated." Another way to say this is that there is a very strong linear relationship between the datasets. It does not imply causality.

```
correlation_coefficients = np.corrcoef(ages, concerts_attended_in_1995)
correlation_coefficients[0, 1]
```

```
0.8480108438760603
```

You may see the correlation coefficient written as ρ instead of *r*. ρ is the Greek letter "rho."

Given two datasets *D* and *E* with values d_0 through d_n and e_0 through e_n, the formulas for covariance Cov(*D*, *E*) and correlation coefficient *r* are shown in Figure 14.3.

$$\text{mean of } D = \mu_D = \frac{d_1 + \cdots + d_n}{n} = \frac{1}{n}\sum_{j=1}^{n} d_j$$

$$\text{mean of } E = \mu_E = \frac{e_1 + \cdots + e_n}{n} = \frac{1}{n}\sum_{j=1}^{n} e_j$$

$$\text{standard deviation of } D = \sigma_D = \sqrt{\frac{(\mu_D - d_1)^2 + \cdots + (\mu_D - d_n)^2}{n}} = \sqrt{\frac{1}{n}\sum_{j=1}^{n} (\mu_D - d_j)^2}$$

$$\text{standard deviation of } E = \sigma_E = \sqrt{\frac{(\mu_E - e_1)^2 + \cdots + (\mu_E - e_n)^2}{n}} = \sqrt{\frac{1}{n}\sum_{j=1}^{n} (\mu_E - e_j)^2}$$

$$\text{covariance of } D \text{ and } E = \text{Cov}(D, E) = \frac{1}{n}\sum_{j=1}^{n} (d_j - \mu_D)(e_j - \mu_E)$$

$$\text{correlation coefficient of } D \text{ and } E = r_{DE} = \frac{1}{n\,\mu_D\,\mu_E}\sum_{j=1}^{n} (d_j - \mu_D)(e_j - \mu_E)$$

Figure 14.3: Formulas for covariance and correlation

Exercise 14.5

Why does Cov(D, E) equal Cov(E, D)? Is the correlation coefficient similarly symmetric?

Exercise 14.6

Write Python functions to compute the covariance and correlation coefficient from their formulas.

14.2 Cats and commas

The City of Greater Dandenong in Australia freely provides data about the cats that have been registered with the city as of January 6, 2021. [CAT] Greater Dandenong is in the state of Victoria and is a suburb of Melbourne. [DAN]

The file is in CSV format, which means that commas separate the values on each line. You may think of the commas as separating columns of data. Each row of data contains the information about one cat.

We can peek at the file by opening it in a text editor or using the *read* function to examine the lines at the beginning of the file. The first line is not data but is a list of the labels or headings for each column.

```
cats = open("src/examples/registered-cats.csv", "rt")
print(cats.readline())

"Locality","Postcode","Animal_Type","Breed_Description",
"Colour_Description","GENDER"
```

There are six columns, and their labels are in double quotes. Not every CSV file includes the column labels as the first line. Some start with data and assume you otherwise know the column definitions.

The second line is the data for the first cat.

```
print(cats.readline())
cats.close()

"DANDENONG NORTH","3175","Cat","DOMSH","TAB","F"
```

The intersection of a column and a row is a *field*. So the data in the Locality field in the first record is DANDENONG NORTH. The data includes codes such as DOMSH for "domestic short hair," "TAB" for "tabby," and "F" for "female."

14.3 pandas DataFrames

A *DataFrame* is a flexible 2-dimensional data structure that represents rows and columns. The rows are records, and the columns are data of a particular *feature*. For example, we will see a DataFrame where the rows correspond to cats, and the columns include their breed and color.

We can access the rows and columns in ways that resemble both lists and dictionaries; that is, we can use numeric indices or string labels/keys. **pandas** is built on **numpy** and **matplotlib**, so you can use their functions and methods for computation and visualization. We discussed those packages in *Chapter 13, Creating Plots and Charts*.

pandas is a very large Python package and has many functions and methods for working with DataFrames. [PAN] You can often get the result you seek by using Pythonic combinations of row and column operations together with indexing and slicing. Other times, you may use loops on lists or arrays. I give examples of both styles of **pandas** coding. For any particular task, there are likely several ways of getting the same result.

14.3.1 Loading a DataFrame

We begin by importing **pandas** as pd and using the *read_csv* function to load our cat data from a file.

```
import pandas as pd

df = pd.read_csv("src/examples/registered-cats.csv")
```

Now we use *info* to tell us about the data:

```
df.info()

<class 'pandas.core.frame.DataFrame'>
RangeIndex: 3485 entries, 0 to 3484
Data columns (total 6 columns):
```

```
 #    Column              Non-Null Count   Dtype
---   ------              --------------   -----
 0    Locality            3485 non-null    object
 1    Postcode            3485 non-null    int64
 2    Animal_Type         3485 non-null    object
 3    Breed_Description   3482 non-null    object
 4    Colour_Description  3483 non-null    object
 5    GENDER              3485 non-null    object
dtypes: int64(1), object(5)
memory usage: 163.5+ KB
```

This display shows us that the new DataFrame has

- Six columns and 3,485 records
- Postcode fields loaded as **numpy** 64-bit integers (dtype **int64**)
- dtype **object** fields for Locality, Animal_Type, Breed_Description, Colour_Description, and GENDER

For our purposes, we can think of the objects as strings.

While a Python **int** number can have an arbitrary number of digits, the **int8**, **int16**, **int32**, and **int64** typed objects use 8, 16, 32, and 64 bits, respectively. For use in **numpy** and **pandas**, use the smallest size that can safely fit your values.

The *head* function shows the first five records in the DataFrame by default. Give it a positive integer argument to show more or fewer.

```
df.head()
```

```
          Locality  Postcode Animal_Type Breed_Description Colour_Description  \
0  DANDENONG NORTH      3175         Cat            DOMSH               TAB
1  DANDENONG NORTH      3175         Cat            DOMLH            BLAWHI
2        DANDENONG      3175         Cat            DOMSH            TABWHI
3       SPRINGVALE      3171         Cat              DOM            TORWHI
4        DANDENONG      3175         Cat              DOM            WHIGRE

  GENDER
0      F
1      M
2      F
3      F
4      M
```

The *tail* function shows the last five records in the DataFrame by default.

```
df.tail()
```

	Locality	Postcode	Animal_Type	Breed_Description	Colour_Description	\
3480	SPRINGVALE	3171	Cat	RAG	SEAL	
3481	NOBLE PARK	3174	Cat	DOMM	BLACK	
3482	NOBLE PARK	3174	Cat	DOMM	WHIGIN	
3483	SPRINGVALE	3171	Cat	RAGX	BLUEP	
3484	SPRINGVALE	3171	Cat	DOMSH	GINGER	

	GENDER
3480	M
3481	F
3482	M
3483	M
3484	M

That display is quite messy. Let's get rid of extraneous data and improve the column names.

14.3.2 Changing columns

In section 3.2.4, we saw how to use del to remove a list item. We can use it to remove a column from our DataFrame. The Animal_Type column is redundant since we know every animal is a cat. Let's remove it:

```
del df["Animal_Type"]

df.head()
```

	Locality	Postcode	Breed_Description	Colour_Description	GENDER
0	DANDENONG NORTH	3175	DOMSH	TAB	F
1	DANDENONG NORTH	3175	DOMLH	BLAWHI	M
2	DANDENONG	3175	DOMSH	TABWHI	F
3	SPRINGVALE	3171	DOM	TORWHI	F
4	DANDENONG	3175	DOM	WHIGRE	M

The display is more concise, but we can still shorten the Breed_Description and Colour_Description column labels. Also, it's odd that the GENDER column is uppercase when none of the others are.

To change the column labels, we create a dictionary with the old and new names, and then call *rename*:

```
new_column_names = {
    "Breed_Description": "Breed",
    "Colour_Description": "Colour",
    "GENDER": "Gender"
}
```

```
df.rename(columns=new_column_names).head()
```

```
        Locality  Postcode  Breed  Colour Gender
0  DANDENONG NORTH     3175  DOMSH     TAB      F
1  DANDENONG NORTH     3175  DOMLH  BLAWHI      M
2        DANDENONG     3175  DOMSH  TABWHI      F
3        SPRINGVALE     3171    DOM  TORWHI      F
4        DANDENONG     3175    DOM  WHIGRE      M
```

That looks good! Let's look at our DataFrame again to see if we want to make any other changes.

```
df.head()
```

```
        Locality  Postcode Breed_Description Colour_Description GENDER
0  DANDENONG NORTH     3175            DOMSH                TAB      F
1  DANDENONG NORTH     3175            DOMLH             BLAWHI      M
2        DANDENONG     3175            DOMSH             TABWHI      F
3        SPRINGVALE     3171              DOM             TORWHI      F
4        DANDENONG     3175              DOM             WHIGRE      M
```

What? The DataFrame doesn't have the new column names. To change the DataFrame itself rather than produce an altered copy of the DataFrame, you must use the `inplace=True` keyword argument to *rename*. This is true for many **pandas** functions.

```
df.rename(columns=new_column_names, inplace=True)
df.head()
```

```
        Locality  Postcode  Breed  Colour Gender
0  DANDENONG NORTH     3175  DOMSH     TAB      F
1  DANDENONG NORTH     3175  DOMLH  BLAWHI      M
2        DANDENONG     3175  DOMSH  TABWHI      F
3        SPRINGVALE     3171    DOM  TORWHI      F
4        DANDENONG     3175    DOM  WHIGRE      M
```

Now we can move on to looking at the data itself. By the way, I use the Australian English spelling "colour" instead of my native US English spelling "color" while we work with this DataFrame.

14.3.3 Examining a DataFrame

The *shape* property of a DataFrame returns a tuple of the number of rows and columns. Note that this is a property and not a method, so we do not use parentheses.

```
df.shape
```

```
(3485, 5)
```

This DataFrame has over three thousand rows, but data scientists routinely use DataFrames with tens or hundreds of thousands of rows. The **pandas** package has many functions that allow you to summarize the contents of rows and columns. For example, the *nunique* function returns the number of unique items in each column.

```
df.nunique()
```

```
Locality      10
Postcode       6
Breed         53
Colour       137
Gender         3
dtype: int64
```

As you can see, the data has relatively few localities, postcodes, and genders but many breeds and colours.

The *value_counts* function tells you how many of each unique item is represented in a column.

```
df["Locality"].value_counts()
```

```
NOBLE PARK          737
KEYSBOROUGH         656
DANDENONG NORTH     651
DANDENONG           549
SPRINGVALE          333
SPRINGVALE SOUTH    263
NOBLE PARK NORTH    224
BANGHOLME            68
DANDENONG SOUTH       3
LYNDHURST             1
Name: Locality, dtype: int64
```

If you use the `normalize=True` keyword argument, *value_counts* gives the result as a *frequency,* which is a percentage divided by 100.

```
df["Locality"].value_counts(normalize=True)
```

```
NOBLE PARK              0.211478
KEYSBOROUGH            0.188235
DANDENONG NORTH       0.186801
DANDENONG             0.157532
SPRINGVALE            0.095552
SPRINGVALE SOUTH      0.075466
NOBLE PARK NORTH      0.064275
BANGHOLME             0.019512
DANDENONG SOUTH       0.000861
LYNDHURST             0.000287
Name: Locality, dtype: float64
```

value_counts sorts the results in descending order by default. Use `ascending=False` to reverse that order, and `sort=False` to turn off sorting.

The indices in this *series* are locality names like BANGHOLME and NOBLE PARK NORTH. While a DataFrame is 2-dimensional, a series is 1-dimensional. For example, we can extract a single column as a series. Use the *sort_index* method to sort the series by its index values.

```
cat_data = df["Locality"].value_counts().sort_index()
cat_data
```

```
BANGHOLME               68
DANDENONG              549
DANDENONG NORTH       651
DANDENONG SOUTH         3
KEYSBOROUGH           656
LYNDHURST               1
NOBLE PARK            737
NOBLE PARK NORTH      224
SPRINGVALE            333
SPRINGVALE SOUTH      263
Name: Locality, dtype: int64
```

Exercise 14.7

How do the `ascending` and `inplace` keyword arguments change the behavior of *sort_index?*

Exercise 14.8

How many different breeds and colours are in the DataFrame, and what are they?

To better understand this data, we now create a bar chart. I won't repeat the discussion of **matplotlib** bar charts from section 13.2; but instead, show you how to use **pandas** and its corresponding visualization functions.

Given a **pandas** DataFrame or series object, use *plot* with the `kind="bar"` keyword argument to create a simple bar chart.

```
import matplotlib.pyplot as plt
plt.clf()
cat_data.plot(kind="bar")

<AxesSubplot:>
<Figure size 432x288 with 1 Axes>

plt.draw()
```

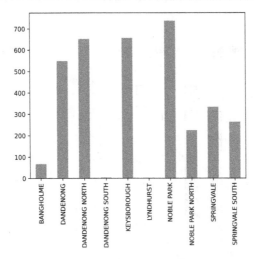

We can also use `cat_data.plot.bar()`.

Now we embellish our plot. First, we rotate the horizontal axis labels by 45 degrees because they are so long. We also make the bars light gray with a black outline and diagonal hatch marks.

```
plt.clf()
```

```
cat_data.plot(kind="bar", color="lightgray", edgecolor="black",
            rot=45, hatch="///")
```

```
<AxesSubplot:>
<Figure size 432x288 with 1 Axes>
```

The horizontal labels are still large, so we reduce their font size.

```
plt.tick_params(axis='x', labelsize=7)
```

```
<Figure size 432x288 with 1 Axes>
```

Finally, we add labels to the axes and a plot title.

```
plt.xlabel("Locality")
plt.ylabel("Number of cats")
plt.title("Number of registered cats in 2021 per locality in the City
of Greater Dandenong")
```

```
Text(0.5, 1.0, 'Number of registered cats in 2021 per locality in the
City of Greater Dandenong')
<Figure size 432x288 with 1 Axes>
```

```
plt.draw()
```

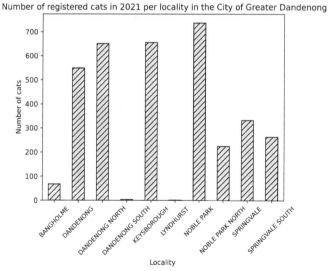

The **pandas** documentation includes a complete guide to its visualization capabilities. [P.V.Z] As in the example, most build on what you can do in **matplotlib**.

Exercise 14.9

Experiment with different strings such as `'/'`, `'\\'`, `'+'`, `'-'`, `'x'`, `'o'`, and `'*'` for the `hatch` keyword argument to *plot*. What changes when you use multiple symbols such as `"//"`, `"///"`, and `"////"`?

14.3.4 Accessing rows

pandas uses two primary properties to access rows: *iloc* and *loc*. The "loc" in each is short for "location." The "i" in *iloc* is short for "integer" and refers to the numeric position of the row in the DataFrame. This is similar to a list index.

Since our cats DataFrame is so large, let's begin by using slicing and *iloc* to extract seven rows. I also make a copy so we don't accidentally change the original DataFrame.

```
small_df = df.iloc[50:57].copy()
small_df
```

```
          Locality  Postcode  Breed  Colour Gender
50  SPRINGVALE SOUTH     3172  DOMSH  BLAWHI      F
51  SPRINGVALE SOUTH     3172   DOMM  WHIBRO      M
52  SPRINGVALE SOUTH     3172  DOMSH  REDCRM      M
53         DANDENONG     3175   DOMM  BLAWHI      F
54         DANDENONG     3175  DOMLH  GREWHI      F
55         DANDENONG     3175   CHIN  WHISIL      M
56         DANDENONG     3175  DOMSH     TRI      F
```

Now we can more easily see the difference between *iloc* and *loc*. The first row is at integer position 0:

```
small_df.iloc[0]
```

```
Locality    SPRINGVALE SOUTH
Postcode                3172
Breed                  DOMSH
Colour                BLAWHI
Gender                     F
Name: 50, dtype: object
```

Take a look at the display of `small_df`. See the numbers running down the left-hand side? While they were originally integer positions in `df`, they are also indices. Think of them as row labels.

The first row in `small_df` has the index 50, though it is at integer position 0.

```
small_df.loc[50]
```

```
Locality      SPRINGVALE SOUTH
Postcode                  3172
Breed                    DOMSH
Colour                  BLAWHI
Gender                       F
Name: 50, dtype: object
```

If you confuse indices and integer positions, you will likely get an error.

```
small_df.loc[0]
```

```
ValueError: 0 is not in range
```

```
small_df.iloc[50]
```

```
IndexError: single positional indexer is out-of-bounds
```

To remove one or more rows, use the *drop* method with a list of the row indices. As usual, specifying `inplace=True` changes the DataFrame itself instead of making a copy.

```
small_df.drop([51], axis=0, inplace=True)
small_df
```

	Locality	Postcode	Breed	Colour	Gender
50	SPRINGVALE SOUTH	3172	DOMSH	BLAWHI	F
52	SPRINGVALE SOUTH	3172	DOMSH	REDCRM	M
53	DANDENONG	3175	DOMM	BLAWHI	F
54	DANDENONG	3175	DOMLH	GREWHI	F
55	DANDENONG	3175	CHIN	WHISIL	M
56	DANDENONG	3175	DOMSH	TRI	F

A DataFrame has two *axes*, not to be confused with **matplotlib** plot axes. Axis 0 refers to the rows, and axis 1 is for the columns. In the case of *drop*, the default axis is 0, but I included it for clarity. You can also use `axis="index"` and `axis="columns"` instead of `axis=0` and `axis=1`.

```
small_df.drop([53, 54], axis="index")
```

	Locality	Postcode	Breed	Colour	Gender
50	SPRINGVALE SOUTH	3172	DOMSH	BLAWHI	F

```
52   SPRINGVALE SOUTH     3172   DOMSH   REDCRM     M
55           DANDENONG    3175   CHIN    WHISIL     M
56           DANDENONG    3175   DOMSH      TRI     F
```

Exercise 14.10

Did the previous example change `small_df`?

14.3.5 Accessing columns

To extract a subset of a DataFrame as a new DataFrame, use column labels within double brackets.

```
small_df[["Gender", "Breed"]]
```

```
     Gender  Breed
50        F  DOMSH
52        M  DOMSH
53        F   DOMM
54        F  DOMLH
55        M   CHIN
56        F  DOMSH
```

Note that I changed the order of the columns from the original DataFrame.

We can combine this with *loc* to extract both columns and rows:

```
small_df.loc[[54, 56], ["Breed", "Colour"]]
```

```
     Breed   Colour
54   DOMLH   GREWHI
56   DOMSH      TRI
```

There is a difference between using double and single brackets when you have one column label:

```
x = small_df[["Breed"]]
x
```

```
     Breed
50   DOMSH
52   DOMSH
```

```
53    DOMM
54   DOMLH
55    CHIN
56   DOMSH
```

```
y = small_df["Breed"]
y
```

```
50    DOMSH
52    DOMSH
53     DOMM
54    DOMLH
55     CHIN
56    DOMSH
Name: Breed, dtype: object
```

In the first case, we have a **DataFrame**, which is 2-dimensional. In the second, we have a 1-dimensional **Series**.

```
type(x), type(y)
```

```
(pandas.core.frame.DataFrame, pandas.core.series.Series)
```

```
x.shape, y.shape
```

```
((6, 1), (6,))
```

Since x is a DataFrame, we use *loc* to access a row:

```
x.loc[54]
```

```
Breed    DOMLH
Name: 54, dtype: object
```

The result is a series:

```
type(x.loc[54])
```

```
pandas.core.series.Series
```

y is a series, so we use single brackets and the index:

```
y[54]
```

```
'DOMLH'
```

The result is a single data item, often called a *scalar*:

```
type(y[54])
```

```
str
```

Exercise 14.11

What does y[[54]] return? What is its type? Why?

Exercise 14.12

Is the result of calling *nunique* a **DataFrame** or a **Series**? How do you access the Locality data?

Exercise 14.13

A DataFrame is 2-dimensional, a series is 1-dimensional, and a scalar is 0-dimensional. Can you devise a rule that connects the dimension of the original object with the dimension of the result when you use double or single brackets?

14.3.6 Creating a DataFrame

We loaded the df DataFrame from a file, but you can create one directly from a dictionary. The column labels are the dictionary keys. Here I combine the two concert-goers data sequences from section 14.1.4:

```
concert_df = pd.DataFrame(
    {
        "Age": ages,
        "1995 Concerts": concerts_attended_in_1995
    }
)

concert_df.iloc[3:9]
```

```
    Age  1995 Concerts
3   35              0
4   42              2
```

5	37	0
6	45	4
7	43	4
8	42	4

If you wish to use a different column order, use the `columns` keyword argument with a list of the column headings in the sequence you wish.

The *describe* method provides a descriptive statistical summary of the numeric columns in the DataFrame.

```
concert_df.describe()
```

```
              Age   1995 Concerts
count  100.000000      100.000000
mean    40.620000        1.880000
std      4.556669        2.275384
min     29.000000        0.000000
25%     38.000000        0.000000
50%     40.000000        0.000000
75%     43.000000        4.000000
max     52.000000        7.000000
```

I now want to give each concert attendee a unique identification number, starting with 1,000. Let's use the *sample* function from the **random** module. We use series notation to insert the identification numbers as a new column:

```
import random

random.seed(23)

num_rows = concert_df.shape[0]
concert_df["Id"] = random.sample(range(1000, 1001 + num_rows),
num_rows)

concert_df.head(5)
```

```
    Age  1995 Concerts    Id
0    40              0  1099
1    41              0  1037
2    39              0  1010
3    35              0  1002
4    42              2  1075
```

Exercise 14.14

I made a mistake and meant for the Id values to be strings and not integers.

```
concert_df.info()

<class 'pandas.core.frame.DataFrame'>
RangeIndex: 100 entries, 0 to 99
Data columns (total 3 columns):
 #   Column         Non-Null Count   Dtype
---  ------         --------------   -----
 0   Age            100 non-null     int32
 1   1995 Concerts  100 non-null     int32
 2   Id             100 non-null     int64
dtypes: int32(2), int64(1)
memory usage: 1.7 KB
```

Delete that column and add a new one with the same name using random numbers converted to **str**.

Use double brackets to reorder the columns. You need to assign the result to the original DataFrame, if you wish, since there is no `inplace` option.

```
concert_df = concert_df[["Id", "Age", "1995 Concerts"]]
```

```
concert_df.head(5)

     Id  Age  1995 Concerts
0  1099   40              0
1  1037   41              0
2  1010   39              0
3  1002   35              0
4  1075   42              2
```

We assigned the Id values randomly, but we can sort the DataFrame by the values in that column:

```
concert_df.sort_values(by="Id", inplace=True)
```

```
concert_df.head(5)

      Id  Age  1995 Concerts
93  1000   35              0
```

```
14   1001    45              6
3    1002    35              0
18   1003    42              4
29   1004    48              5
```

 Exercise 14.15

Sort the DataFrame in place by age.

14.3.7 Saving a DataFrame

In section 14.3.1, we used *read_csv* to read a DataFrame from a file. To write a CSV file, use *to_csv*.

```
concert_df.to_csv("work/concert-goers.csv", index=False)
```

This is a text file, so we can peek at it with *readline*. Here are the header and the first five rows of data in the file:

```
with open("work/concert-goers.csv", "rt") as csv_file:
    for _ in range(6):
        print(csv_file.readline(), end="")
Id,Age,1995 Concerts
1000,35,0
1001,45,6
1002,35,0
1003,42,4
1004,48,5
```

pandas supports several file data formats with methods to read and write them:

- *read_csv* and *to_csv* for CSV files
- *read_excel* and *to_excel* for Microsoft .xlsx files
- *read_json* and *to_json* for JSON files (see section 8.8.2)
- *read_pickle* and *to_pickle* for pickle files (see section 8.8.1)

This expression writes our concert-goers DataFrame to a Microsoft Excel spreadsheet:

```
concert_df.to_excel("work/concert-goers.xlsx", index=False)
```

Figure 14.4 is a screenshot of the result. Note that I included index=False so that **pandas** would only write the data to the Excel sheet. Other keyword arguments to the read and write methods allow you to specify columns, rows, and sheets.

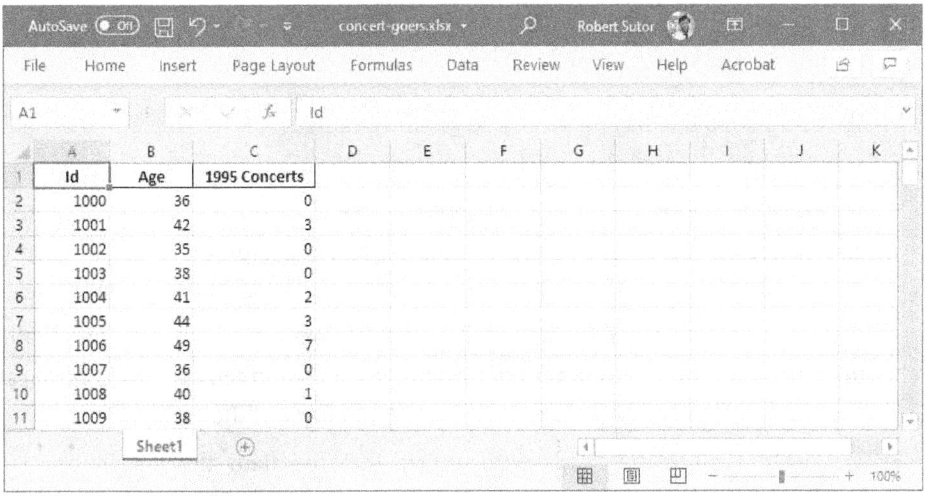

Figure 14.4: Microsoft Excel with the spreadsheet generated by "to_excel"

14.4 Data cleaning

Cleaning data is an important topic, and authors have written dozens of books, chapters, and papers on the subject. [CLD] What do you do when data is wrong or missing?

I was surprised when I first looked at the cats DataFrame and discovered that the GENDER column had three codes: F, M, and U. Presumably, the last stands for "unknown."

```
df['Gender'].value_counts()

F    1863
M    1616
U       6
Name: Gender, dtype: int64
```

We use a conditional expression to filter the rows to see only those with U for gender:

```
df[df["Gender"] == "U"]

          Locality  Postcode  Breed  Colour Gender
259  DANDENONG NORTH      3175    DOM  UNKNOW      U
```

```
611          SPRINGVALE     3171  DOMSH   WHITE    U
690     NOBLE PARK NORTH    3174  DOMSH  SILTAB    U
1273         NOBLE PARK     3174  DOMSH     TAB    U
1697        KEYSBOROUGH     3173  DOMSH   BLACK    U
1750    DANDENONG NORTH     3175    DOM  BROWHI    U
```

What do we do with these rows?

- We could go to the data provider and see if we can get correct values that are either F or M. This action might be easy, hard, or impossible.

- We could delete the rows. This is simple but could throw off the statistical analysis elsewhere. What if one of these rows corresponded to a unique cat of a given breed or colour? What if it was for the lone cat that lives in a locality or postcode area?

- We could replace the fields with values that maintain the statistical distribution of the entire column. There is roughly the same number of female cats as male cats, so we could randomly assign three of the cats F and three of the cats M.

- We could replace the fields with values that maintain the distribution within locality, postcode, breed, or colour. Which do we choose? See section 14.8 for an interesting case connecting gender and colour.

Whatever you do, write it down, and be sure to note all changes to the original dataset in any exercise, paper, study, or report you write. Failure to do so will bring into doubt your results and your reputation. Editors of prestigious journals have withdrawn previously published papers when subsequent analysis showed that the authors ignored or modified significant data.

Suppose we find out that the gender of that cat in the row with the index 611 should be F. We change it using *iloc* on the left of an assignment:

```
df.loc[611, ["Gender"]] = 'F'

df[df["Gender"] == "U"]
```

```
              Locality  Postcode  Breed  Colour Gender
259     DANDENONG NORTH     3175    DOM  UNKNOW    U
690    NOBLE PARK NORTH     3174  DOMSH  SILTAB    U
1273         NOBLE PARK     3174  DOMSH     TAB    U
1697        KEYSBOROUGH     3173  DOMSH   BLACK    U
1750    DANDENONG NORTH     3175    DOM  BROWHI    U
```

If we have no other changes and decide to remove all other records with U for gender, we can do so in one line:

```
df = df[df["Gender"].str.contains("F|M")]
df['Gender'].value_counts()
```

```
F    1864
M    1616
Name: Gender, dtype: int64
```

 Exercise 14.16

Why is there a vertical bar "|" in the argument to *contains*? Hint: see section 12.2.3.

There are now gaps in the index names compared to the integer positions. Use *reset_index* to set them to the same values based on the new number of rows.

```
df.reset_index(drop=True, inplace=True)
```

In the last example, we had data in a column, but we decided it was incorrect. What happens if the data is missing? **pandas** uses np.nan as the value for any missing data. Test for this "not a number" value using *isna*.

```
df[df["Breed"].isna()]
```

	Locality	Postcode	Breed	Colour	Gender
437	DANDENONG NORTH	3175	NaN	GINGER	F
2461	KEYSBOROUGH	3173	NaN	TAB	F
2566	DANDENONG	3175	NaN	TABWHI	F

You can use *isnull* as an alias for *isna*.

If you look at the original file from which we loaded the DataFrame, you will see that the missing values are empty strings between commas. The Colour column is missing some data, but the Locality column is complete.

```
df[df["Colour"].isna()]
```

	Locality	Postcode	Breed	Colour	Gender
1205	SPRINGVALE	3171	DOMLH	NaN	F
2462	KEYSBOROUGH	3173	TAB	NaN	F

```
df[df["Locality"].isna()]
```

```
Empty DataFrame
Columns: [Locality, Postcode, Breed, Colour, Gender]
Index: []
```

Now that we know where the missing values are, we can replace them as we did before or remove the offending rows with *dropna*.

```
df.dropna(inplace=True)
df[df["Breed"].isna()]
```

```
Empty DataFrame
Columns: [Locality, Postcode, Breed, Colour, Gender]
Index: []
```

```
df[df["Colour"].isna()]
```

```
Empty DataFrame
Columns: [Locality, Postcode, Breed, Colour, Gender]
Index: []
```

Exercise 14.17

The methods *notna* and *notnull* are the negated forms of *isna* and *isnull*, respectively.

dropna removes any row with any missing values. Use *notna* in an expression that removes only the rows that are missing values in the Breed column.

14.5 Statistics with pandas

pandas implements versions of all the statistical functions we saw in section 14.1. The *describe* method summarizes many of these for the columns with numeric data. It returns a DataFrame.

```
concert_df.describe()
```

	Id	Age	1995 Concerts
count	100.00000	100.000000	100.000000
mean	1049.69000	40.620000	1.880000
std	29.28084	4.556669	2.275384
min	1000.00000	29.000000	0.000000
25%	1024.75000	38.000000	0.000000
50%	1049.50000	40.000000	0.000000

| 75% | 1074.25000 | 43.000000 | 4.000000 |
| max | 1100.00000 | 52.000000 | 7.000000 |

14.5.1 Means, and medians, and deviations, oh my

pandas provides methods to compute each of these statistics. *mean* operates on a **DataFrame** object and returns the computed averages as a **Series**:

```
concert_df.mean()
```

```
Id               1049.69
Age                40.62
1995 Concerts       1.88
dtype: float64
```

Given a series, *mean* returns a scalar.

```
concert_df["Age"].mean()
```

```
40.62
```

Exercise 14.18

Explain the syntax of these two examples and why they successfully compute the mean of the Age column:

```
concert_df.describe().loc["mean", "Age"]
```

```
40.62
```

```
concert_df.mean()["Age"]
```

```
40.62
```

The next several examples compute the standard deviation, variance, first quartile, median, and third quartile of the Age column.

```
concert_df.std()["Age"]
```

```
4.556668994258012
```

```
concert_df.var()["Age"]
```

```
20.76323232323232
```

```
concert_df.quantile(q=0.25)["Age"]
```

38.0

```
concert_df.median()["Age"]
```

40.0

```
concert_df.quantile(q=0.75)["Age"]
```

43.0

Exercise 14.19

Use the *min* and *max* methods to compute the minimum and maximum values in the `1995 Concerts` column.

14.5.2 Covariance and correlation, again

As we saw in section 14.1.4, covariance and correlation measure how two sequences of data vary together. Typically, we want to see if there is a linear relationship between the two and whether one increases or decreases as the other increases or decreases. The Pearson correlation coefficient is the covariance divided by the standard deviations of the sequences.

In the language of **pandas**, the data sequences are usually **Series** objects. The *cov* and The *corr* methods compare each numeric column with every numeric column, including itself. The results are tables or, more mathematically speaking, *matrices*.

```
concert_df.cov()
```

	Id	Age	1995 Concerts
Id	857.367576	0.729495	3.255354
Age	0.729495	20.763232	8.792323
1995 Concerts	3.255354	8.792323	5.177374

```
cc = concert_df.corr(method="pearson")
cc
```

	Id	Age	1995 Concerts
Id	1.000000	0.005468	0.048861
Age	0.005468	1.000000	0.848011
1995 Concerts	0.048861	0.848011	1.000000

Look at the diagonal of numbers going from the upper left to the lower right. These compare columns to themselves, and the columns must be highly correlated!

We can ignore any row or column that includes the Id because we assigned these values randomly.

The interesting positions in the table are where one of the rows is Age, and the other is 1995 Concerts.

```
cc.loc["Age", "1995 Concerts"]
```

```
0.8480108438760597
```

This number is positive and much closer to 1.0 than 0.0. Therefore, the two columns are highly correlated, and one increases as the other does. I interpret this to mean that the older you are, the more concerts you likely attended in 1995.

See where I used ["Age", "1995 Concerts"] to get the correlation coefficient? This indexing retrieved the value from a specific row and column. That's why in section 14.1.4, I used [0, 1] to get the values from the first rows, second positions of the covariance and correlation matrices.

14.6 Converting categorical data

The GENDER column contains *categorical data* rather than numeric. The items in the column belong to a fixed set of values, which are usually strings. In this case, the values are 'F' and 'M'. While we can check if an item is equal to one of these, it is often easier to convert the categorical column to multiple numeric "dummy columns" containing 0 and 1.

Here are the first two rows of df:

```
df.head(2)
```

```
         Locality  Postcode  Breed  Colour Gender
0  DANDENONG NORTH     3175  DOMSH     TAB      F
1  DANDENONG NORTH     3175  DOMLH  BLAWHI      M
```

and this is what we get when we use *get_dummies* on the GENDER column:

```
pd.get_dummies(df, columns=["Gender"]).head(2)
```

```
         Locality  Postcode  Breed  Colour  Gender_F  Gender_M
0  DANDENONG NORTH     3175  DOMSH     TAB         1         0
1  DANDENONG NORTH     3175  DOMLH  BLAWHI         0         1
```

pandas removed the GENDER column and replaced it with two new columns, Gender_M and Gender_F. By default, *get_dummies* inserts the original column name as the prefix for the new names and an underscore as the separator. Use the prefix keyword argument to start the new column names with the string of your choice:

```
pd.get_dummies(df, columns=["Gender"], prefix="G").head(2)
```

```
        Locality  Postcode  Breed  Colour  G_F  G_M
0  DANDENONG NORTH     3175  DOMSH     TAB    1    0
1  DANDENONG NORTH     3175  DOMLH  BLAWHI    0    1
```

You can remove the prefix and the separator entirely by using empty strings with prefix and prefix_sep.

```
df = pd.get_dummies(df, columns=["Gender"], prefix="", prefix_sep="")
df.head(2)
```

```
        Locality  Postcode  Breed  Colour  F  M
0  DANDENONG NORTH     3175  DOMSH     TAB  1  0
1  DANDENONG NORTH     3175  DOMLH  BLAWHI  0  1
```

Since the new columns are numeric, *describe* returns statistics about them:

```
df.describe()
```

```
          Postcode            F            M
count  3475.000000  3475.000000  3475.000000
mean   3173.969209     0.534964     0.465036
std      13.652882     0.498848     0.498848
min    3171.000000     0.000000     0.000000
25%    3173.000000     0.000000     0.000000
50%    3174.000000     1.000000     0.000000
75%    3175.000000     1.000000     1.000000
max    3975.000000     1.000000     1.000000
```

Exercise 14.20

What do the mean row entries for F and M say about the relative percentages of each? Why?

Exercise 14.21

Why might you not want to create dummy columns from the `Breed` and `Colour` columns?

14.7 Cats by gender in each locality

Our goal in this section is to create and plot a DataFrame containing the number of female and male cats per locality. We begin with a version of building the DataFrame via comprehensions.

The *unique* function returns a **numpy** array of the unique values in a **pandas** column.

```
localities = df["Locality"].unique()
localities
```

```
array(['DANDENONG NORTH', 'DANDENONG', 'SPRINGVALE', 'SPRINGVALE SOUTH',
       'KEYSBOROUGH', 'NOBLE PARK NORTH', 'NOBLE PARK', 'BANGHOLME',
       'DANDENONG SOUTH', 'LYNDHURST'], dtype=object)
```

```
type(localities)
```

```
numpy.ndarray
```

To make it easier for users to understand the results, we put the localities in alphabetical order. The **numpy** *sort* method sorts an array in place.

```
localities.sort()
localities
```

```
array(['BANGHOLME', 'DANDENONG', 'DANDENONG NORTH', 'DANDENONG SOUTH',
       'KEYSBOROUGH', 'LYNDHURST', 'NOBLE PARK', 'NOBLE PARK NORTH',
       'SPRINGVALE', 'SPRINGVALE SOUTH'], dtype=object)
```

Now I use two list comprehensions to create lists of female and male cat counts per locality. Note how I filter the rows of the DataFrame by localities within each comprehension.

```
number_of_male_cats = [
    df[df["Locality"] == locality]['M'].sum()
        for locality in localities
]
```

```
number_of_male_cats

[36, 263, 295, 1, 301, 1, 337, 109, 144, 129]

number_of_female_cats = [
    df[df["Locality"] == locality]['F'].sum()
        for locality in localities
]

number_of_female_cats

[32, 285, 353, 2, 352, 0, 399, 114, 188, 134]
```

Let's also change the locality names so that only the first letters in each word are capitalized. This is called "title case."

```
localities = [locality.title() for locality in localities]
localities

['Bangholme',
 'Dandenong',
 'Dandenong North',
 'Dandenong South',
 'Keysborough',
 'Lyndhurst',
 'Noble Park',
 'Noble Park North',
 'Springvale',
 'Springvale South']
```

Exercise 14.22

What is the difference between the **str** methods *capitalize* and *title*?

Exercise 14.23

localities is now a **list**. How would you convert it to a **numpy** array?

We have the pieces, and now we create the DataFrame:

```
df_cats = pd.DataFrame(
    {
        "Locality": localities,
        'F': number_of_female_cats,
        'M': number_of_male_cats
    }
)

df_cats
```

```
              Locality    F    M
0            Bangholme    32   36
1            Dandenong   285  263
2      Dandenong North   353  295
3      Dandenong South     2    1
4          Keysborough   352  301
5            Lyndhurst     0    1
6           Noble Park   399  337
7     Noble Park North   114  109
8           Springvale   188  144
9     Springvale South   134  129
```

That was straightforward, but we could have done most of it in one step using *groupby*. This function collects data into groups and then applies some aggregation or transformation function within each group. Here we group by Locality and sum the numeric values in each column within each group:

```
df_cats = df.groupby(["Locality"]).sum()
df_cats
```

```
                   Postcode      F      M
Locality
BANGHOLME            215900   32.0   36.0
DANDENONG           1739900  285.0  263.0
DANDENONG NORTH     2057400  353.0  295.0
DANDENONG SOUTH        9525    2.0    1.0
KEYSBOROUGH         2071969  352.0  301.0
LYNDHURST              3975    0.0    1.0
NOBLE PARK          2336064  399.0  337.0
```

```
NOBLE PARK NORTH     707802  114.0  109.0
SPRINGVALE          1052772  188.0  144.0
SPRINGVALE SOUTH     834236  134.0  129.0
```

That's not quite right. We do not need the sum of the `Postcode` column, so we specify the two columns we do want:

```
df_cats = df.groupby(["Locality"])[['F', 'M']].sum()
df_cats
```

```
                       F      M
Locality
BANGHOLME           32.0   36.0
DANDENONG          285.0  263.0
DANDENONG NORTH    353.0  295.0
DANDENONG SOUTH      2.0    1.0
KEYSBOROUGH        352.0  301.0
LYNDHURST            0.0    1.0
NOBLE PARK         399.0  337.0
NOBLE PARK NORTH   114.0  109.0
SPRINGVALE         188.0  144.0
SPRINGVALE SOUTH   134.0  129.0
```

The localities are the index, which you can see on the left-hand side of the DataFrame. The *reset_index* method makes the localities a column and creates a new numeric index.

```
df_cats.reset_index(inplace=True)
df_cats
```

```
             Locality      F      M
0           BANGHOLME   32.0   36.0
1           DANDENONG  285.0  263.0
2     DANDENONG NORTH  353.0  295.0
3     DANDENONG SOUTH    2.0    1.0
4         KEYSBOROUGH  352.0  301.0
5           LYNDHURST    0.0    1.0
6          NOBLE PARK  399.0  337.0
7    NOBLE PARK NORTH  114.0  109.0
8          SPRINGVALE  188.0  144.0
9    SPRINGVALE SOUTH  134.0  129.0
```

Use *astype* to change the type of the values in a column or row. Here we set the F and M columns to be **numpy** 8-bit integers.

```
df_cats = df_cats.astype({'F': "int8", 'M': "int8"})
```

Our final transformation is to title case the Locality column:

```
df_cats["Locality"] = df_cats["Locality"].str.title()
df_cats
```

```
                 Locality    F     M
0               Bangholme   32    36
1               Dandenong   29     7
2         Dandenong North   97    39
3         Dandenong South    2     1
4             Keysborough   96    45
5               Lyndhurst    0     1
6              Noble Park -113    81
7        Noble Park North  114   109
8               Springvale  -68 -112
9         Springvale South -122 -127
```

Exercise 14.24

What does df_cats["Locality"].str.title() return? What is its type?

From here, it is simple to generate a side-by-side bar graph:

```
plt.clf()

df_cats.plot.bar(
    x="Locality",
    y=['F', 'M'],
    color=["gray", "black"],
    title="Number of female and male cats per locality",
    xlabel="Locality",
    ylabel="Count"
)
```

```
<AxesSubplot:title={'center':'Number of female and male cats per
locality'}, xlabel='Locality', ylabel='Count'>
```

```
<Figure size 432x288 with 0 Axes>
<Figure size 432x288 with 1 Axes>
```

```
plt.draw()
```

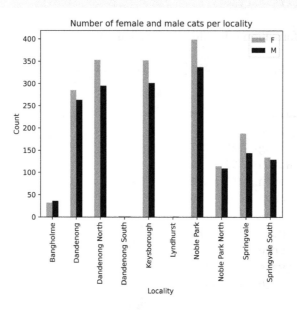

Add the `stacked=True` keyword argument to show stacked bars as in Figure 14.5.

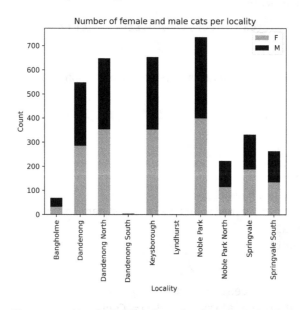

Figure 14.5: Number of female and male cats per locality

14.8 Are all tortoiseshell cats female?

Our DataFrame contains cats of many colours. Here's some Python to print the colour codes in eight neat columns:

```python
colours = sorted(df["Colour"].unique())

for count, colour in enumerate(colours, start=1):
    print(f"{colour:<9}", end=("\n" if count % 8 == 0 else ""))
```

AGOUTI	AMB	APR	APRWHI	BEIBLA	BEIGE	BEIWHI	BLAAPR
BLABRI	BLABRO	BLACK	BLACRE	BLAGIN	BLAGRE	BLAORA	BLATAN
BLATOR	BLAWHI	BLONDE	BLUBI	BLUCRE	BLUE	BLUEP	BLUGRE
BLUMIT	BLUSPO	BLUTAB	BLUTOR	BLUWHI	BONE	BRIBRO	BRIGRE
BRIND	BROBEI	BROBLA	BROCRE	BROGRE	BROMER	BROSPOT	BROTAB
BROWHI	BROWN	BRSPTAB	CALI	CHAMP	CHAR	CHARTAB	CHAWHI
CHMWHI	CHOC	CHOCP	CHOCRE	CHOWHI	CINN	CREAM	CREAPOI
CREBEI	CREBRO	CREGIN	CREGRE	CREWHI	DKGREY	DTORT	FAWN
GINBLA	GINBRO	GINGER	GINTAB	GINWHI	GOLD	GREBLA	GREBLU
GREBRO	GRECRE	GREGIN	GREGOL	GRESIL	GRETAB	GRETAN	GREWHI
GREY	LILAC	LILBI	LILMIT	LILWHI	LIVWHI	MARB	MINK
MOTT	ORABLA	ORAGRE	ORANGE	ORAWHI	PEACH	RED	REDCRM
REDGIN	REDPOINT	REDTAB	SABLE	SAND	SBROW	SEAL	SEALM
SEALTORT	SEPIA	SILBLA	SILGRE	SILTAB	SILVER	SILWHI	SPOT
TAB	TABBLA	TABGREY	TABPOI	TABWHI	TANWHI	TORGRY	TORT
TORTTAB	TORWHI	TRI	UNKNOW	WHIAPR	WHIBLA	WHIBLU	WHIBRO
WHICRE	WHIGIN	WHIGRE	WHIORA	WHIRED	WHISIL	WHITAN	WHITE
WHITOR							

Presumably, a TANWHI cat is more tan than white when compared to a WHITAN cat. I'm interested in the cats that have TOR in their colour codes. TOR is an abbreviation for "tortoiseshell," and I had always heard that such cats are always female. Does our data support this?

There are nine colour codes containing TOR:

```python
[colour for colour in colours if "TOR" in colour]
```

```python
['BLATOR',
 'BLUTOR',
```

```
    'DTORT',
    'SEALTORT',
    'TORGRY',
    'TORT',
    'TORTTAB',
    'TORWHI',
    'WHITOR']
```

and we can subset the full DataFrame to just those rows for tortoiseshell cats:

```
tor_df = df[df["Colour"].str.contains("TOR")]
```

groupby and *sum* give us the counts by gender:

```
tor_df.groupby("Colour")[['M', 'F']].sum()
```

```
              M    F
Colour
BLATOR        0   11
BLUTOR        0    2
DTORT         0    2
SEALTORT      0    1
TORGRY        0   18
TORT         10  224
TORTTAB       0    7
TORWHI        2   51
WHITOR        1   15
```

Except for the TORT row, it appears that almost all the tortoiseshell cats are female. Genetically, this is generally the case, and the few male tortoiseshell cats are usually sterile.

The larger TORT and TORWHI row sums for M may come from incorrect information in the original file. If you wanted to use this DataFrame for a serious study, you should validate the gender information. This is part of the data cleaning process we discussed in section 14.4.

We can do the same analysis for calico cats in one line of Python **pandas** code:

```
df[df["Colour"].str.contains("CALI")].groupby("Colour")[['M',
  'F']].sum()
```

```
        M   F
Colour
CALI    0  20
```

Exercise 14.25

Repeat the preceding analysis to determine if ginger cats (those with GIN in their colour) are predominantly male or female.

Exercise 14.26

Create a horizontal bar chart that shows how many female ginger cats are in each postcode area.

14.9 Cats in trees and circles

Though I have focused on **matplotlib** as the primary visualization engine for data, several other Python packages may suit your advanced needs. I especially suggest you look at **seaborn** and **plotly**. [SEA] [PLY]

I end this chapter by using our cat data to create treemaps and Venn diagrams.

14.9.1 Treemaps

Now that we have the colour data, we can visualize it in many different ways. The **squarify** module creates *treemaps*, which are rectangles broken down into sub-rectangles based on relative data counts.

First, we install **squarify** from the operating system command line: [SQU]

```
pip install squarify
```

and import it:

```
import matplotlib.pyplot as plt
import squarify
```

In the last section, we created the list `colours` containing all the unique cat colour codes in our DataFrame. We now create three other lists:

```
grey_colours = [colour for colour in colours if "GRE" in colour]
print(*grey_colours)
```

```
BLAGRE BLUGRE BRIGRE BROGRE CREGRE DKGREY GREBLA GREBLU GREBRO GRECRE
GREGIN GREGOL GRESIL GRETAB GRETAN GREWHI GREY ORAGRE SILGRE TABGREY
WHIGRE
```

```
tan_colours = [colour for colour in colours if "TAN" in colour]
print(*tan_colours)
```

```
BLATAN GRETAN TANWHI WHITAN
```

```
white_colours = [colour for colour in colours if "WHI" in colour]
print(*white_colours)
```

```
APRWHI BEIWHI BLAWHI BLUWHI BROWHI CHAWHI CHMWHI CHOWHI CREWHI GINWHI
GREWHI LILWHI LIVWHI ORAWHI SILWHI TABWHI TANWHI TORWHI WHIAPR WHIBLA
WHIBLU WHIBRO WHICRE WHIGIN WHIGRE WHIORA WHIRED WHISIL WHITAN WHITE
WHITOR
```

Exercise 14.27

Explain what is happening in the expression `print(*grey_colours)`.

The lists are the color codes used for cats that are at least partially grey, tan, or white. Here's how we draw a treemap plot of the relative sizes of these lists and the total number of colours.

```
squarify.plot(
    sizes=[len(colours), len(grey_colours),
           len(tan_colours), len(white_colours)],
    label=["total", "Grey Cats", "Tan Cats", "White Cats"],
    color=["#DDDDDD", "#BBBBBB", "#999999", "white"])
```

```
<AxesSubplot:>
<Figure size 432x288 with 1 Axes>
```

```
plt.draw()
```

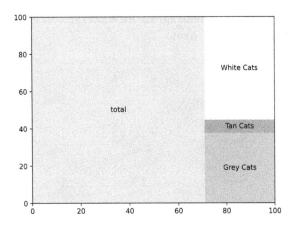

If you omit the `color` keyword argument, **squarify** randomly chooses the rectangle shades for you. These may change every time you call *plot*.

Exercise 14.28

Extend this example to include ginger cats.

14.9.2 Venn diagrams

Many of us learned about Venn diagrams in school. They use overlapping circles to show how sets of data intersect.

To draw them with Python, we install **matplotlib-venn** from the operating system command line and import it: [VEN]

```
pip install matplotlib-venn
```

```
import matplotlib_venn
```

This module can draw Venn diagrams with two and three circles. I use the lists of grey, tan, and white colours from the last section as the data for a 3-circle diagram:

```
matplotlib_venn.venn3(
    subsets=(set(grey_colours), set(tan_colours), set(white_colours)),
    set_labels=("Grey", "Tan", "White")
)
```

```
<matplotlib_venn._common.VennDiagram at 0x227fb9a02e0>
<Figure size 432x288 with 1 Axes>
```

```
plt.draw()
```

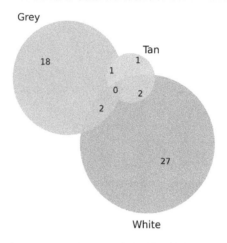

Exercise 14.29

What do the numbers 18 and 27 in the "Grey" and "White" circles mean?

14.10 Summary

In this chapter, we learned the fundamental concepts of descriptive statistics and saw how to use **numpy** and **matplotlib** for computation and visualization. We then moved on to **pandas**, a highly sophisticated Python package for reading, writing, and analyzing DataFrames. We revisited statistics to show the ease with which **pandas** allows us to examine and summarize large and small datasets.

The next chapter concludes this book, but for the appendices, references, and indices. I will provide a brief survey of using Python for machine learning. This knowledge will enable you, I hope, to move on to more advanced books and articles to understand the full breadth of Python tools for artificial intelligence.

15

Learning, Briefly

Learning is the only thing the mind never exhausts,
never fears, and never regrets.

—Leonardo da Vinci

Machine learning is not new, but it and its sub-discipline, deep learning, are now being used extensively for many applications in artificial intelligence (AI). There are hundreds of academic and practical coding books about machine learning.

This final chapter introduces machine learning and neural networks primarily through the scikit-learn **sklearn** module. Consider this a jumping-off point where you can use the Python features you've learned in this book to go more deeply into these essential AI areas if they interest you.

Topics covered in this chapter

15.1 What is machine learning?

Consider my complete product browsing and purchase history with an online retailer. This data represents my experience of buying from the seller and their experience selling to me. How can the data and my new purchases help the retailer sell me additional products and services?

Machine learning is a large set of techniques where computer algorithms learn from existing data rather than having hardcoded decisions built into them. Such an algorithm improves its accuracy as it receives new data. It might also get better when humans or processes intervene to judge the correctness and quality of previous results.

In *unsupervised learning*, an algorithm looks for patterns in the data that may help users gain insight. An unsupervised algorithm may help the retailer see that I have increased interests in cooking, travel, or home repair during certain seasons. The seller can then preemptively give me product recommendations several weeks before I would typically purchase them.

In *supervised learning*, the data also has among its features the "answer" to a question we might ask about it. Suppose the online retailer also has a travel bureau. I bought recommended books and clothes after I purchased the trip. In that case, the retailer can label the book and clothes data "purchased for travel." Given new clothing purchases, the retailer's machine learning algorithm might strongly suggest that I am looking to take a vacation and offer possible flights and hotels.

In *reinforcement learning*, the algorithm adds some kind of reward or penalty based on the outcome. If I purchase and then return a product, the algorithm might insert a negative score as a feature in the data associated with the deduced result. This feature could help the algorithm make a better future choice between two possible actions. Perhaps I am taking a trip, and I want to buy new shirts. If I purchase and then return some of them, the retailer can update the data to note that I don't like long-tailed shirts with buttons when going to a warm climate.

Another way to get reinforcement data is to ask the user directly. If the retailer website shows you a product and then asks, "Do you like this recommendation?" the seller can use a "yes" or "no" response to tune its recommendation algorithm. Of course, if you purchase the product, that should produce positive reinforcement.

Exercise 15.1

How might you use reinforcement learning to teach a self-driving car how fast it should be traveling when going around a curve in the road?

The information a machine learning algorithm uses to "learn" is called the *training set*. For supervised learning, the training set has one or more *label* features. When we get new unlabeled data, we run the algorithm and hope it matches the expected results. When we *fit* the training data with an algorithm to make predictions, we create a machining learning *model*.

A common method to test the machine learning algorithm is to select 50% of all the labeled data randomly and designate it to the training set. Then, take the remaining data, ignore the labels, and see how well the algorithm predicts the results in this *test set*. Keep adjusting the algorithm to improve accuracy, if possible. You may need to change which features you use, normalize the features (section 15.3), reduce the number of features (section 15.4), or switch to an entirely different technique.

15.2 Cats again

We again employ the data about cats from the City of Greater Dandenong that I introduced in section 14.2. As before, we clean the data and create gender dummy columns.

```python
import pandas as pd

df = pd.read_csv("src/examples/registered-cats.csv")
del df["Animal_Type"]
del df["Postcode"]

new_column_names = {
    "Breed_Description": "Breed",
    "Colour_Description": "Colour",
    "GENDER": "Gender"
}

df.rename(columns=new_column_names, inplace=True)

df.loc[611, ["Gender"]] = 'F'
df = df[df["Gender"].str.contains("F|M")]
df.reset_index(drop=True, inplace=True)
```

```
df.dropna(inplace=True)
df.reset_index(drop=True, inplace=True)
df = pd.get_dummies(df, columns=["Gender"], prefix="", prefix_sep="")
df = df.astype({"F": "int8", "M": "int8"})
df.info()

<class 'pandas.core.frame.DataFrame'>
RangeIndex: 3475 entries, 0 to 3474
Data columns (total 5 columns):
 #   Column    Non-Null Count   Dtype
---  ------    --------------   -----
 0   Locality  3475 non-null    object
 1   Breed     3475 non-null    object
 2   Colour    3475 non-null    object
 3   F         3475 non-null    int8
 4   M         3475 non-null    int8
dtypes: int8(2), object(3)
memory usage: 88.4+ KB
```

I deleted the Postcode column since it is closely related to Locality.

For our purposes in this chapter, we need all our column entries to be numeric. Our strategy is this: for each of the Locality, Breed, and Colour columns, collect and sort the unique entries. Assign an **int** code to each unique entry. Build dictionaries to go from the **str** names to the **int** codes.

I begin with the breeds:

```
import numpy as np

breed_dict = dict()

for count, breed in enumerate(np.sort(df["Breed"].unique())):
    breed_dict[breed] = count

breed_dict["DOMSH"]

21
```

I do the same for colours:

```
colour_dict = dict()

for count, colour in enumerate(np.sort(df["Colour"].unique())):
    colour_dict[colour] = count
```

Now for the localities:

```
locality_dict = dict()

for count, locality in enumerate(np.sort(df["Locality"].unique())):
    locality_dict[locality] = count
```

Now we are ready to update the entries. I iterate over the DataFrame rows and pull out the **str** breed, colour, and locality values as a tuple. I then look up the **int** codes and put those in the DataFrame:

```
for n in range(df.shape[0]):
    breed, colour, locality = df.loc[n, ["Breed", "Colour", "Locality"]
]
    df.at[n, "Breed"] = breed_dict[breed]
    df.at[n, "Colour"] = colour_dict[colour]
    df.at[n, "Locality"] = locality_dict[locality]
```

Note how I used *at* to work with a single value in each row and column.

Exercise 15.2

Is *at* working with an index or numeric row position value? What if I told you there is also a function called *iat*?

We finish by converting the Breed, Colour, and Locality column types to **int16**:

```
df = df.astype({"Breed": "int16", "Colour": "int16", "Locality": "int16"})
df.info()

<class 'pandas.core.frame.DataFrame'>
RangeIndex: 3475 entries, 0 to 3474
Data columns (total 5 columns):
 #   Column    Non-Null Count  Dtype
---  ------    --------------  -----
 0   Locality  3475 non-null   int16
 1   Breed     3475 non-null   int16
 2   Colour    3475 non-null   int16
 3   F         3475 non-null   int8
 4   M         3475 non-null   int8
dtypes: int16(3), int8(2)
memory usage: 27.3 KB
```

```
df.head()
```

```
   Locality  Breed  Colour  F  M
0         2     21     112  1  0
1         2     19      17  0  1
2         1     21     116  1  0
3         8     18     121  1  0
4         1     18     130  0  1
```

The following sentences are equivalent:

- df has 5 columns.
- df has 5 *features.*
- df has 5 *dimensions.*
- df is 5-dimensional.

Since we will be using and seeing many floating-point numbers, I now use a **pandas** feature to show floats in standard, non-scientific notation:

```
pd.set_option("display.float_format", lambda x: "%.4f" % x)
```

The format notation is similar to what we use in f-strings but comes from an earlier time in Python history. It says, "substitute x into the string as a **float**, and display 4 digits after the decimal point." We will still see computational round-off errors where Python displays numbers that are mathematically supposed to be 0.0 and 1.0 as close but not equal to those values.

We do something similar for **numpy**:

```
np_float_formatter = "{:.4f}".format
np.set_printoptions(formatter={'float_kind':np_float_formatter})
```

15.3 Feature scaling

Since we now have dimensions, it makes sense that geometry will enter the picture. If I have two points in the Cartesian plane, $v = (v_1, v_2)$ and $w = (w_1, w_2)$, then the *Euclidean distance*, or simply *distance*, between them is

$$\|v - w\| = \sqrt{(v_1 - w_1)^2 + (v_2 - w_2)^2}$$

For example, the distance between (-3, -1) and (2, 1.5) is approximately 5.6. Note the double bars around the difference of the points.

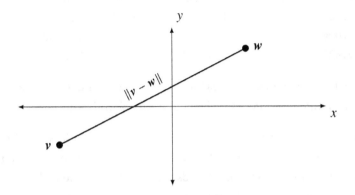

In words, we say that the distance between any two points is the square root of the sum of the squares of the differences between corresponding coordinates. In three dimensions, the distance formula is

$$\|v - w\| = \sqrt{(v_1 - w_1)^2 + (v_2 - w_2)^2 + (v_3 - w_3)^2}$$

for points (v_1, v_2, v_3) and (w_1, w_2, w_3). Note the convention of using a bold italic font for the point when it has two or more dimensions: $v = (v_1, v_2, v_3)$.

These are generalizations of the formula for the distance between two numbers, v and w, on a line, which is the absolute value.

$$|v - w| = \sqrt{(v - w)^2}$$

Our DataFrame has 5 dimensions, and each column entry is numeric. We can treat each row as a point and measure the distance between them. This might help us know when two cats, each represented by a row, are "similar." The values vary quite a bit across all the rows and columns.

```
df.describe()
```

	Locality	Breed	Colour	F	M
count	3475.0000	3475.0000	3475.0000	3475.0000	3475.0000
mean	4.4518	21.7706	77.2702	0.5350	0.4650

std	2.6689	7.0630	43.0268	0.4988	0.4988
min	0.0000	0.0000	0.0000	0.0000	0.0000
25%	2.0000	19.0000	23.0000	0.0000	0.0000
50%	4.0000	21.0000	80.0000	1.0000	0.0000
75%	6.0000	21.0000	116.0000	1.0000	1.0000
max	9.0000	52.0000	136.0000	1.0000	1.0000

The minimum and maximum values for Locality are 0 and 9. For Breed, they are 0 and 52, and so on. The ranges of values in each column vary significantly from column to column.

Suppose we have three cats, which we represent by (locality, breed) pairs as (0, 0), (0, 1), and (0, 52). The breeds do not represent ordered and continuous data: there is no reason to think breeds 0 and 52 are "closer" than breeds 0 and 1. Nevertheless, the "distance" between the first two cats is 1, and the second two cats is 51. This second and larger number might sway our calculations so that we think those cats are much more different from the first two.

To make the values more comparable across columns, we will *normalize* or *scale* the values in each column. Some algorithms require normalization, and others run much faster if you normalize beforehand. There are several techniques for normalizing data, and I next highlight three of them.

15.3.1 Min-max normalization

Min-max scaling or *normalization* produces values between 0 and 1, inclusive. Effectively, we subtract the minimum column value from each number in a column and then divide by the new maximum. You would not do this if the minimum and maximum were both zero.

Mathematically, this is

$$x_{new} = \frac{x_{old} - x_{min}}{x_{max} - x_{min}}$$

for a column value x_{old}. After computation, we replace x_{old} with x_{new} in the column.

Exercise 15.3

What is one way to adjust this so that the range of values is between −1 and 1?

To compute this for the `Locality` column, we extract the column items using the *values* property and then call *minmax_scale* from **sklearn**:

```
from sklearn.preprocessing import minmax_scale

new_locality = minmax_scale(df["Locality"].values)
new_locality[:10]

array([0.2222, 0.2222, 0.1111, 0.8889, 0.1111, 0.1111, 0.2222, 0.2222,
       0.2222, 0.2222])
```

Though I do not do this now, I could set the `Locality` column to have these new scaled values by executing `df["Locality"] = new_locality`.

Exercise 15.4

Verify that the minimum and maximum values in `new_locality` are 0.0 and 1.0.

Exercise 15.5

Show that min-max normalization leaves the F and M columns unchanged.

Exercise 15.6

Write a Python function to perform min-max normalization for a named numeric column in a DataFrame. You may use *min* and *max* from **pandas** but no functions or methods from **numpy** or **sklearn**.

15.3.2 Standardization

Let μ be the mean of a column and let σ be its non-zero standard deviation. The *standardization* of a column value x_{old} is

$$x_{new} = \frac{x_{old} - \mu}{\sigma}$$

If $x_{old} = \mu + \sigma$, then $x_{new} = 1$. Similarly, if $x_{old} = \mu - \sigma$, then $x_{new} = -1$.

Exercise 15.7

Can standardized values be greater than 1.0 or less than −1.0?

Exercise 15.8

What is the mean of the column with the standardized values? What is its standard deviation?

sklearn has many classes, methods, and functions to allow you to transform your data in different ways. For example, the following code uses the **StandardScaler** class to compute the standardization of our Locality column data:

```
from sklearn.preprocessing import StandardScaler

new_locality = StandardScaler().fit_transform(df[["Locality"]])
pd.Series(new_locality.flatten()).describe()
```

```
count    3475.0000
mean       -0.0000
std         1.0001
min        -1.6683
25%        -0.9188
50%        -0.1693
75%         0.5802
max         1.7044
dtype: float64
```

We can also do this using **numpy** methods. First, we compute the mean and standard deviation:

```
locality = df["Locality"].values
mu = locality.mean()
sigma = locality.std()
```

Because **numpy** is so flexible in working with arrays, the computation of the standardized array is simple:

```
new_locality = (locality - mu)/sigma
pd.Series(new_locality).describe()
```

```
count    3475.0000
mean       -0.0000
std         1.0001
min        -1.6683
25%        -0.9188
50%        -0.1693
75%         0.5802
max         1.7044
dtype: float64
```

Exercise 15.9

Use **pandas** methods instead of **numpy** methods to standardize the Locality column.

15.3.3 Mean normalization

Let μ be the mean of a column. The *mean normalization* of a column value x_{old} is

$$x_{new} = \frac{x_{old} - \mu}{x_{max} - x_{min}}$$

We assume the denominator is not zero.

We again use **numpy** methods to compute this:

```
max_minus_min = locality.max() - locality.min()

assert max_minus_min != 0.0

new_locality = (locality - mu) / max_minus_min
pd.Series(new_locality).describe()
```

```
count    3475.0000
mean       -0.0000
std         0.2965
min        -0.4946
25%        -0.2724
50%        -0.0502
75%         0.1720
max         0.5054
dtype: float64
```

Exercise 15.10

Use **pandas** methods instead of **numpy** methods to perform mean normalization on the Locality column.

15.4 Feature selection and reduction

Recall that our df DataFrame has 5 features/columns/dimensions. Do we need all these features for our analysis, or are any of them redundant?

In our df DataFrame, we do not need both the F and M columns. When a column entry in one is 0, the entry in the other is 1, and vice versa. This is easy to see, but other and more subtle relationships may exist among columns.

For example, if *a*, *b*, and *c* are floating-point numbers and *X*, *Y*, and *Z* are columns, we might have $z = a x + b y + c$ for each value *x* in *X*, *y* in *Y*, and *z* in *Z* in the same row. In column notation, $Z = a X + b Y + c$.

Exercise 15.11

Show that $F = -M + 1$.

Exercise 15.12

Interpret this correlation coefficient matrix for the F and M columns:

```
df[['F', 'M']].corr(method="pearson")

          F        M
F   1.0000  -1.0000
M  -1.0000   1.0000
```

For further analysis of the cats DataFrame, we can and should drop one of the F and M columns.

Principal Component Analysis (PCA) is a technique to reduce the number of features in our data while maintaining many of its statistical properties.

Notice that I said, "reduce the number of features." I did not say, "select a subset of existing features." In general, PCA creates fewer and different features by mathematically combining existing data. If you are familiar with linear algebra, PCA computes a linear transformation matrix that maps our data to lower dimensions. In general, we will lose information, but we may retain enough to have fast enough computation and useful results.

Before we perform PCA, we need to normalize or standardize our data. I choose to standardize (section.15.3.2) each of the columns so that they have the same mean and standard deviation, subject to round-off errors.

```python
def standardize_DataFrame_column(df, column_label):
    data = df[column_label].values
    mu = data.mean()
    sigma = data.std()
    assert sigma != 0  # don't forget this!
    df[column_label] = (data - mu)/sigma
```

```python
for label in df.columns.values:
    standardize_DataFrame_column(df, label)
```

```python
df.describe()
```

	Locality	Breed	Colour	F	M
count	3475.0000	3475.0000	3475.0000	3475.0000	3475.0000
mean	-0.0000	-0.0000	-0.0000	-0.0000	-0.0000
std	1.0001	1.0001	1.0001	1.0001	1.0001
min	-1.6683	-3.0828	-1.7961	-1.0726	-0.9324
25%	-0.9188	-0.3923	-1.2615	-1.0726	-0.9324
50%	-0.1693	-0.1091	0.0635	0.9324	-0.9324
75%	0.5802	-0.1091	0.9003	0.9324	1.0726
max	1.7044	4.2806	1.3652	0.9324	1.0726

Exercise 15.13

What does `df.columns.values` return? What is its type?

Now we compute a PCA for our DataFrame. Instead of reducing the number of features, let's see what the statistics of the result tell us if we maintain five dimensions.

```
from sklearn.decomposition import PCA

pca = PCA(n_components=5)
pca.fit(df)

pca.explained_variance_ratio_

array([0.4045, 0.2060, 0.1977, 0.1917, 0.0000])
```

PCA creates the new first feature by trying to capture the maximum amount of variance. That is, the first feature contains as much of the information as we can transform into one feature. The first item in pca.explained_variance_ratio_ is 0.4045, which means that the first feature includes approximately 40.45% of the original information in the DataFrame.

The second feature is "orthogonal" to the first, which means that it includes other information not represented in that first new feature. The second feature represents another dimension compared to the first. It captures an additional 20.6% of the information in the DataFrame.

The PCA algorithm then continues, building out the dimensions up to the number you specify in the n_components keyword argument.

Note that something strange happens by the time we get to the end of pca.explained_variance_ratio_. Subject to round-off error, we have captured 100% of the original information in only four new features. The fifth adds nothing at all.

This should not surprise us because we already know that we only need one of F and M! All our useful information is in four features.

If we compute the PCA but ask for three components, we get this:

```
pca = PCA(n_components=3)
pca.fit(df)

pca.explained_variance_ratio_

array([0.4045, 0.2060, 0.1977])
```

The values are the same as the first three in the previous computation. Other **sklearn** functions, methods, and properties give us the transformed lower-dimensional data that we can use in a new DataFrame for further processing.

For now, I drop the redundant M column.

```
del df['M']
df.head(2)
```

```
   Locality   Breed  Colour      F
0   -0.9188 -0.1091  0.8073  0.9324
1   -0.9188 -0.3923 -1.4010 -1.0726
```

If I use or develop a supervised machine learning to determine a cat's gender from its Locality, Breed, and Colour, the F column is the label in the training set.

Exercise 15.14

Optional: I cover linear algebra in Chapter 5 of *Dancing with Qubits.* [D.W.Q] If you are familiar with that, research how PCA is related to the eigenvectors and eigenvalues of the DataFrame covariance matrix.

15.5 Clustering

Suppose I have a CSV dataset containing 75 (x, y) geometric coordinates. I load these into the xy_df **pandas** DataFrame and look at its descriptive statistical summary:

```
xy_df = pd.read_csv("src/examples/clustering-xy.csv")
xy_df.describe()
```

```
              x        y
count  75.0000  75.0000
mean    7.5733   4.5401
std     4.0102   2.1265
min     1.9796   1.1947
25%     3.4182   2.8896
50%     7.0173   3.6819
75%    12.2170   6.9615
max    13.5643   8.2785
```

Here is the usual sample of the first five points:

```
xy_df.head()
```

```
         x       y
0  13.4832  3.2657
1   7.6388  7.0170
2   2.9279  2.9603
3   7.4514  6.4439
4   3.3011  2.4642
```

How are these points spread out geometrically? Are they uniformly distributed within their minimum and maximum ranges?

A scatter plot would help us see the distribution because we are in two dimensions, but let's try to collect or *cluster* the points into *k* groups first. Here, *k* is a positive integer. Clustering is an example of an unsupervised machine learning algorithm. For this, we use the **KMeans** class from **sklearn**.

```
from sklearn.cluster import KMeans
k_means = KMeans(n_clusters=3, random_state=0).fit(xy_df)
```

I stated that I wanted *k* = 3 clusters via the n_clusters=3 keyword argument. I also passed in the random_state=0 keyword argument. This argument gives a seed to any random process happening within *KMeans* and makes the calculation reproducible. We saw random seeds in section 5.7.

The *labels_* property gives the cluster number for each point. Since we asked for three clusters, **sklearn** numbered them 0, 1, and 2.

```
k_means.labels_
```

```
array([1, 0, 2, 0, 2, 2, 2, 0, 1, 1, 1, 0, 1, 1, 1, 1, 2, 2, 2, 1, 2, 1,
       2, 1, 2, 0, 2, 2, 2, 0, 1, 0, 1, 1, 0, 0, 0, 1, 1, 1, 2, 2, 0, 2,
       0, 2, 0, 1, 1, 2, 0, 0, 0, 1, 0, 0, 2, 0, 2, 1, 2, 0, 0, 1, 0, 2,
       0, 1, 1, 0, 2, 2, 1, 0, 2])
```

The center of each cluster is called its *centroid*, and the *k_means.cluster_centers_* property returns those:

```
k_means.cluster_centers_
```

```
array([[7.0397, 7.3487],
       [12.6463, 2.9810],
       [3.0339, 3.2906]])
```

A baseball team on defense has three players positioned in the outfield. The points in our data are the geometric coordinates of where the outfielders of a fictitious team were standing as 25 successive players on the opposing team came up to bat.

Figure 15.1: Photo of a baseball field

The photo in Figure 15.1 shows a professional baseball stadium and field. (See appendix section E.1 for attribution.) The outfielders are standing toward the top in the photo, and I've placed circles around their positions.

So far, I have just given you the points, and we did some calculations. I delayed creating a scatter plot, but we draw that now in Figure 15.2.

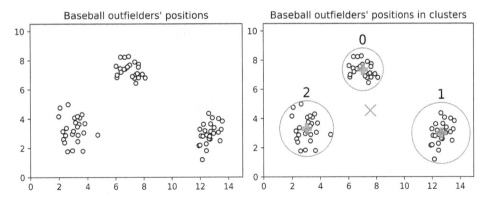

Figure 15.2: Data clustering on two features

I've drawn the locations as circles in Figure 15.2. I marked the centroids with large "+" signs. I drew the circles around the cluster centroids to enclose the points in each cluster. Moving from left to right, we have cluster 2 for the left fielder, cluster 0 for the center fielder, and cluster 1 for the right fielder.

From the scatter plot, I can see that the point (14, 3) is in the right-most cluster, number 1. The *predict* function computes this:

```
k_means.predict([[14, 3]])
```

```
array([1])
```

Stated in the language of machine learning, given the CSV file as the training set for the k-means clustering algorithm with 3 clusters, we can predict that the baseball player standing at (14, 3) is most likely the right fielder.

Exercise 15.15

The point in the scatter plot that I marked with an "x" is at the geometric average of the three centroids. Its coordinates are:

```
center_point = np.mean(k_means.cluster_centers_, axis=0)
center_point
```

```
array([7.5733, 4.5401])
```

where I specified `axis=0` to average along the columns of the 2-dimensional **numpy** array `k_means.cluster_centers_`.

How do these coordinates compare with the means reported previously in `xy_df.describe()`?

In which predicted cluster does **sklearn** place this point? How much do you need to add to or subtract from the x and y coordinates to land in each of the other clusters?

Let's now return to k-means clustering, where k is the number of clusters. Remember that while clustering plots are in three or fewer dimensions, nothing restricts you from using more numeric features and hence more dimensions.

Here is an outline of the algorithm:

1. If your features represent different kinds of data or value ranges, normalize or standardize them as we did in section 15.3.

2. Number the n points as x_0 to x_{n-1}. We must have $k \le n$.

3. Choose k **different** points, c_0 to c_{k-1}, at random as your initial guesses for the centroids. This is why the *KMeans* function has the optional random_state argument for the seed.

4. We now generate our first guesses for the clusters. We call the clusters C_0 to C_{k-1}. Starting with x_0, compute the **squares** of the distances between x_0 and each centroid. If the centroid c_j has the smallest squared distance and j is the smallest integer with this value, put x_0 in C_j. For example, if x_0 has the same squared distance from c_2, c_3, and c_7, put x_0 in C_2 because $2 < 3 < 7$.

5. Repeat the last step for every point. We have now placed every point in a cluster.

6. Recalculate each centroid as the mean of all the points in its cluster.

7. Repeat the assignments of points to the k clusters.

8. Are we done yet? Recalculate the centroids, and if they have not changed since the last time, we are finished. Exit the algorithm with the computed centroids and clusters. If the centroids did change, go back to the previous step and assign the points to clusters.

Exercise 15.16

Why do we know the clusters each contain at least one point after step 4?

Unfortunately, this algorithm is not always guaranteed to converge to a unique set of clusters. It may keep bouncing back and forth between different centroids. You may want to limit the last steps to a fixed number of iterations, such as 10 or 50. Experiment!

In the final step, I said, "if they have not changed since the last time." You might want to change this to something like "if they have not changed by more than 0.1 since the last iteration." That is, settle for close but not perfect.

Not all datasets and subsets of features cluster well. It is worth trying clustering as a first technique to understand if there is some apparent structure to the information.

Though I drew circles around the clusters to visualize them, clusters may take many geometric forms. We can also use different functions to measure the "distance" from the centroids. I show two such examples in Figure 15.3.

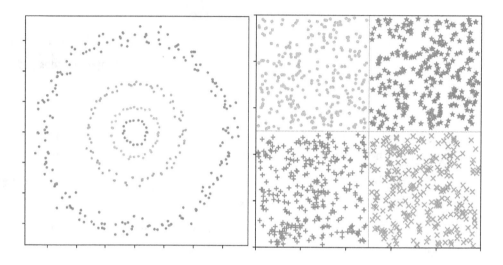

Figure 15.3: Two additional forms of clustering

Exercise 15.17

Discuss how you could use clustering to recommend books based on your past buying habits and those of people like you. What could "people like you" mean here?

Exercise 15.18

You apply to be hired by a company. How might the company's Human Resources department match you to an appropriate role within the organization using clustering?

Exercise 15.19

How could a search engine use clustering and NLP (section 12.3) to show you news articles that are similar to each other?

15.6 Classification

Classification provides answers to questions like the following:

- Wear the blue, white, or yellow shirt?

- Buy or lease the car?

- Hire or don't hire the person?

- Watch or don't watch the movie?

- Take or don't take the class?

- Read the email or move it to the spam folder?

- Issue the credit card or not?

Classification looks at an *observation* and places it in one of several *classes*. In the first question, the observation is the situation for which I am choosing the shirt and my past behavior. There are three possible shirts, and so three classes. The other questions require *binary classification* because they each have only two possible answers, "yes" or "no."

The general classification discipline of statistics and machine learning does not limit itself to two classes, though binary classification is very common. The *classifier* assigns the observation to a class by using a model previously built using a training set. Classification is an example of supervised machine learning.

A classifier may produce the probability of an observation being in a class instead of the class label. Consider the information about the gender of cats in section 14.8. If you present me with a cat that is either calico or tortoiseshell, I can place it in the F class with very high probability. For cats of other colors, the chances are much closer to 50–50 that the cat is female or male.

Before we look at some classification techniques, let's load a DataFrame with the outfielder position coordinates and computed cluster labels from section 15.5. We only consider points lying in cluster 1 or cluster 2, and I show these in Figure 15.4.

```
df = pd.read_csv("src/examples/clustering-xy-labels.csv")
df = df[df["label"] != 0]
df.head()
```

```
        x       y  label
0  13.4832  3.2657      1
2   2.9279  2.9603      2
4   3.3011  2.4642      2
```

```
5  3.4519 4.0798      2
6  3.3845 3.6431      2
```

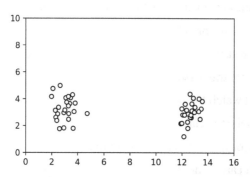

Figure 15.4: Positions of left and right fielders

We now look at support vector machines for binary classification and *k*-nearest neighbors for multi-class classification. You will encounter other techniques if you learn more about machine learning, including logistic regression and decision trees.

15.6.1 Support vector machines (SVM)

Given data with two features, as in Figure 15.4, can we use the data and the labels to draw a line between them to help us make predictions? In this case, the points are the initial field locations for two baseball players, a left fielder and a right fielder. Given a new point, if it lies to the left of the line, then it is where the left fielder is standing. Similarly, if it is to the right of the line, it is the location of the right fielder. We are performing binary classification.

There are many such lines, and I show five of them in Figure 15.5. Is there a "best" line that has nice mathematical properties?

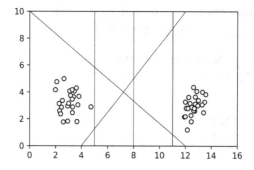

Figure 15.5: Lines separating clusters of points

Finding that best line is the task of a *support vector machine (SVM)* performing *linear support vector classification (SVC)*. We need some terminology before I show you how to compute it using Python and **sklearn**.

From your mathematics classes in school, you know what a "plane" is. It could be the Cartesian plane I have been using for 2-dimensional plots or any one of the infinite planes in three dimensions. For example, the set of all points $(x, y, 0)$ is the plane in 3D that lies flat with all vertical coordinates equal to 0.

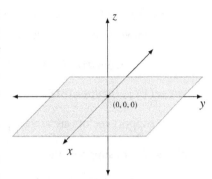

A *hyperplane* is a linear object that has one less dimension than the space in which it lives. A plane is 2-dimensional, so a plane in three dimensions is a hyperplane. A line has one dimension, so a line in a plane is a hyperplane. A point on a line is a hyperplane. If we have 16 features, a 15-dimensional linear subspace within it is a hyperplane.

SVC finds the best hyperplane to separate data, if it exists. The definition of the SVC hyperplane ensures that the hyperplane maximizes the distance from it to the closest point in either group.

Figure 15.6 shows two groups of points marked by white circles on the left and black-filled circles on the right. There is a gap between them, and so the groups are *linearly separable*. The *support vectors* are 5, the maximum of the left-most points, and 7, the minimum of the right-most points. The hyperplane is the point 6, which is halfway between the support vectors.

Figure 15.6: Linear support vector classification in one dimension

Returning to our baseball players and their 2-dimensional locations, we use the **SVC** class from **sklearn** to compute the support vectors.

```
from sklearn import svm

classifier = svm.SVC(kernel="linear")
classifier.fit(df[['x', 'y']], df["label"])

SVC(kernel='linear')
```

SVC creates a classifier object. I specify that I want to perform a linear classification via the kernel keyword argument. The first argument to *fit* is the DataFrame of x and y coordinates, and the second is the label series.

Now that we have a classifier, we can use *predict* to see to which label we should assign [14, 3]. As with the clustering example, it belongs in group 1.

```
classifier.predict([[14, 3]])
```

```
array([1], dtype=int64)
```

The support vectors are always points in the training set, so let's retrieve them:

```
classifier.support_vectors_
```

```
array([[11.9108,  2.1622],
       [11.9830,  3.2777],
       [ 4.7231,  2.8998]])
```

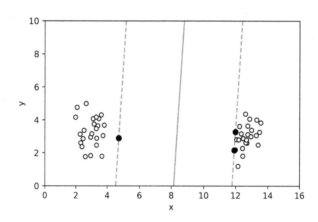

Figure 15.7: Linear support vector classification in two dimensions

In this example, the first two support vectors belong to the same group. We can compute the slope of the line connecting them:

```
a = classifier.support_vectors_[0]
b = classifier.support_vectors_[1]
slope = (a[1] - b[1]) / (a[0] - b[0])
slope
```

```
15.450138504155019
```

Exercise 15.20

From geometry, what is the formula for the slope of the line between two points? Explain my calculation of `slope`.

Now that you know the slope, compute the equation of the line passing through the support vector in the group on the right.

What is the equation of the parallel line passing through the support vector in the other group?

The hyperplane we seek is halfway between the parallel lines you just computed. What is its equation?

In practice, we want the classifier function to make predictions for new data, and don't need the hyperplane's equation. It is possible to get the equation for the hyperplane from the attributes of the classifier. This method requires deeper discussion than I can provide in this high-level survey of machine learning in Python.

What happens if the data is not linearly separable? Can we apply some function, called a *kernel*, to separate and classify the data? For example, a polynomial defines a curve, and a kernel using that polynomial might separate the points. Can we use more features to increase the number of dimensions, as in Figure 15.8? One technique, called "the kernel trick," lifts our data to higher dimensions where it may become separable.

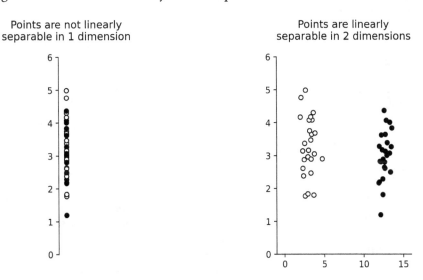

Figure 15.8: Using two features instead of one to allow separability

15.6.2 K-nearest neighbors

In New York City, there are two major league baseball teams: the Yankees and the Mets. At a fictional "Fans of New York Baseball" party, 150 fans have come together to talk about the sport and good-naturedly argue about which team is better.

Let's load a **pandas** DataFrame with the x and y coordinates where the fans are standing in the room one hour after the party began.

```
import pandas as pd

df = pd.read_csv("src/examples/baseball-xy.csv")
df = df.astype({"label": "int16"})

df.info()
<class 'pandas.core.frame.DataFrame'>
RangeIndex: 150 entries, 0 to 149
Data columns (total 3 columns):
 #   Column  Non-Null Count  Dtype
---  ------  --------------  -----
 0   x       150 non-null    float64
 1   y       150 non-null    float64
 2   label   150 non-null    int16
dtypes: float64(2), int16(1)
memory usage: 2.8 KB

df.iloc[2]
x        5.8227
y        7.9895
label    0.0000
Name: 2, dtype: float64

df.iloc[63]
x        13.1151
y         7.9477
label     1.0000
Name: 63, dtype: float64
```

The Mets fans have label 0 and the Yankees fans have label 1.

Exercise 15.21

How many of each type of fan is at the party?

Figure 15.9 shows these positions. The black dots are Yankees fans, and the "×" markers are Mets fans.

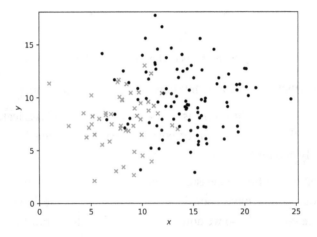

Figure 15.9: Locations of New York baseball fans at a party

There are more Yankees fans at the party than there are Mets fans. The Yankee fans are more common in the upper right, while the Mets fans are more common in the lower left.

Three friends are new to the city and wander into the party. They don't know much about baseball, but they want to fit in. They position themselves near people at the party. Each friend adopts a simple rule: ask the nearest fan their favorite team, the Yankees or the Mets. Whatever the fan answers is the friend's new favorite team.

The circles in Figure 15.10 show where the friends are standing in the room. I have labeled the friends and their positions as **A**, **B**, and **C**.

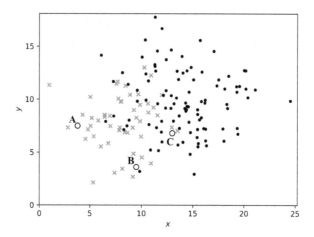

Figure 15.10: Locations of the three friends at the party

Visually, I guess that **A** and **C** are Mets fans, and **B** is a Yankees fan.

The *k-nearest neighbors* algorithm is a supervised learning classification technique that looks at the *k* points closest to an observed point that we wish to classify. Here, *k* is a positive integer and is usually odd for binary classification.

The most common method of measuring "nearness" is via the Euclidean distance formula. In this example, our points have coordinates measured from the lower left. The *x* and *y* coordinates have the same scale, so we don't need normalize them, nor the observed point.

Once we have the *k* nearest points or neighbors, we ask each, "to which class do you belong?". We sum the "votes," and the class with the highest count wins. We classify the observed point as being in the class to which the majority of the *k* neighbors belong. In this example, *k* = 1, and we say we are performing "nearest neighbor classification."

Now let's use **sklearn** to confirm our guesses. The locations of **A**, **B**, and **C** are position_A, position_B, and position_C.

```
position_A = [3.8, 7.5]
position_B = [9.5, 3.6]
position_C = [13, 6.8]
```

We import the **KNeighborsClassifier** class and fit a classifier instance to our data for one neighbor:

```
from sklearn.neighbors import KNeighborsClassifier
```

```
k_neighbors = KNeighborsClassifier(n_neighbors=1)
k_neighbors.fit(df[['x', 'y']], df["label"])

KNeighborsClassifier(n_neighbors=1)
```

Does **sklearn** predict **A** is a Mets fan, as we did?

```
k_neighbors.predict([position_A])

array([0], dtype=int16)
```

It does, since the result is an array with the single value 0, and 0 corresponds to the Mets. Since there is a unique closest neighbor, the probability of correct classification is 1.0, or 100%.

```
k_neighbors.predict_proba([position_A])

array([[1.0000, 0.0000]])
```

Exercise 15.22

What is the probability of there being two points in different classes that are equidistant from our observed point? In which class does **sklearn** place our point?

Here is a Python function that makes the results easier to read:

```
def fan_predictor(name, position):
    teams = ("Mets", "Yankees")
    team = k_neighbors.predict([position])[0]
    probability = k_neighbors.predict_proba([position])[0]

    print(f"'{name}' is a {teams[team]} fan with " +
          f"probability {probability[team]}")

fan_predictor('A', position_A)

'A' is a Mets fan with probability 1.0

fan_predictor('B', position_B)

'B' is a Yankees fan with probability 1.0

fan_predictor('C', position_C)

'C' is a Mets fan with probability 1.0
```

Go back to Figure 15.10 and **B**. Although the closest neighbor is a Yankees fan, they are in a group of Mets fans. What happens if we increase *k* to 5?

```
k_neighbors = KNeighborsClassifier(n_neighbors=5)
k_neighbors.fit(df[['x', 'y']], df["label"])

KNeighborsClassifier()

fan_predictor('A', position_A)
fan_predictor('B', position_B)
fan_predictor('C', position_C)
```

```
'A' is a Mets fan with probability 1.0
'B' is a Mets fan with probability 0.8
'C' is a Yankees fan with probability 0.6
```

B and **C** have switched teams!

The default value for n_neighbors is 5, but you should experiment with various values to validate if the *k*-nearest neighbors classification algorithm provides insight and reliable prediction power from your data.

Exercise 15.23

Explain the following results:

```
for j in range(0, 120, 20):
    k_neighbors = KNeighborsClassifier(n_neighbors=j + 1)
    k_neighbors.fit(df[['x', 'y']], df["label"])
    fan_predictor('B', position_B)
```

```
'B' is a Yankees fan with probability 1.0
'B' is a Mets fan with probability 0.6666666666666666
'B' is a Mets fan with probability 0.6097560975609756
'B' is a Mets fan with probability 0.5901639344262295
'B' is a Mets fan with probability 0.5185185185185185
'B' is a Yankees fan with probability 0.5346534653465347
```

Exercise 15.24

What happens when you set *k* equal to the number of rows in your DataFrame? How does classification relate to the size of each labeled class in the training set?

15.7 Linear regression

Linear regression is a technique we use to fit a hyperplane to a training set of independent variables and one dependent variable.

A common use is determining the "best" line that best approximates the points in a 2-dimensional scatter plot. Here, the x coordinate is the independent variable, and the y coordinate is the dependent variable. This is *simple linear regression*.

If we have more than one independent variable, we are performing *multiple linear regression*. In this section, we look at a simple case and the ordinary least squares algorithm.

Given a new independent observation, x, the line's equation allows us to compute an estimated value for its dependent y. Let x_{min} and x_{max} be the minimum and maximum values of x in the training set. If

$$x_{min} \leq x \leq x_{max}$$

then the prediction of y is *interpolation*. Otherwise, the prediction is *extrapolation* or *forecasting*.

Exercise 15.25

Why is linear regression a supervised technique?

The equation for a line in the Cartesian xy-plane is

$$y = m\,x + b$$

where m is the slope, and b is the y-intercept because $y = b$ when $x = 0$. Statisticians call m and b the *unknown parameters* until we fit the line to the training data.

Exercise 15.26

How many different points do you need to define a line in the plane? How many different points do you need to determine a hyperplane?

Relate your answers to these questions to the minimum number of rows you need in your testing set DataFrame.

When we have two points, say (2.0, 2.0) and (8.0, 5.0), it is trivial to compute the line through them. In this case, $m = 0.5$ and $b = 1.0$, and Figure 15.11 shows the plot.

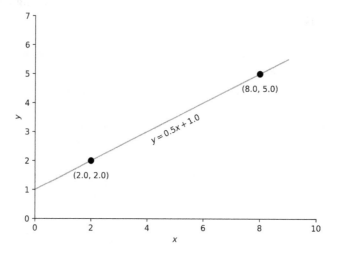

Figure 15.11: Simple linear regression on two points

What happens when we add the point (4.0, 5.0)? There is no line that passes through all three points, so we seek a line that somehow represents the data better than any other line.

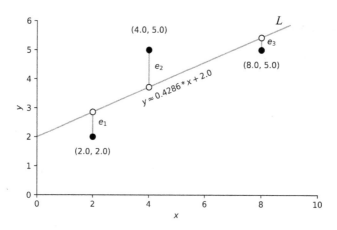

Figure 15.12: Simple linear regression on three points

I've labeled the line we want as L in Figure 15.12. The three initial points are the black-filled circles. We need to compute the equation of L so that we minimize the sum of the squares of the vertical distances from our points to L. The white circles are the points on L closest to our initial points, and their vertical distances are e_1, e_2, and e_3.

Think of the "e" in e_0 as the "error" between the data and the ideal line L. You may see some people use the Greek letter epsilon = ε instead of e.

We call the following technique the *linear least squares algorithm* because we want to minimize

$$e_1^2 + e_2^2 + e_3^2$$

More generally, this sum divided by the number of observations is called the *Mean Squared Error* or *MSE*:

$$MSE = (e_1^2 + e_2^2 + e_3^2) / 3$$

This is an example of a *cost function*. We define and use these frequently in machine learning and, more generally, in optimization problems. Minimizing a cost function's output is often the goal in best fitting a machine learning model to training data.

Another cost function is *Mean Absolute Error,* or *MAE*: where we take the absolute value of the e_j instead of squaring them.

$$MAE = (|e_1| + |e_2| + |e_3|) / 3$$

We considered absolute value versus squaring in section 14.1.3 when we defined absolute deviation and standard deviation.

We have three points

$$(x_1, y_1) = (2.0, 2.0)$$
$$(x_2, y_2) = (4.0, 5.0)$$
$$(x_3, y_3) = (8.0, 5.0)$$

Let $y = m x + b$ be the equation of the line L. We wish to find m and b so that

$$(y_1 - m x_1 - b)^2 + (y_2 - m x_2 - b)^2 + (y_3 - m x_3 - b)^2$$

is as small as possible.

The math to compute this is beyond the scope of this book, but if we let μ_x be the mean of the x coordinates and μ_y be the mean of the y coordinates of our original points, then

$$m = \frac{(x_1 - \mu_x)(y_1 - \mu_y) + (x_2 - \mu_x)(y_2 - \mu_y) + (x_3 - \mu_x)(y_3 - \mu_y)}{(x_1 - \mu_x)^2 + (x_2 - \mu_x)^2 + (x_3 - \mu_x)^2}$$

$$b = \mu_y - m\mu_x$$

Using summation notation and generalizing to n points, we have

$$m \ = \ \frac{\sum_{j=1}^{n}(x_j - \mu_x)(y_j - \mu_y)}{\sum_{j=1}^{n}(x_j - \mu_x)^2} \ = \ \frac{n\sum_{j=1}^{n}x_j y_j - \left(\sum_{j=1}^{n}x_j\right)\left(\sum_{j=1}^{n}y_j\right)}{n\sum_{j=1}^{n}x_j^2 - \left(\sum_{j=1}^{n}x_j\right)^2}$$

$$b \ = \ \mu_y - m\mu_x$$

Admittedly, that looks complicated. The second definition of m is easy to translate to a Python function using **numpy**:

```python
def simple_linear_regression(xs, ys):
    # return the slope and y-intercept of the simple
    # linear regression line

    n = xs.size
    assert n == ys.size

    numer = n * (xs * ys).sum() - xs.sum() * ys.sum()
    denom = n * (xs * xs).sum() - (xs.sum())**2
    slope = numer / denom
    y_intercept = ys.mean() - slope * xs.mean()
    return slope, y_intercept
```

Note how easy it is to express the summations in **numpy**. Now we compute m and b, and hence L:

```python
xs = np.array([2.0, 4.0, 8.0])
ys = np.array([2.0, 5.0, 5.0])

simple_linear_regression(xs, ys)

(0.42857142857142855, 2.0)
```

Exercise 15.27

Why do I have the `assert` in the definition of *simple_linear_regression*?

Exercise 15.28

Does the point (μ_x, μ_y) lie on L?

Exercise 15.29

Re-express the general formula for *m* in terms of the standard deviations and covariance of the sets of *x* and *y* coordinates. See section 14.1.4.

sklearn makes it easy to perform linear regressions. When we *fit* the data to a **LinearRegression** object, the first argument is a 2-dimensional array of the values of the independent variables or a DataFrame. The second argument is the array of dependent variable values.

```
from sklearn.linear_model import LinearRegression

lin_reg = LinearRegression()
lin_reg.fit(pd.DataFrame({'x': xs}), ys)

LinearRegression()
```

Now we can extract *m*:

```
lin_reg.coef_

array([0.4286])
```

and the *y*-intercept *b*:

```
lin_reg.intercept_

1.9999999999999996
```

Exercise 15.30

Explain how we get the interpolated result

```
lin_reg.predict([[6]])

array([4.5714])
```

and the extrapolated result

```
lin_reg.predict([[10]])

array([6.2857])
```

As we add more observations, we can update our linear regression model. That is, we can *learn* from further experience. Figure 15.13 shows a line fitted to 100 points.

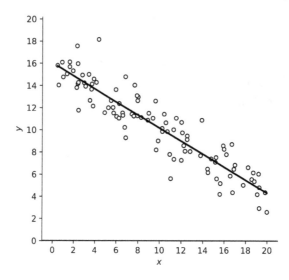

Figure 15.13: Simple linear regression on 100 points

Exercise 15.31

Compute the simple linear regression line for the data shown in Figure 14.2 in section 14.1.4.

In the general multi-dimensional case, we have more than one input, which I have called *x*. In four dimensions, the equation of a hyperplane is

$$y = w_1 x_1 + w_2 x_2 + w_3 x_3 + b$$

where the w_j and *b* are constant numbers, and the x_j are the independent variables. You may see other letters used instead of *w*: *a* and *c* are common. The value *b* is a constant and is frequently 0. Within machine learning, we call *b* "the bias."

In addition to prediction, we can use linear regression for binary classification, as in Figure 15.14. For example, if $y \geq 0$, return 1 if the new observation is "true." Otherwise, return 0 for "false."

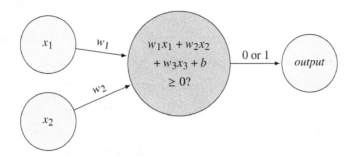

Figure 15.14: Linear regression model with inputs and output

Exercise 15.32

Another binary classification is *logistic regression*. Research this and why it might or might not be better than linear regression in this example.

15.8 Concepts of neural networks

Now let's generalize what we just saw. We keep the three input *nodes* and call this the *input layer*. These feed into the *hidden layer*, and the results are in the *output layer*.

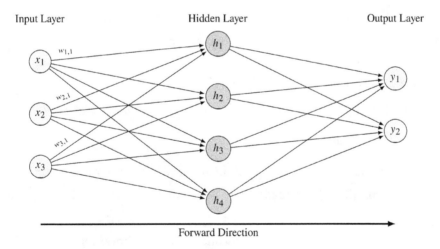

Figure 15.15: Neural network with 3 inputs, 4 hidden nodes, and 2 outputs

Processing in a neural network moves from left to right, and we call this the "forward direction."

Figure 15.15 shows the three input nodes, four nodes in the hidden layer, and two output nodes. I've also shown the *weights* $w_{i,1}$ for the input nodes going to the first hidden node h_1. We also have weights relating the input nodes to h_2, h_3, and h_4, and relating those to the output nodes. I have not shown them to keep the diagram less cluttered.

Another name for a node is *neuron*, and Figure 15.15 shows a *neural network*. More precisely, this is an *artificial neural network (ANN)* because of its similarity to what we believe happens in the brain. As we shall see, we can do machine learning with such networks.

Important questions include:

- Can we have more inputs? (Yes)
- Can we have more hidden nodes? (Yes)
- Can we have more outputs? (Yes)
- What happens in the nodes in the hidden layer?
- Can we have more hidden layers?

I answered the first three, so let's look inside a hidden node to answer the fourth question. I use h_1 as an example. As with linear regression, we compute

$$w_{1,1}\, x_1 + w_{2,1}\, x_2 + w_{3,1}\, x_3 + b_1$$

where I have adjusted the subscripts to indicate we are working with h_1. I call this result z_1. What now? What do we send on to y_1 and y_2 in the output layer?

15.8.1 Activation functions

Inside h_1, we compute z_1 and then apply an *activation function*, which I call *af*.

$$af(z_1) = af(w_{1,1}\, x_1 + w_{2,1}\, x_2 + w_{3,1}\, x_3 + b_1)$$

We send the result of *af* to y_1 and y_2. If the result we send is 0, we are saying, "this node has nothing to contribute." Alternatively, a non-zero result says, "this node/neuron has activated."

If *af* is a linear function, this neural network implements linear regression and so is not very interesting as a new technique. Nevertheless, machine learning practitioners sometimes use the identity function $af(x) = x$ as an activation function. Its effect is to take z_1 and pass it forward to y_1 and y_2 without change.

This passing or feeding values forward from left to right gives our neural network a longer name: a *feed-forward* artificial neural network.

Non-linear activation functions give neural networks much of their distinguishing power. There are many such functions, and I describe four of the most commonly used.

We define the non-linear *binary step* activation function by

$$af(x) = H(x) = 0 \text{ if } x < 0$$
$$af(x) = H(x) = 1 \text{ if } x \geq 0$$

At $x = 0$, this function jumps from 0 to 1. Historically, this was called the *Heaviside Step Function H* after Oliver Heaviside, a 19th-century British English physicist and mathematician.

The *rectified linear unit (ReLU)* activation function is

$$af(x) = ReLU(x) = 0 \text{ if } x \leq 0$$
$$af(x) = ReLU(x) = x \text{ if } x > 0$$

At $x = 0$, this function moves from flat to rising linearly.

Because of the presence of the bias b_1 value in

$$af(w_{1,1} \, x_1 + w_{2,1} \, x_2 + w_{3,1} \, x_3 + b_1)$$

we can adjust where the step occurs in *H*, or the linear transition occurs in *ReLU*.

I have shown these two activation functions in the *xy*–plane in Figure 15.16.

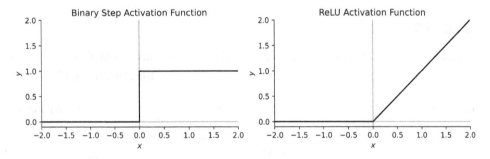

Figure 15.16: The binary step and *ReLU* activation functions

The *sigmoid* activation function returns values greater than 0 and less than 1. We define it by

$$af(x) = sigmoid(x) = 1 / (1 + e^{-x}) = e^x / (1 + e^x)$$

The *hyperbolic tangent* = *tanh* activation function returns values greater than −1 and less than 1. We define it by

$$af(x) = tanh(x) = (e^x - e^{-x}) / (e^x + e^{-x})$$

I plot these two S-shaped functions in Figure 15.17.

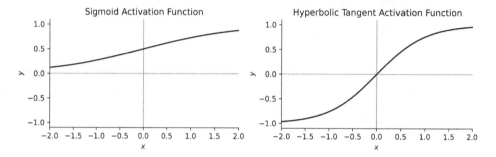

Figure 15.17: The *sigmoid* and *tanh* activation functions

Exercise 15.33

Show that *tanh*(x) = 2 *sigmoid* (2 x) − 1.

If you have studied calculus, note that the binary step and *ReLU* functions are not differentiable at $x = 0$. The *sigmoid*, *tanh*, and linear functions are continuously differentiable.

15.8.2 An example

Suppose I am considering whether I should go to the grocery store to buy food or order pizza for delivery. The output will be 0 for "go to the grocery store" or 1 for "order pizza."

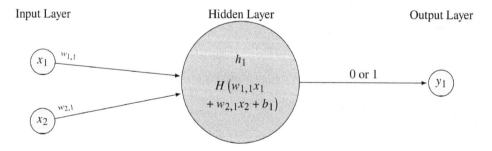

Figure 15.18: Neural network for pizza versus grocery store

Within h_1, we compute

$$w_{1,1} x_1 + w_{2,1} x_2 + b_1$$

and then apply the binary step activation function. We must define the input variables, the weights, and the bias.

x_1 is a binary-valued variable. It has the value 0 if you do **not** have two hours to go to the grocery store, come home, and prepare dinner. It has the value 1 if you do. When x_1 is 0, I am ordering pizza because I don't have time to go shopping.

x_2 is a **float** from 0.0 to 1.0. The value is 1.0 if you very much want pizza and is 0.0 if you don't want to eat pizza. I assume that you can eat pizza, but 0.0 means you very much prefer to eat something else.

Consider

$$z_1 = -10 \, x_1 + 20 \, x_2 + 1$$

after which I apply the binary step function H. When x_1 is 0, z_1 is positive for all allowable values of x_2. Therefore, we pass 1 forward to y_1: we will order pizza.

When x_1 is 1, meaning that I do have time to go shopping,

$$z_1 = -10 \times 1 + 20 \, x_2 + 1 = 20 \, x_2 - 9$$

When $x_2 = 0.45$, $z_1 = 0.0$. When $x_2 = 0.5$, $z_1 = 1.0$. This implies that when I have time and my desire for pizza is greater than 0.45, I will order pizza. Otherwise, I will go to the store.

In practice, we would have more input and hidden nodes. We would collect many example numeric inputs from people about the features they consider essential. Then we would split all this data into training and testing sets. Remember that for the training set, we know if people ordered pizza or not: this is structured learning.

After running a training example, we mathematically adjust the weights, and then process the next example. We successively tune the weights and bias. If we have enough data and it is linearly separable, we will stabilize eventually. Then we use the testing set to confirm that we have a model that correctly predicts the output a large percentage of the time.

This is only a sketch of what machine learning software and practitioners do with simple neural networks, but I hope it is enough to give you an idea.

15.8.3 Deep learning

When we have more than one hidden layer, we are performing *deep learning*. People vary in their views of the minimum layers you need for deep learning, but, for simplicity, I say it is two. Figure 15.19 shows a network with three hidden layers.

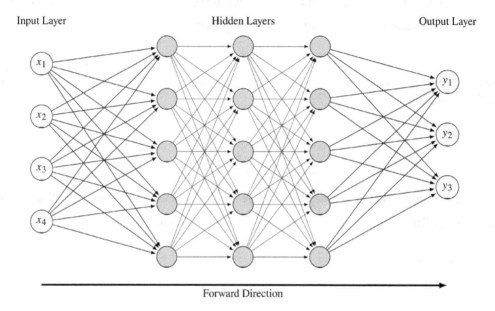

Figure 15.19: Deep neural network with 4 inputs, 3 hidden layers, and 3 outputs

Note that it is not always the case that every hidden layer has the same number of nodes as every other. We may not have every node in one layer connected to every node in the next. When we do, the network is *fully connected*.

What happens within each level? A fully connected layer may multiply its inputs by weights and then add a bias value. Another layer may have the sole function to apply one of the activation functions we saw previously.

A *normalization layer* applies a normalization or standardization transformation we discussed in section 15.3 or another transformation appropriate to the data type. This activity might be part of other layers, such as the input layer.

Many examples of deep learning networks involve image recognition. Here, you might imagine that the input is a 3-dimensional array of pixels. With two of those dimensions, we represent the rectangular layout of the image pixels, and in the third, we hold the RGB values for each pixel.

If a photo has a resolution of 4,032 by 3,024, it has 1,2192,768 pixels. If we multiply this by 3, we get 3,6578,304. That's a lot of input! If the first hidden layer has 100 nodes and we fully connect the input to this hidden layer, we have over 36 million weights. This leads to a lot of computation.

Given this input, we might have a *pooling layer* that looks at square blocks of pixels and produces the mean or maximum RGB values of the pixels in each square. In this way, we can reduce the image size, albeit while losing resolution.

Another type of layer performs *convolution*. Think of convolution as a process that filters or concentrates aspects of the input. For example, given an input photo, we can use convolution to detect the edges of objects.

A *cropping layer* cuts off parts of the image, possibly creating a smaller image. For example, we might want to remove extra whitespace around an image.

We move from our 36 million inputs to image recognition by connecting and applying these and other kinds of layers. Yes, we might say, "here are the three cats in the photo" or "this is a dog and not a fish," but we can also possibly recognize tumors in medical images.

Deep learning has driven extraordinary advances in Natural Language Processing (NLP), a topic we discussed in section 12.3. Practitioners use deep neural networks for sentiment analysis, named entity resolution, text and speech generation, and implementing chatbots and other applications.

We train a deep learning network by inputting data, running it through the layers, and comparing the computed output with what we expected. Remember that this is supervised learning, so we know the output for data in the training set. Using linear algebra and calculus, we work backward from the difference between the expected and received output to adjust the weights and other internal parameters in the layers. This technique is called *backpropagation*.

To learn about the mathematics and techniques of neural networks and deep learning, I highly recommend the book *Deep Learning* by Ian Goodfellow, Yoshua Bengio, and Aaron Courville. [DLG]

15.8.4 Python neural network and deep learning libraries

Python has a rich collection of libraries for machine learning, neural networks, and deep learning. In this chapter, we used **sklearn** for PCA, normalization, SVC, regression, and clustering. You can also build neural networks with it, but the **sklearn** implementors warn

> *This implementation is not intended*
> *for large-scale applications. In particular,*
> *scikit-learn offers no GPU support.* [**SKLN**]

It may seem strange that there is a connection between a GPU, a **G**raphical **P**rocessing **U**nit, and neural networks. GPUs are used in computers to accelerate on-screen graphics for video games, photo editing, and design work. It turns out that the math you need for those applications, linear algebra, is also required for neural networks and deep learning. Without GPU acceleration, training times for large production datasets might take many weeks or months of computation.

PyTorch is a good choice among the open source Python packages for machine and deep learning. [**TOR**] Like **sklearn**, **pytorch** builds on **numpy**. It has good support from cloud providers, and there are tools for NLP, computer vision, and image recognition. Facebook contributed PyTorch to the open source community.

TensorFlow is a comprehensive library for machine learning and is a popular choice among many AI practitioners. [**TFL**] *Keras* builds on top of TensorFlow and provides a "user-friendly" set of programming interfaces. [**KER**] Like PyTorch, TensorFlow and Keras have strong cloud provider support and many books and web references on applying them for applications. Google contributed TensorFlow and Keras to open source.

15.9 Quantum machine learning

As I write this, "quantum machine learning" is in the early stages of theory, research, and development. Why are we interested in techniques connecting quantum computing and AI?

As we have seen in this chapter, machine learning involves significant mathematical computation. We believe that when quantum computers get large and powerful enough, they may perform some essential AI calculations much faster than we can now.

Next, we double the dimension of our quantum computational space every time we add a qubit because of entanglement. If we have 10 qubits, the dimension of the space in which the quantum states reside is $2^{10} = 1,024$. With 64 qubits, the dimension is $2^{65} = 18,446,744,073,709,551,616$, or eighteen quintillion, four hundred forty-six quadrillion, seven hundred forty-four trillion, seventy-three billion, seven hundred nine million, five hundred fifty-one thousand, six hundred sixteen.

While we will never need that many features in a dataset, researchers are looking at how we can use the "good" growth of dimensions in quantum computing to control the "bad" growth of features and computational structures like matrices in AI.

The techniques we saw in this chapter were developed over time by many people, and there are many more activation functions, algorithms, hidden layer types, and methods than I have discussed. Can the fundamentally different ideas in quantum computing like superposition and entanglement give us new ways to match patterns? Can we create better kernels with less training for classification?

Quantum computers are not "big data machines." They cannot ingest gigabytes of data and then process it. What we can do, though, is look at how to combine the best classical and new quantum ideas to create better AI for the benefit of society.

15.10 Summary

In this chapter, we took a brief tour of machine learning, neural networks, and deep learning. We focused on the core ideas behind these technologies, and I showed you how to get started with several **sklearn** coding examples in Python.

But for the appendices and back matter, this chapter concludes this introduction to Python. Together, we explored classical programming techniques and saw how quantum computing is a natural complement and extension of those. The future of coding is the integration of classical and quantum objects, constructs, controls, and algorithms. I wish you much success on your personal journey to that exciting future!

Appendices

A
Tools

It's one thing to read a book about writing software, and it's quite another to install the tools you need and then use them. In this appendix, I discuss how to get and install Python, install modules and packages, start and use the Python interpreter, use integrated development environments, install and use Qiskit on your machine and online, and check your code via "linting."

Topics covered in this appendix

A.1 The operating system command line

At several points in this book, I ask you to enter and execute a statement from "the operating system command line." For example, in section 10.1.3, I ask you to run `pytest unipoly.py` from the command line.

The way to get a command line varies by the operating system you are using:

- For Microsoft Windows, open **Windows PowerShell** or **Windows Command Prompt** from the **Start** menu or by typing "PowerShell" or "Prompt" in the search box.
- For Apple macOS, start the **Terminal** application from **Launchpad** or the `Application/Utilities` folder.
- For Linux, if you are in a graphical environment, try typing Ctrl-Alt-T to open the **terminal**. Otherwise, look for a "Terminal" application in the system menus. If you are not in a graphical environment, you are in a terminal!

The command line is also called "the shell prompt" in several operating systems.

A.2 Installing Python

You may already have Python 3 installed on your system. Try running

```
python -VV
```

or

```
python3 -VV
```

from the operating system command line. If you see a version and description of the Python 3 application, it is already installed. Some systems may have Python version 2 installed, and the `python -VV` command may start and show information about that. Try `python3 -VV` to see if Python 3 is also installed.

This book uses Python 3.9, and some examples may not work for versions less than that. The process to upgrade to a newer version of Python may require you to have special permissions for installing software on your computer. For current information about upgrading your system, do a web search for one of

```
windows upgrade python
mac upgrade python
linux upgrade python
```

and follow the directions.

To perform a new Python installation, go to

`https://www.python.org/downloads/`

and download the latest version for your system. For Windows, make sure you download the 64-bit version, as you need this for **qiskit**.

For macOS and Windows, run the software installer you downloaded. For Windows, choose the options to install for all users and to update `PATH`.

Linux requires you to do more of the installation work and is not necessarily for new users:

- After downloading the source "tarball," use `tar -xvf` followed by the file name to extract the Python files.
- Go to the folder with the files and issue `./configure` on the command line.
- Run `make` on the command line.
- Run `make install` on the command line.

Once Python is installed, use either `python` or `python3` to start it. The name may vary by your operating system or installation type. I refer to `python` in what follows.

A.3 Installing Python modules and packages

PyPI is the "Python Package Index" and contains over 300,000 modules and packages that you can download and `import`. You can browse and search this catalog at

`https://pypi.org/`

The **pip** application downloads and manages PyPI packages on your system. Use either `pip` or `pip3` from the command line to start the application. I refer to `pip` in what follows to demonstrate several of its capabilities. When you run these commands, your output will likely be different from what I show here because I have other packages installed. If you are curious, I list the Python packages on my machine in *Appendix F, Production Notes*.

Show help information about pip

```
pip --help
```

```
Usage:
  pip <command> [options]
```

```
Commands:
  install                 Install packages.
  download                Download packages.
  uninstall               Uninstall packages.
  freeze                  Output installed packages in requirements
format.
  list                    List installed packages.
  ...
```

To see detailed help for a specific command, insert its name between `pip` and `--help`.

```
pip list --help
```

```
Usage:
  pip list [options]

Description:
  List installed packages, including editables.

Packages are listed in a case-insensitive sorted order.

List Options:
  -o, --outdated          List outdated packages
  -u, --uptodate          List uptodate packages
  -e, --editable          List editable projects.
  -l, --local             If in a virtualenv that has
                          global access, do not list
                          globally-installed packages.
  ...
```

List installed Python modules and packages

```
pip list
```

```
Package                     Version
--------------------------- ---------
alabaster                   0.7.12
appdirs                     1.4.4
argon2-cffi                 20.1.0
astroid                     2.5.6
async-generator             1.10
atomicwrites                1.4.0
attrs                       21.2.0
  ...
```

To see the packages that are out of date, add the `--outdated` option:

```
Package            Version Latest Type
------------------ ------- ------ -----
click              7.1.2   8.0.1  wheel
decorator          4.4.2   5.0.9  wheel
h5py               3.1.0   3.2.1  wheel
idna               2.10    3.1    wheel
isort              4.3.21  5.8.0  wheel
...
```

Install a package

We install the **treemap** module:

```
pip install treemap
```

```
Collecting treemap
  Downloading treemap-1.05.tar.gz (5.4 kB)
Building wheels for collected packages: treemap
  Building wheel for treemap (setup.py) ... done
  Created wheel for treemap: filename=treemap-1.5-py3-none-any.whl
size=6115

sha256=82cda8fc01b5159ea028900ec198c11a593edd0944223ba636a69850da03cab9
  Stored in directory: c:
\users\rssut\appdata\local\pip\cache\wheels\ef\c0\38\d343a9431900820517
18405c15277e4cf5977d0689b7fa345f
Successfully built treemap

Installing collected packages: treemap
Successfully installed treemap-1.5
```

You may install multiple packages by listing them all after `pip install`.

Upgrade a previously installed package

We add the `--upgrade` option to `pip` to upgrade a previously installed package. In the case of **treemap**, we just installed it, so there is nothing to upgrade.

```
pip install --upgrade treemap
```

```
Requirement already satisfied: treemap in d:\python\python39\lib\site-
packages (1.5)
```

The **flake8** module does need upgrading. In the output, you can see how `pip` checks packages on which **flake8** depends. If they need upgrading, `pip` will do so.

```
pip install --upgrade flake8
```

```
Collecting flake8
  Using cached flake8-3.9.2-py2.py3-none-any.whl (73 kB)
Requirement already satisfied: pyflakes<2.4.0,>=2.3.0 in
     d:\python\python39\lib\site-packages (from flake8) (2.3.1)
Requirement already satisfied: pycodestyle<2.8.0,>=2.7.0 in
     d:\python\python39\lib\site-packages (from flake8) (2.7.0)
Requirement already satisfied: mccabe<0.7.0,>=0.6.0 in
     d:\python\python39\lib\site-packages (from flake8) (0.6.1)

Installing collected packages: flake8
  Attempting uninstall: flake8
    Found existing installation: flake8 3.8.4
    Uninstalling flake8-3.8.4:
      Successfully uninstalled flake8-3.8.4
Successfully installed flake8-3.9.2
```

Uninstall a package

I don't need **treemap** on my system, so I uninstall it:

```
pip uninstall treemap
```

```
Found existing installation: treemap 1.5
Uninstalling treemap-1.5:
  Would remove:
    d:\python\python39\lib\site-packages\treemap-1.5.dist-info\*
    d:\python\python39\lib\site-packages\treemap\*
    d:\python\python39\scripts\treemap_coverage.exe
    d:\python\python39\scripts\treemap_demo.exe
Proceed (y/n)? y
  Successfully uninstalled treemap-1.5
```

Note how **pip** prompted me to confirm if I wanted to remove **treemap**. I responded with "y" for "yes." Had I included the `--yes` option to `pip`, it would not have asked me for confirmation.

Update pip

Though **pip** is installed with Python, its developers update it more frequently than Python itself. You should update it regularly to have the latest version:

```
python -m pip install -U pip
```

```
Requirement already satisfied: pip in d:\python\python39\lib\site-
packages (21.1.2)
```

In this case, the application is current, but I check once a week.

A.4 Installing a virtual environment

In *Appendix F, Production Notes*, I list the Python modules I have installed on my computer. There are almost 200 of them.

I wish I could say that each of these modules works flawlessly with every other, but that's not the case. When a developer updates one module, it could break other code that depends on it.

Instead of having one big Python environment of your system, you can create smaller *virtual environments*. A virtual environment holds the Python interpreter, tools, standard libraries, and only the compatible **pip**-installed modules and packages you need for your project. This feature is especially useful when you work on several projects that require different sets of modules because you can set up a virtual environment for each. [PYV]

Once you have installed Python, as I discussed in section A.2, you can install a virtual environment. First, decide where you want to put it. A common choice is to place it in a subdirectory of your home directory and call it **.venv**. On a command line for Linux, Mac, or Windows PowerShell, issue the following:

```
cd ~
python -m venv .venv
```

These commands move you to your home directory and install the core Python tools you require. To use this virtual environment, enter

```
.venv\Scripts\activate.bat
```

on Windows or

```
source .venv/bin/activate
```

on Linux or Mac.

Now you can go ahead and use **pip** as in section A.3 to install whatever modules and packages you need.

Instead of using `.venv` under your home directory, you can specify any valid directory path for your virtual environment files. For example, on Windows, I could issue:

```
python -m venv c:\my-qiskit-env
c:\my-qiskit-env\Scripts\activate.bat
pip install qiskit
```

A.5 Installing the Python packages used in this book

Issue the following commands to download and install the packages used in this book:

```
pip install wheel coverage flake8 flashtext jupyter
pip install matplotlib matplotlib-venn numpy pandas
pip install pylint pytest qiskit sklearn spacy squarify sympy
python -m spacy download en_core_web_sm
```

A.6 The Python interpreter

The Python interpreter is a classic example of a command-line-driven **Read-Evaluate-Print-Loop (REPL)**.

Figure A.1 shows how I start Python from the operating system command line.

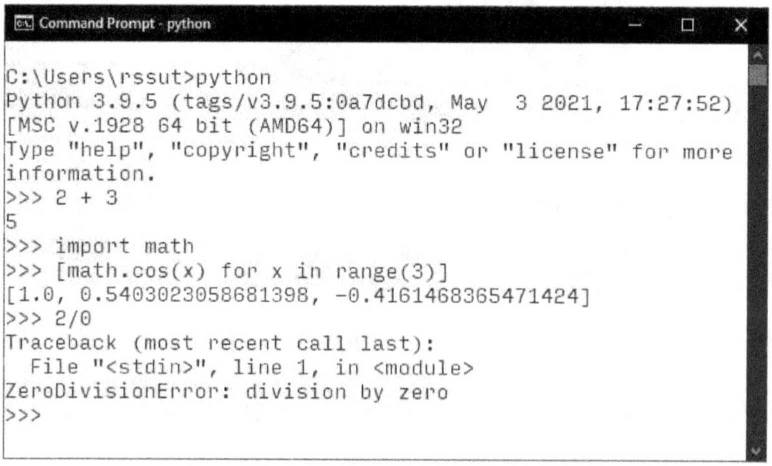

```
C:\Users\rssut>python
Python 3.9.5 (tags/v3.9.5:0a7dcbd, May  3 2021, 17:27:52)
[MSC v.1928 64 bit (AMD64)] on win32
Type "help", "copyright", "credits" or "license" for more
information.
>>> 2 + 3
5
>>> import math
>>> [math.cos(x) for x in range(3)]
[1.0, 0.5403023058681398, -0.4161468365471424]
>>> 2/0
Traceback (most recent call last):
  File "<stdin>", line 1, in <module>
ZeroDivisionError: division by zero
>>>
```

Figure A.1: The Python interpreter

After displaying version information, Python displays a ">>>" prompt. I enter the first command: 2 + 3.

- Python *reads* 2 + 3 and *parses* it into an internal data structure it can manipulate.
- It then determines that it must *evaluate* the expression by applying "+" to **int** 2 and **int** 3. It does so and produces the result 5.
- Since the result is not **None**, Python prints 5.
- It then displays another ">>>" and waits for input.

The next expression is **import math**. Python reads and evaluates this code and loads the **math** module into your workspace. It produces no output.

The input in Figure A.1 is a list comprehension, and Python shows the computed result.

The final expression raises a **ZeroDivisionError** exception. Python shows the traceback to the code that caused the problem.

The Python interpreter has limited editing capabilities. You can use the keyboard arrow keys to move backward and forward in what you type, and you can delete and enter characters. Cut-and-paste also works. For more sophisticated editing and to work with files, you need a more sophisticated Python environment.

A.7 IDLE

IDLE is the Python Integrated **D**evelopment and **L**earning **E**nvironment. It is a windowed application and ships with Python. Figure A.2 shows **IDLE** in edit mode for the file in *Appendix C* containing the **UniPoly** class.

Figure A.2: Python IDLE

You can start **IDLE** from the command line:

```
python -m idlelib
```

or from an applications menu or folder. Figure A.3 shows **IDLE** in my Microsoft Windows Start menu.

Figure A.3: The Windows Start menu for Python IDLE

IDLE operates either in interpreter mode with enhanced code editing or in file edit mode, as in Figure A.2. **IDLE** highlights syntax elements like comments, strings, punctuation, and keywords in different colors. It also automatically indents your code.

One important feature is *debugging*. **IDLE** allows you to walk step by step through your code.

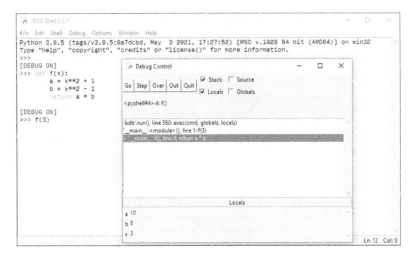

Figure A.4: Python IDLE in debug mode

In Figure A.4, I first turn on debug mode from the menu. I define a function *f* and then call f(3). By repeatedly clicking the "Step" button, I move through my code. At the bottom of the Debug Control window, you can see the values of all local variables.

This form of debugging helps you understand if your code is doing what you expect it to do and, if not, see where the problem is occurring. You can also set a *breakpoint*, a location in your code at which debug execution will stop. Rather than run your code line by line, you can say, "execute everything up to this point and then stop so I can look around." [IDL]

A.8 Visual Studio Code

An *IDE* is an Integrated Development Environment. These tools are sometimes dedicated to one programming language, but I think the best offer uniform and consistent support for several languages. **Visual Studio Code** is the IDE I use now, and it is my favorite by far. I wrote the content for this book using **Visual Studio Code**. [VSC]

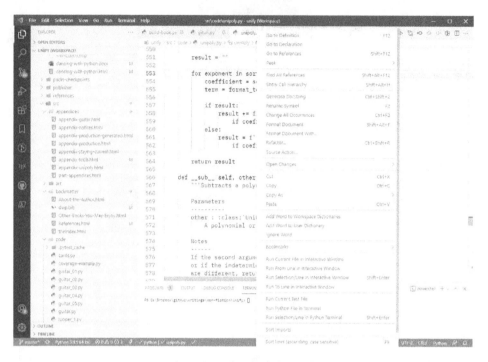

Figure A.5: Visual Studio Code

Figure A.5 is a screenshot of **Visual Studio Code** editing the file in Appendix C. The text has color syntax highlighting, and the editor has features like code completion and automatic indentation. You can see a directory listing of some of the content files for this book on the left. The editor displayed the context menu when I right-clicked on sorted in the code.

From within **Visual Studio Code**, you can create, run, and debug your code. The "integrated" part of "IDE" means that the tool offers support for version control systems like **git** and brings together advanced features at the project level. You can group files into a *workspace*, and **Visual Studio Code** can jump to the definitions of your functions, classes, and methods. At the bottom of the image shown in Figure A.5 is an integrated command prompt, so you can do almost all your work without leaving the IDE.

For me, the most exciting parts of **Visual Studio Code** are the user-contributed extensions. Need to start using a new programming language? Someone has probably written an extension that fully supports it. Want the environment to look different? Someone has probably written a "theme" that has the colors and fonts that you like. Want to sort a JSON file by key? I use an extension for that. Want to reformat your HTML code? There are several extension choices for that.

I use **Visual Studio Code** to work with Python, HTML, CSS, XML, Markdown, LaTeX, JSON, C++, PHP, and JavaScript.

If you want to explore other editors and IDEs, I suggest you look at **PyCharm**, **Komodo**, **Sublime Text**, **Eclipse/PyDev**, and **Kite**.

A.9 Jupyter notebooks

Jupyter is a browser-based interactive environment mainly written in Python. While you can use it to edit and run Python immediately, it also supports other installable runtime environments called *kernels* and is widely used by data scientists. Developers have written over one hundred kernels and these support languages like R, Scala, Go, Java, Ruby, Rust, and even Fortran. [JUP]

Jupyter creates *notebooks*, which are combinations of text, images, code, and code execution results. A notebook is persistent, which means that you can save it and restart it later. By default, **Jupyter** uses the Markdown description language for formatting text.

Install and start **Jupyter** from the command line:

```
pip install jupyter
jupyter notebook
```

You may get a windowed prompt asking you which browser you would like to use. Pick one, and in the upper right-hand corner, click "New". Choose "Python 3". You should now see a browser window similar to Figure A.6.

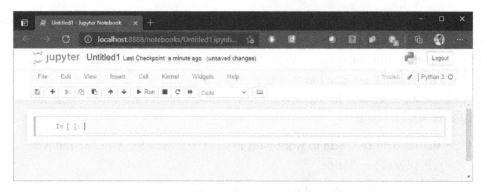

Figure A.6: New Jupyter session

A notebook is a sequence of *cells* from top to bottom. You can run and rerun individual cells or execute all commands in a notebook. Here is what you do to create the notebook shown in Figure A.7:

1. By default, new cells are for Python input. Click on the box for the first cell after "In [ ]:". Then go to the menu where there is a dropdown list that says "Code". Click it and change the cell type to "Markdown". Go back to the cell and type: "## Drawing a parabola". Press the **Run** button on the menu. Your text now appears as a heading, and **Jupyter** displays a new input cell.

2. Enter the Python code `import matplotlib.pyplot as plt` in the cell. Now either click the **Run** button or enter the **Shift + Enter** key combination. You get a new cell.

3. Change the new cell to Markdown and enter the text in Figure A.7 starting with "Now we define the points". Run this cell (and I'm going to stop telling you to do this).

4. Enter the Python code to define `x_values` and `y_values`. Note how your input can be more than one line of code.

5. Enter the Markdown text starting with "Finally, we create".

6. The last step is to create the plot. Enter `plt.plot(x_values, y_values)`. **Jupyter** displays the plot.

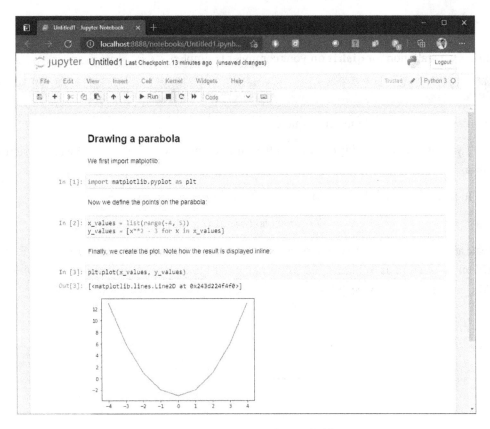

Figure A.7: Jupyter session with matplotlib output

I encourage you to explore the other commands provided within the **Jupyter** window. You can run the application in a non-interactive mode from Python: I executed the Python commands in this book and collected their output via **Jupyter** notebooks that did not use the browser interface.

You can store your notebooks on your computer or on the cloud via several commercial services. Some online tools like the IBM Quantum Lab use **Jupyter** on the web and extend it via custom widgets.

A.10 Installing and setting up Qiskit

The basic installation for **qiskit** on your computer is simple:

```
pip install wheel 'qiskit[visualization]'
```

You may or may not need the single quotes.

If you plan to use Qiskit on Microsoft Windows, be sure to install the 64-bit version of Python.

To access the cloud-based IBM Quantum hardware systems, visit

$$https://quantum-computing.ibm.com/$$

Create an account if necessary, and once you are on the IBM Quantum home page, look for the "API token" section as in Figure A.8. You need an API token to authenticate yourself with the IBM Quantum cloud service. [**QIS**]

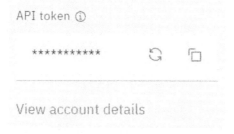

Figure A.8: IBM Quantum API token

If the field is blank instead of filled with asterisks, click the button with the circular arrows. Then click the button to its right, which has a square overlapping another square. This button copies your API token to your clipboard.

You can also retrieve your API token directly from

$$https://quantum-computing.ibm.com/account$$

Now start Python and enter the following, cutting and pasting the API token as a substitute for TOKEN into the call to *save_account*:

```
from qiskit import IBMQ

IBMQ.save_account("TOKEN")
```

Note that *save_account* expects a string as its argument. To test that you correctly set up your account, enter the Python command:

```
IBMQ.load_account()
<AccountProvider for IBMQ(hub='ibm-q', group='open', project='main')>
```

You should see information about your account provider.

A.11 The IBM Quantum Composer and Lab

IBM Quantum provides an online drag-and-drop circuit editor and execution environment called the IBM Quantum Composer. You can access it by visiting

https://quantum-computing.ibm.com/composer/

From this web page, you can read the documentation and interactive tutorials, design circuits, run circuits on hardware and in simulators, and visualize the execution results in several ways.

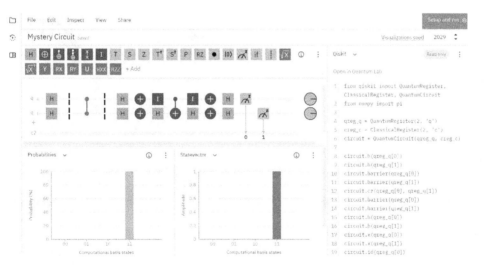

Figure A.9: IBM Quantum Composer

Figure A.9 shows a circuit I created. On the right-hand side, the Composer displays the **qiskit** code you can use in Python to build the same circuit. The visualizations on the bottom show static analyses of the expected results from the circuit.

Exercise A.1

Figure A.9 shows which quantum circuit from *Chapter 11, Searching for the Quantum Improvement*?

The IBM Quantum Lab is an online **Jupyter**-based environment for coding and visualizing quantum circuits using **qiskit** and Python. Access it from:

https://quantum-computing.ibm.com/lab/

Figure A.10 shows the IBM Quantum Lab with a basic 3-gate circuit. When you create a new Lab file, it automatically imports the most-used Python **qiskit** and IBM Quantum modules. It then loads your account information from the cloud.

In this file, I define a `QuantumCircuit` object and then add the gates to it. The circuit *draw* method displays its output within the **Jupyter** window. Since this is a **Jupyter** notebook, you can save your work.

Figure A.10: IBM Quantum Lab

The IBM Quantum Composer is a powerful tool that runs in the cloud. You need not download anything to your machine if you wish to use Python and **qiskit** together to create quantum circuits and programs.

A.12 Linting

A *linter* is an application that scans your code looking for errors or variations from stylistic conventions. Many IDEs like **Visual Studio Code** have built-in support for one or more linters. [VSL] A linter does *static analysis* in that it does not run your code and only looks at what you have written.

A linter usually has three classes of messages:

- **Informational** – "You might want to know about this, but you will likely get annoyed if I keep telling you."
- **Warning** – "You did something here that might cause problems, or you are not following formatting conventions. Don't procrastinate, and take a look now."
- **Error** – "Fix this. Now."

A good Python linter to start with is **pylint**. Install it via:

```
pip install pylint
```

Once you have it installed, issuing `pylint --help` from the operating system command line will display several screens of information about the options.

If we store the incorrect code

```
def f(x=1, y)
  return x+y
```

in the file **src\code\linting\lint-01.py**, or **src/code/linting/lint-01.py** on non-Windows systems, and then run

```
pylint src\code\linting\lint-01.py
```

from the operating system command line, we see the output

```
************* Module lint-01
src\code\linting\lint-01.py:1:13: E0001: non-default argument follows
default
argument (<unknown>, line 1) (syntax-error)
```

Since the message code is "E0001" and starts with an "E", this is an error.

flake8 is another Python linting application. Install it via:

```
pip install flake8
```

Now I place another file in the directory, this time called **src\code\linting\lint-02.py**, or **src/code/linting/lint-02.py** on non-Windows systems. It contains the code:

```
def f(x, y):
    #useless comment
    return x+y
```

Running **flake8** on the **directory**,

```
flake8 src\code\linting
```

we see:

```
src\code\linting\lint-01.py:1:13: E999 SyntaxError: non-default
argument follows default argument
src\code\linting\lint-01.py:2:3: E111 indentation is not a multiple of
four
src\code\linting\lint-01.py:2:3: E113 unexpected indentation
src\code\linting\lint-01.py:2:13: W292 no newline at end of file
src\code\linting\lint-02.py:2:5: E265 block comment should start with
'# '
src\code\linting\lint-02.py:3:15: W292 no newline at end of file
```

We get messages about both **lint-01.py** and **lint-02.py**.

I find the "W292" warning annoying, so I turn it off:

```
flake8 --ignore=W292 src\code\linting
```

and get

```
src\code\linting\lint-01.py:1:13: E999 SyntaxError: non-default
argument follows default argument
src\code\linting\lint-01.py:2:3: E111 indentation is not a multiple of
four
src\code\linting\lint-01.py:2:3: E113 unexpected indentation
src\code\linting\lint-01.py:2:11: E226 missing whitespace around
arithmetic operator
src\code\linting\lint-02.py:2:5: E265 block comment should start with
'# '
```

```
src\code\linting\lint-02.py:3:13: E226 missing whitespace around
arithmetic operator
```

Exercise A.2

What new error message in the code do we see from **flake8**? Can you find the problem in the code? [PEP008]

Other linters you should investigate and consider using include **prospector**, **pycodestyle**, **pydocstyle**, and **pylama**. Some of the linters integrate other linting applications within their processing. Also consider **mccabe** for evaluating code complexity and **bandit** for security checks.

I do not run linters from the command line and only use them within IDEs. **Visual Studio Code** supports all the linters I mentioned via its Python extension. Figure A.11 shows the messages and the locations of the problems in **lint-02.py** within the editor.

```
unify > src > code > linting > lint-02.py >
1 v def f x, y :
2       #useless comment    block comment should start with '# '
3       return x+y    missing whitespace around arithmetic operator
```

Figure A.11: Visual Studio Code linting example

Within **Visual Studio Code**, you configure the linters via the preferences and settings system. Figure A.12 shows some of these settings.

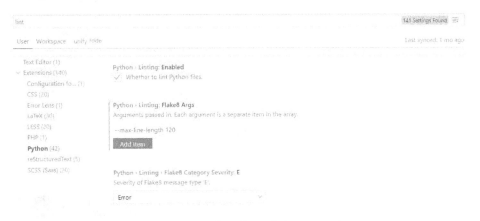

Figure A.12: Visual Studio Code linting settings

B

Staying Current

Any book about a technical subject is a snapshot of what was available at the time of publication. In this appendix, we point you to the primary sources of information about Python and the tools we have discussed.

Topics covered in this appendix

B.1 python.org

The foremost site for the Python programming language and environment is

```
https://www.python.org/
```

There you will find the latest downloadable versions for Linux, Windows, macOS, AIX, z/OS, and other operating systems.

For documentation, begin at

```
https://docs.python.org/3/
```

and browse the "What's new" document every few weeks.

B.2 qiskit.org

The home of the Qiskit open source quantum computing software development kit (SDK) is

https://qiskit.org/

The canonical source for learning about **qiskit** is the Qiskit Textbook at

https://qiskit.org/textbook/preface.html

It features a review of quantum computing along with interactive Python examples. [LQCQ]

The main web page for the Qiskit documentation is

https://qiskit.org/documentation/index.html

The release notes for all versions are at

https://qiskit.org/documentation/release_notes.html

You can see your installed version of **qiskit** and its constituent modules by issuing the following in Python:

```
import qiskit

qiskit.__qiskit_version__

{'qiskit-terra': '0.18.0', 'qiskit-aer': '0.8.2', 'qiskit-ignis':
'0.6.0', 'qiskit-ibmq-provider': '0.16.0', 'qiskit-aqua': '0.9.4',
'qiskit': '0.28.0', 'qiskit-nature': None, 'qiskit-finance': None,
'qiskit-optimization': None, 'qiskit-machine-learning': None}
```

This is the version referenced in the Release Notes:

```
qiskit.__qiskit_version__["qiskit"]

'0.28.0'
```

B.3 Python expert sites

There are dozens of websites where you can find information about doing this or that in Python. Many of them are correct, and many of them use Python 3. Look for articles written in the last five years. If you see code that uses *print* without a following open parenthesis, stop reading. They are discussing Python 2, not Python 3.

I think **Real Python** at

```
https://realpython.com/
```

has consistently helpful and comprehensive articles about most aspects of coding with Python. You can only see so many articles at no charge before they ask you to join and pay. (I have no commercial relationship with them.)

The Python Tutorial at **W3Schools**

```
https://www.w3schools.com/python/
```

is a great place to find Python examples. These might remind you of syntax or the functions or methods that accomplish a particular task.

I recommend you visit and use the websites for four of the major Python packages we used in this book: **matplotlib**, **numpy**, **pandas**, and **sklearn**:

```
https://matplotlib.org/
https://numpy.org/
https://pandas.pydata.org/
https://sklearn.org/
```

B.4 Asking questions and getting answers

When you do a web search about how to do something in Python, there is an excellent chance you will end up at **Stack Overflow**:

```
https://stackoverflow.com/questions/tagged/python
```

You will see questions and answers to straightforward and challenging problems. If you sign in and participate, over time, you can up-vote and down-vote questions and answers. Many answers are helpful, but some are incorrect or posted by non-experts. In any case, you will likely find some code that you can use or modify for your project.

You can learn a lot by browsing the questions people ask, even if you don't have an immediate need for the code. You will learn what others are doing with Python. You can ask your own questions but **always** do a thorough web search first. It is frustrating to experts if you do not respect their time by doing your basic research first.

C

The Complete UniPoly Class

This appendix contains the complete and final definition of the **UniPoly** class for univariate polynomials that we developed in *Chapter 7, Organizing Objects into Classes*.

```
"""The unipoly module contains the UniPoly class for polynomials.
"""


# The next line is for the spell-checker in Visual Studio Code
# cspell:ignore radd rsub rmul


class UniPoly:
    """Polynomial with one variable and integer coefficients.

    UniPoly creates a univariate polynomial with a single term
    and an integer coefficient. Polynomials may use different
    variables, here called 'indeterminates', but their names
    must each be a single alphabetic character. These polynomials
    are immutable.

    Parameters
    ----------
    coefficient : int, optional
        The coefficient of the term. Default value is 1.
    indeterminate : str, optional
        The "variable." Must be a single alphabetic character.
        Default value is 'x'.
    exponent : int, optional
```

```
            The exponent of the term. Default value is 1.

    Raises
    ------
    ValueError
        If the coefficient is not an int.
    ValueError
        If the exponent is not a non-negative int.
    ValueError
        If the indeterminate is not a single
        alphabetic character.
    """


    # --------------------------------------------------------------
    # Initialization
    # --------------------------------------------------------------

    def __init__(self,
                 coefficient=1,
                 indeterminate='x',
                 exponent=0):

        # Instance variables
        # ------------------
        # __indeterminate : str
        #     The "variable" of the polynomial.
        # __terms : dict
        #     The terms of the polynomials with the exponents
        #     as keys and values being coefficients.

        if not isinstance(coefficient, int):
            raise ValueError(
                "The coefficient for a UniPoly must "
                "be an int.")

        if not isinstance(exponent, int) or exponent < 0:
            raise ValueError(
                "The exponent for a UniPoly must "
                "be a non-negative int.")

        if (not isinstance(indeterminate, str) or
                len(indeterminate) != 1 or
```

```
              not indeterminate[0].isalpha()):
          raise ValueError(
              "The indeterminate for a UniPoly must "
              "be an alphabetic str of length 1.")

      self.__indeterminate = indeterminate

      if coefficient != 0:
          self.__terms = {exponent: coefficient}
      else:
          self.__terms = dict()

  # -------------------------------------------------------------------
  # Magic methods
  # -------------------------------------------------------------------

  def __add__(self, other):
      """Adds a polynomial to a polynomial or an integer.

      Parameters
      ----------
      other : :class:`UniPoly` or int
          A polynomial or integer.

      Notes
      ------
      If the second argument is not a polynomial or integer,
      or if the indeterminates in non-constant polynomials
      are different, returns NotImplemented.

      Returns
      -------
      :class:`UniPoly` or NotImplemented
          The sum of self and other, or NotImplemented.
      """

      if not self.__terms:
          return other

      # if other is an integer, create a constant polynomial
      if isinstance(other, int):
          other = UniPoly(other, self.__indeterminate, 0)
```

```
            # both objects are polynomials

        if isinstance(other, UniPoly):
            # if other == 0, return self
            if not other.__terms:
                return self

            if self.__indeterminate != other.__indeterminate:
                if self.is_constant and other.is_constant:
                    return UniPoly(self.constant_coefficient +
                                   other.constant_coefficient,
                                   self.__indeterminate,
                                   0)
                return NotImplemented

            new_poly = UniPoly(0, self.__indeterminate, 0)

            for exponent, coefficient in self.__terms.items():
                if exponent not in other.__terms:
                    new_poly.__terms[exponent] = coefficient
                else:
                    sum = self.__terms[exponent] \
                        + other.__terms[exponent]
                    if sum:
                        new_poly.__terms[exponent] = sum

            for exponent, coefficient in other.__terms.items():
                if exponent not in self.__terms:
                    new_poly.__terms[exponent] = coefficient

            return new_poly

        return NotImplemented

    def __radd__(self, other):
        """Compute other + self.

        Returns
        -------
        :class:`UniPoly`
            The sum of other and self.
```

```
    """

        # Addition is commutative, so we call __add__.

        return self.__add__(other)

    def __bool__(self):
        """Returns True if the polynomial is non-zero, False otherwise.

        Returns
        -------
        bool
        """

        return bool(self.__terms)

    def __call__(self, x):
        """Evaluates a polynomial by substituting `x` for the
indeterminate.

        Python calls this when it sees a polynomial being used
        in function application. If `p` is a polynomial, then
        `p(3.4)` substitutes 3.4 for the indeterminate and
        does the math as expressed by the polynomial.

        Parameters
        ----------
        x : any object that supports -, +, *, **
            The value to be substituted for the indeterminate.

        Returns
        -------
        Probably `type(x)`
        """
        result = 0

        for exponent, coefficient in self.__terms.items():
            result += coefficient * x**exponent

        return result

    def __copy__(self):
```

```
        """Copies a polynomial, returning a new polynomial.

        This creates a deep copy of a polynomial, even if the
        original was zero.

        See Also
        --------
        `deepcopy`

        Returns
        -------
        :class:`UniPoly`
            A complete copy of the polynomial.
        """
        new_poly = UniPoly(0, self.__indeterminate)
        for exponent in self.__terms:
            new_poly.__terms[exponent] = self.__terms[exponent]

        return new_poly

    def __deepcopy__(self, memo):
        """Copies a polynomial, returning a new polynomial.

        This creates a deep copy of a polynomial, even if the
        original was zero.

        See Also
        --------
        `copy`

        Returns
        -------
        :class:`UniPoly`
            A complete copy of the polynomial.
        """
        return self.__copy__()

    def __eq__(self, other):
        """Tests for equality between a polynomial and polynomial or
`int`.

        Parameters
```

```
        ----------
        other : :class:`UniPoly` or `int`
            The object to be be tested for equality

        Returns
        -------
        bool
            True if the objects are mathematically equal.
        """

        # If other is an `int`, turn it into an polynomial.
        if isinstance(other, int):
            other = UniPoly(other, self.__indeterminate, 0)

        if isinstance(other, UniPoly):
            # Look for a reason why they are not equal.

            # Do they have different indeterminates?
            if self.__indeterminate != other.__indeterminate:
                return False

            # Do they have different numbers of terms?
            if len(self.__terms) != len(other.__terms):
                return False

            # Are there terms of equal exponent and coefficient?
            for exponent in self.__terms:
                if exponent not in other.__terms:
                    # an exponent in self is missing from other
                    return False
                if self.__terms[exponent] != other.__terms[exponent]:
                    # the coefficients of terms of equal exponent
                    # are different
                    return False

            # They must be equal.
            return True

        return NotImplemented

    def __getitem__(self, exponent):
        """Retrieves the coefficient of the term with the exponent.
```

```
        Supports square bracket access to coefficients of terms in
        the polynomial. If there is no term with the given exponent,
        returns 0.

        Parameters
        ----------
        exponent : int
            A non-negative integer representing an exponent
            in a polynomial term.

        Returns
        -------
        int or NotImplemented
            The coefficient of the term with the given exponent.

        Raises
        ------
        ValueError
            Raised if the exponent is an integer but less than zero.
        """
        if not isinstance(exponent, int):
            return NotImplemented
        if exponent < 0:
            raise ValueError("polynomial exponents are integers >= 0")

        if not self.__terms:
            return 0
        return self.__terms.get(exponent, 0)

    def __hash__(self):
        """Computes a hash code for a polynomial.

        If p and q are equal polynomials, then their hash codes
        are equal.

        Returns
        -------
        int
            An integer hash code.
        """
        h = hash(self.__indeterminate)
```

```python
        for exponent_coefficient in self.__terms:
            h += hash(exponent_coefficient)
        return h

    def __int__(self):
        """Returns an `int` if the polynomial is constant.

        Returns
        -------
        int
            The constant term of the constant polynomial.

        Raises
        ------
        ValueError
            Raised if self is not constant.
        """
        if not self.__terms:
            return 0
        if len(self.__terms) == 1 and 0 in self.__terms:
            return self.__terms[0]
        raise ValueError("non-constant polynomial for int()")

    def __iter__(self):
        """Iterator returning exponent-coefficient tuple pairs.

        Notes
        -----
        The order of the pairs is not guaranteed to be sorted.

        Returns
        -------
        tuple:
            (exponent, coefficient) in polynomial.
        """

        return iter(self.__terms.items())

    def __mul__(self, other):
        """Multiplies a polynomial to a polynomial or an integer.

        Parameters
```

```
          ----------
      other : :class:`UniPoly` or int
          A polynomial or integer.

      Notes
      ------
      If the second argument is not a polynomial or integer,
      or if the indeterminates in non-constant polynomials
      are different, returns NotImplemented.

      Returns
      -------
      :class:`UniPoly` or NotImplemented
          The product of self and other, or NotImplemented.
      """


      # if self == 0, return it
      if not self.__terms:
          return self


      # if other is an integer, create a constant polynomial
      if isinstance(other, int):
          other = UniPoly(other, self.__indeterminate, 0)


      # both objects are polynomials


      if isinstance(other, UniPoly):
          # if other == 0, return other
          if not other.__terms:
              return other


          # We now have two non-zero polynomials.


          if self.__indeterminate != other.__indeterminate:
              if self.is_constant and other.is_constant:
                  return UniPoly(self.constant_coefficient *
                                 other.constant_coefficient,
                                 self.__indeterminate,
                                 0)
              return NotImplemented


          new_poly = UniPoly(0, self.__indeterminate, 0)
```

```python
        for exponent_1, coefficient_1 in self.__terms.items():
            for exponent_2, coefficient_2 in other.__terms.items():
                new_poly += UniPoly(coefficient_1 * coefficient_2,
                                    self.__indeterminate,
                                    exponent_1 + exponent_2)
        return new_poly

    return NotImplemented

def __rmul__(self, other):
    """Computes other * self.

    Returns
    -------
    :class:`UniPoly`
        The product of other and self.
    """

    # Multiplication is commutative, so we call __mul__.

    return self.__mul__(other)

def __neg__(self):
    """Negates a polynomial, returning a new polynomial.

    Returns
    -------
    :class:`UniPoly`
        A new polynomial.
    """
    # negate the polynomial term-wise unless it is already 0
    if not self.__terms:
        return self

    new_poly = UniPoly(0, self.__indeterminate)
    for exponent in self.__terms:
        new_poly.__terms[exponent] = -self.__terms[exponent]

    return new_poly

def __pow__(self, exponent):
```

```
        """Raises a polynomial to a non-negative integer exponent.

        `__pow__` used by both `**` and `pow`.

        Parameters
        ----------
        exponent : int
            The non-negative exponent to which self is raised.

        Notes
        -----
        If the exponent is not a non-negative integer, this method
        returns NotImplemented. This allows another class to implement
        quotients of polynomials and expressions like `p**(-1)`.

        Returns
        -------
        :class:`UniPoly` or NotImplemented
            The polynomial `self` raised to `exponent`, or
NotImplemented.
        """

        # Validate the exponent.
        if not isinstance(exponent, int) or exponent < 0:
            return NotImplemented

        # Handle the simple cases
        if exponent == 0:
            return UniPoly(1, self.__indeterminate, 0)

        if exponent == 1:
            return self

        if exponent == 2:
            return self * self

        # Compute the exponent by recursive repeated squaring.
        power = self**(exponent // 2)
        power *= power
        if exponent % 2 == 1:
            power *= self
```

```
        return power

def __repr__(self):
    """Returns the human-readable representation of a polynomial.

    Notes
    -----
    For simplicity, we are returning the result of `__str__`. We
    should be returning something like
    UniPoly(2, 'x', 6) + UniPoly(1, 'x', 3) + UniPoly(-1, 'x', 0)

    Returns
    -------
    str
    """
    return str(self)

def __str__(self):
    """Creates a human-readable string representation.

    This returns forms that look like 2x**6 + x**3 - 1.

    Returns
    -------
    str
        Mathematical human-readable form of the polynomial.
    """
    if not self.__terms:
        return '0'

    def format_term(coefficient, exponent):
        """Format a single term in the polynomial.

        This function formats a term and handles the special
        cases when the coefficient is +/- 1 or the exponent is
        0 or 1.

        Parameters
        ----------
        coefficient : int
            The coefficient of the term.
        exponent : int
```

```
            The exponent of the term.

        Returns
        -------
        str
            The human-readable representation of a polynomial term.
        """
        coefficient = abs(coefficient)

        if exponent == 0:
            return str(coefficient)

        if exponent == 1:
            if coefficient == 1:
                return self.__indeterminate
            return f"{coefficient}*{self.__indeterminate}"

        if coefficient == 1:
            return f"{self.__indeterminate}**{exponent}"

        return f"{coefficient}*{self.__indeterminate}**{exponent}"

    # Collect the formatted and sorted terms in result, taking
    # case to handle leading or middle negative signs.

    result = ""

    for exponent in sorted(self.__terms, reverse=True):
        coefficient = self.__terms[exponent]
        term = format_term(coefficient, exponent)

        if result:
            result += f" - {term}" \
                if coefficient < 0 else f" + {term}"
        else:
            result = f"-{term}" \
                if coefficient < 0 else term

    return result

def __sub__(self, other):
    """Subtracts a polynomial or an integer from a polynomial.
```

```
Parameters
----------
other : :class:`UniPoly` or `int`
    A polynomial or integer.


Notes
------
If the second argument is not a polynomial or integer,
or if the indeterminates in non-constant polynomials
are different, returns NotImplemented.


Returns
-------
:class:`UniPoly` or NotImplemented
    The difference of self and other, or NotImplemented.
"""
if not self.__terms:
    return -other


# if other is an integer, create a constant polynomial
if isinstance(other, int):
    other = UniPoly(other, self.__indeterminate, 0)


# both objects are polynomials

if isinstance(other, UniPoly):
    # if other == 0, return self
    if not other.__terms:
        return self

    if self.__indeterminate != other.__indeterminate:
        if self.is_constant and other.is_constant:
            return UniPoly(self.constant_coefficient -
                           other.constant_coefficient,
                           self.__indeterminate,
                           0)
        return NotImplemented

    new_poly = UniPoly(0, self.__indeterminate, 0)

    for exponent, coefficient in self.__terms.items():
```

```
            if exponent not in other.__terms:
                new_poly.__terms[exponent] = coefficient
            else:
                sum = self.__terms[exponent] \
                    - other.__terms[exponent]
                if sum:
                    new_poly.__terms[exponent] = sum

        for exponent, coefficient in other.__terms.items():
            if exponent not in self.__terms:
                new_poly.__terms[exponent] = -coefficient


        return new_poly

    return NotImplemented

def __rsub__(self, other):
    """Computes other - self.

    Returns
    -------
    :class:`UniPoly`
        The difference of other and self.
    """

    # other - self == -(self - other)
    # For efficiency, this should be computed like __sub__
    # so we create fewer intermediate polynomials.

    return self.__sub__(other).__neg__()

# ---------------------------------------------------------------
# Properties
# ---------------------------------------------------------------

@property
def constant_coefficient(self):
    """The coefficient of the term with exponent 0."""
    if self.__terms and 0 in self.__terms:
        return self.__terms[0]
    return 0
```

```python
    @property
    def degree(self):
        """Returns the largest exponent in the polynomial."""
        if self.__terms:
            max_degree = None
            for exponent in self.__terms:
                if max_degree is None or exponent > max_degree:
                    max_degree = exponent
        else:
            return 0


    @property
    def indeterminate(self):
        """The indeterminate (or variable) in the polynomial."""
        return self.__indeterminate


    @property
    def is_constant(self):
        """True if the polynomial only contains a term of exponent
0."""

        if not self.__terms:
            return True
        if len(self.__terms) != 1:
            return False
        return next(iter(self.__terms)) == 0


    @property
    def minimum_degree(self):
        """The smallest exponent among the non-zero terms, or 0."""
        if self.__terms:
            min_degree = None
            for exponent in self.__terms:
                if min_degree is None or exponent < min_degree:
                    min_degree = exponent
        else:
            # Return 0 for the zero polynomial
            return 0


    # ------------------------------------------------------------------
    # Instance methods
    # ------------------------------------------------------------------
```

```python
    def derivative(self):
        """Computes the derivative of a polynomial.

        Returns
        -------
        :class:`UniPoly`
            The derivative.
        """

        # Initialize the result to the zero polynomial
        # with the correct indeterminate.
        result = UniPoly(0, self.__indeterminate, 0)

        # Compute the term-by-term derivative, and sum.
        if self.__terms:
            for exponent, coefficient in self.__terms.items():
                if exponent > 0:
                    result += UniPoly(exponent * coefficient,
                                      self.__indeterminate,
                                      exponent - 1)
        return result

# Sample tests for pytest


def test_good_addition():
    """Tests that an addition is correct."""
    x = UniPoly(1, 'x', 1)
    p = x + 1
    q = x - 1
    assert p + q == 2 * x


def test_bad_multiplication():
    """Tests that a multiplication fails."""
    x = UniPoly(1, 'x', 1)
    p = x + 1
    q = x - 1
    assert p * q == x * x + 1
```

D

The Complete Guitar Class Hierarchy

This appendix contains the complete and final definition of the **Guitar** class for guitars that we developed in *Chapter 7, Organizing Objects into Classes*.

```
"""The guitar module implements the class hierarchy from Chapter 7.

Wooden          MusicalInstrument (ABC)
  |               |
  |               +-- Clarinet (ABC)
  |               |
  |               +-- Guitar
  |                     |
  |                     +-- ElectricGuitar
  |                     |
  +---------------+-- AcousticGuitar
"""
from abc import ABC, abstractmethod
from enum import Enum


class MusicalInstrument(ABC):
    """Abstract base class for musical instruments.

    Parameters
    ----------
    brand : str
```

```
        Manufacturer brand name. E.g., "Fender".
    model : str
        Instrument model name. E.g., "Stratocaster".
    year_built : str
        Year the instrument was built.
    """
    def __init__(self, brand, model, year_built):
        self._brand = brand
        self._model = model
        self._year_built = year_built

    @abstractmethod
    def __str__(self):
        """Abstract method -
        creates a human-readable string representation.

        Returns
        -------
        :class:`NotImplementedType`
            NotImplemented
        """
        return NotImplemented


class Guitar(MusicalInstrument):
    """Guitar.

    Parameters
    ----------
    brand : str
        Manufacturer brand name. E.g., "Fender".
    model : str
        Instrument model name. E.g., "Stratocaster".
    year_built : str
        Year the instrument was built.
    strings : int, optional
        Number of strings. Default is 6.

    Also See
    --------
    :class"`MusicalInstrument`
    """
```

```python
    def __init__(self, brand, model, year_built, strings=6):
        super().__init__(brand, model, year_built)
        self._strings = strings

    def __str__(self):
        """Creates a human-readable string representation.

        Returns
        -------
        str
            Description of the guitar.
        """
        return f"{self._strings}-string " \
            + f"{self._year_built} " \
            + f"{self._brand} {self._model}"


class Pickup(Enum):
    """Enumeration type of pickups for electric guitars.
    """
    UNKNOWN = "unknown"
    SINGLECOIL = "single coil"
    HUMBUCKER = "Humbucker"
    ACTIVE = "active"
    GOLDFOIL = "gold foil"
    TOASTER = "toaster"


class ElectricGuitar(Guitar):
    """Electric guitar.

    Parameters
    ----------
    brand : str
        Manufacturer brand name. E.g., "Fender".
    model : str
        Instrument model name. E.g., "Stratocaster".
    year_built : str
        Year the instrument was built.
    strings : int, optional
        Number of strings. Default is 6.
```

```
        Also See
        --------
        :class"`Guitar`
        """

    def __init__(self, brand, model, year_built, strings=6):
        super().__init__(brand, model, year_built, strings)
        self._pickup = Pickup.UNKNOWN


    def __str__(self):
        """Creates a human-readable string representation.

        Returns
        -------
        str
            Description of the electric guitar.
        """
        return f"{self._strings}-string " \
            + f"{self._year_built} " \
            + f"{self._brand} {self._model} " \
            + f"with {self._pickup.value} pickup"

    @property
    def pickup(self):
        """Style of pickup used in an electric guitar."""
        return self._pickup

    @pickup.setter
    def pickup(self, value):
        if not isinstance(value, Pickup):
            raise ValueError("The pickup must be of "
                             "enumeration type Pickup.")
        self._pickup = value


class Wooden:
    """Object made of wood

    Parameters
    ----------
    wood_name : str
        Name of wood species in object.
    """
```

```python
    def __init__(self, wood_name):
        self._wood_name = wood_name

    def __str__(self):
        """Creates a human-readable string representation.

        Returns
        -------
        str
            Name of the wood used.
        """
        return self._wood_name


class AcousticGuitar(Guitar, Wooden):
    """Acoustic guitar.

    Parameters
    ----------
    brand : str
        Manufacturer brand name. E.g., "Fender".
    model : str
        Instrument model name. E.g., "Stratocaster".
    year_built : str
        Year the instrument was built.
    soundboard_wood : str
        Kind of wood used in the guitar soundboard.
    strings : int, optional
        Number of strings. Default is 6.

    Also See
    --------
    :class"`Guitar`
    :class"`Wooden`
    """
    def __init__(self, brand, model, year_built,
                 soundboard_wood, strings=6):
        super().__init__(brand, model, year_built, strings)
        Wooden.__init__(self, soundboard_wood)

    def __str__(self):
        """Creates a human-readable string representation.
```

```
        Returns
        -------
        str
            Description of the acoustic guitar.
        """
        return f"{self._strings}-string " \
            + f"{self._year_built} " \
            + f"{self._brand} {self._model} " \
            + f"with {self._wood_name} soundboard"


class Clarinet(MusicalInstrument):
    """Abstract base class derived from MusicalInstrument.

    See Also
    --------
    MusicalInstrument
    """

    pass
```

E
Notices

In this appendix, we document the trademarks and legal notices associated with the software we mention in the book.

Topics covered in this appendix

E.1 Photos, images, and diagrams

Unless otherwise noted, all photographs, images, and diagrams were created by the author, Robert S. Sutor. They share the same copyright as the book itself.

The photo of Gus the cat in section 14.2 is by Katie Sutor and is used with permission.

The photo of the baseball field in section 15.5 is licensed under the Creative Commons Attribution-ShareAlike 4.0 International License:

```
https://creativecommons.org/licenses/by-sa/4.0/deed.en
```

The author retrieved it from Wikimedia Commons on May 22, 2021:

```
https://commons.wikimedia.org/wiki/File:2009_World_Baseball
_Classic_Canada_versus_USA_Rogers_Centre_Toronto.jpg
```

The author converted the photo to grayscale, cropped the upper portion, blurred brand names, and placed circles around the positions of the outfielders.

E.2 Data

The dataset "Registered Cats – Greater Dandenong" `registered-cats.csv` is made available by the City of Greater Dandenong under the Creative Commons Attribution 3.0 Australia (CC BY 3.0 AU): [CAT]

https://creativecommons.org/licenses/by/3.0/au/

No changes were made to the original database other than those described in *Chapter 14, Analyzing Data*, to demonstrate techniques of **pandas** DataFrame manipulation and data cleaning.

E.3 Trademarks

IBM®, Qiskit®, AIX®, z/OS®, and IBM Watson® are registered trademarks of IBM Corporation.

Alexa® and Echo® are registered trademarks of Amazon.com, Inc.

Excel®, Windows®, Visual Studio®, and Cortana® are registered trademarks of Microsoft Corporation.

Google Home and Google Sheets are trademark of Google LLC.

The Jupyter Trademark is registered with the U.S. Patent & Trademark Office.

Linux® is the registered trademark of Linus Torvalds in the U.S. and other countries.

Python is copyright © 2001-2021 Python Software Foundation.

The Python logo is a trademark of the Python Software Foundation: https://www.python.org/community/logos/.

PyTorch, the PyTorch logo, and any related marks are trademarks of Facebook, Inc.

Siri® and macOS® are registered trademarks of Apple Inc.

STAR TREK® is the registered trademark of CBS Studios Inc.

TensorFlow, the TensorFlow logo, and any related marks are trademarks of Google Inc.

E.4 Python 3 license

Python 3 is made available under the license at

https://docs.python.org/3/license.html.

At the time of the writing of this book, the license was for version 3.9.5:

1. This LICENSE AGREEMENT is between the Python Software Foundation ("PSF"), and the Individual or Organization ("Licensee") accessing and otherwise using Python 3.9.5 software in source or binary form and its associated documentation.

2. Subject to the terms and conditions of this License Agreement, PSF hereby grants Licensee a nonexclusive, royalty-free, world-wide license to reproduce, analyze, test, perform and/or display publicly, prepare derivative works, distribute, and otherwise use Python 3.9.5 alone or in any derivative version, provided, however, that PSF's License Agreement and PSF's notice of copyright, i.e., "Copyright © 2001-2021 Python Software Foundation; All Rights Reserved" are retained in Python 3.9.5 alone or in any derivative version prepared by Licensee.

3. In the event Licensee prepares a derivative work that is based on or incorporates Python 3.9.5 or any part thereof, and wants to make the derivative work available to others as provided herein, then Licensee hereby agrees to include in any such work a brief summary of the changes made to Python 3.9.5.

4. PSF is making Python 3.9.5 available to Licensee on an "AS IS" basis. PSF MAKES NO REPRESENTATIONS OR WARRANTIES, EXPRESS OR IMPLIED. BY WAY OF EXAMPLE, BUT NOT LIMITATION, PSF MAKES NO AND DISCLAIMS ANY REPRESENTATION OR WARRANTY OF MERCHANTABILITY OR FITNESS FOR ANY PARTICULAR PURPOSE OR THAT THE USE OF PYTHON 3.9.5 WILL NOT INFRINGE ANY THIRD PARTY RIGHTS.

5. PSF SHALL NOT BE LIABLE TO LICENSEE OR ANY OTHER USERS OF PYTHON 3.9.5 FOR ANY INCIDENTAL, SPECIAL, OR CONSEQUENTIAL DAMAGES OR LOSS AS A RESULT OF MODIFYING, DISTRIBUTING, OR OTHERWISE USING PYTHON 3.9.5, OR ANY DERIVATIVE THEREOF, EVEN IF ADVISED OF THE POSSIBILITY THEREOF.

6. This License Agreement will automatically terminate upon a material breach of its terms and conditions.

7. Nothing in this License Agreement shall be deemed to create any relationship of agency, partnership, or joint venture between PSF and Licensee. This License Agreement does not grant permission to use PSF trademarks or trade name in a trademark sense to endorse or promote products or services of Licensee, or any third party.

8. By copying, installing or otherwise using Python 3.9.5, Licensee agrees to be bound by the terms and conditions of this License Agreement.

Production Notes

I wrote the content for this book primarily in HTML. Additional content included:

- Static JPG images.
- LaTeX files using packages including `tikz`, `circuitikz`, and `quantikz`. I processed these with **pdflatex** and then the **convert** utility from **ImageMagick** to create JPG images.
- Python files using **matplotlib** for plots and charts, especially in *Chapter 13, Creating Plots and Charts*. I ran these in batch mode to generate JPG images.
- An SVG file created by **spacy** for the NLP dependency visualization in *Chapter 12, Searching and Changing Text*.

For many images, especially those I generated, I used **convert** from **ImageMagick** to remove extra surrounding whitespace.

I wrote several Python scripts to generate the book and features within it, such as index marking. For example, one script processed HTML chapter files and extracted the Python code. The script then ran the code through **Jupyter** in non-interactive mode to capture all Python output.

I used Python and the **python-docx** package to create a Microsoft Word docx file for proofing the book. I especially want to thank the developers of **python-docx** along with those who published fixes and extensions on the web.

The following sections list the details of my Python and LaTeX development environment during the book creation, though not all modules and packages were used for production.

Python version

Python 3.9.6 (tags/v3.9.6:db3ff76, Jun 28 2021, 15:26:21)
[MSC v.1929 64 bit (AMD64)]

Python modules and packages used

alabaster 0.7.12, **appdirs** 1.4.4, **argon2-cffi** 20.1.0, **astroid** 2.6.5, **async-generator** 1.10, **atomicwrites** 1.4.0, **attrs** 21.2.0, **autopep8** 1.5.7, **Babel** 2.9.1, **backcall** 0.2.0, **bandit** 1.7.0, **beautifulsoup4** 4.9.3, **black** 21.7b0, **bleach** 3.3.1, **blis** 0.7.4, **Brotli** 1.0.9, **cairocffi** 1.2.0, **CairoSVG** 2.5.2, **catalogue** 2.0.4, **certifi** 2021.5.30, **cffi** 1.14.6, **chardet** 4.0.0, **charset-normalizer** 2.0.4, **click** 7.1.2, **colorama** 0.4.4, **coverage** 5.5, **cryptography** 3.4.7, **cssselect2** 0.4.1, **cycler** 0.10.0, **cymem** 2.0.5, **Cython** 0.29.24, **debugpy** 1.4.1, **decorator** 5.0.9, **defusedxml** 0.7.1, **dill** 0.3.4, **dlx** 1.0.4, **dnspython** 1.16.0, **docopt** 0.6.2, **docplex** 2.21.207, **docutils** 0.16, **dodgy** 0.2.1, **en-core-web-sm** 3.0.0, **entrypoints** 0.3, **fastdtw** 0.3.4, **fastjsonschema** 2.15.1, **fett** 0.3.2, **flake8** 3.8.4, **flake8-polyfill** 1.0.2, **flashtext** 2.7, **fonttools** 4.25.2, **frosted** 1.4.1, **furo** 2021.7.31b41, **future** 0.18.2, **gitdb** 4.0.7, **GitPython** 3.1.18, **h5py** 3.2.1, **html5lib** 1.1, **idna** 3.2, **imagesize** 1.2.0, **inflection** 0.5.1, **iniconfig** 1.1.1, **ipykernel** 6.0.3, **ipython** 7.26.0, **ipython-genutils** 0.2.0, **ipywidgets** 7.6.3, **isort** 4.3.21, **jedi** 0.18.0, **Jinja2** 3.0.1, **joblib** 1.0.1, **jsonschema** 3.2.0, **jupyter** 1.0.0, **jupyter-client** 6.1.12, **jupyter-console** 6.4.0, **jupyter-core** 4.7.1, **jupyterlab-pygments** 0.1.2, **jupyterlab-widgets** 1.0.0, **kiwisolver** 1.3.1, **latexcodec** 2.0.1, **lazy-object-proxy** 1.4.3, **lxml** 4.6.3, **MarkupSafe** 2.0.1, **matplotlib** 3.4.2, **matplotlib-inline** 0.1.2, **matplotlib-venn** 0.11.6, **mccabe** 0.6.1, **mistune** 0.8.4, **more-itertools** 8.8.0, **mpmath** 1.2.1, **multitasking** 0.0.9, **murmurhash** 1.0.5, **mypy** 0.910, **mypy-extensions** 0.4.3, **nbclient** 0.5.3, **nbconvert** 6.1.0, **nbformat** 5.1.3, **nest-asyncio** 1.5.1, **networkx** 2.6.2, **notebook** 6.4.0, **ntlm-auth** 1.5.0, **numpy** 1.21.1, **packaging** 21.0, **pandas** 1.3.1, **pandocfilters** 1.4.3, **parso** 0.8.2, **pathspec** 0.9.0, **pathy** 0.6.0, **pbr** 5.6.0, **pep8** 1.7.1, **pep8-naming** 0.10.0, **pickleshare** 0.7.5, **pies** 2.6.7, **Pillow** 8.3.1, **pip** 21.2.2, **pluggy** 0.13.1, **ply** 3.11, **preshed** 3.0.5, **prometheus-client** 0.11.0, **prompt-toolkit** 3.0.19, **prospector** 1.3.1, **psutil** 5.8.0, **py** 1.10.0, **pybind11** 2.7.0, **pycodestyle** 2.7.0, **pycparser** 2.20, **pydantic** 1.8.2, **pydata-sphinx-theme** 0.6.3, **pydocstyle** 6.1.1, **pydot** 1.4.2, **pydyf** 0.0.3, **pyflakes** 2.3.1, **Pygments** 2.9.0, **pylama** 7.7.1, **pylatexenc** 2.10, **pylint** 2.9.6, **pylint-celery** 0.3, **pylint-django** 2.4.4, **pylint-flask** 0.6, **pylint-plugin-utils** 0.6, **pymongo** 3.11.4, **pyparsing** 2.4.7, **pyphen** 0.11.0, **pyroma** 3.2, **pyrsistent** 0.18.0, **pytest** 6.2.4, **python-constraint** 1.4.0, **python-dateutil** 2.8.2, **python-docs-theme** 2021.5, **python-docx** 0.8.11, **python-jsonrpc-server** 0.3.4, **python-pptx** 0.6.19, **pytz** 2021.1, **pywin32** 301, **pywinpty** 1.1.3,

PyYAML 5.4.1, **pyzmq** 22.1.0, **qiskit** 0.28.0, **qiskit-aer** 0.8.2, **qiskit-aqua** 0.9.4, **qiskit-ibmq-provider** 0.16.0, **qiskit-ignis** 0.6.0, **qiskit-terra** 0.18.0, **qtconsole** 5.1.1, **QtPy** 1.9.0, **Quandl** 3.6.1, **regex** 2021.7.6, **requests** 2.26.0, **requests-ntlm** 1.1.0, **requirements-detector** 0.7, **retworkx** 0.9.0, **rstcheck** 3.3.1, **scikit-learn** 0.24.2, **scipy** 1.7.1, **seaborn** 0.11.1, **Send2Trash** 1.7.1, **setoptconf** 0.2.0, **setuptools** 57.2.0, **six** 1.16.0, **smart-open** 5.1.0, **smmap** 4.0.0, **snooty-lextudio** 1.11.0.dev0, **snowballstemmer** 2.1.0, **soupsieve** 2.2.1, **spacy** 3.1.1, **spacy-legacy** 3.0.8, **Sphinx** 4.1.2, **sphinx-rtd-theme** 0.5.2, **sphinxcontrib-applehelp** 1.0.2, **sphinxcontrib-devhelp** 1.0.2, **sphinxcontrib-htmlhelp** 2.0.0, **sphinxcontrib-jsmath** 1.0.1, **sphinxcontrib-qthelp** 1.0.3, **sphinxcontrib-serializinghtml** 1.1.5, **squarify** 0.4.3, **srsly** 2.4.1, **stevedore** 3.3.0, **sty** 1.0.0rc1, **sympy** 1.8, **terminado** 0.10.1, **testpath** 0.5.0, **thinc** 8.0.8, **threadpoolctl** 2.2.0, **tinycss2** 1.1.0, **toml** 0.10.2, **tomli** 1.2.0, **tornado** 6.1, **tqdm** 4.62.0, **traitlets** 5.0.5, **tweedledum** 1.1.0, **typed-ast** 1.4.3, **typer** 0.3.2, **typing-extensions** 3.10.0.0, **ujson** 4.0.2, **urllib3** 1.26.6, **vulture** 2.3, **wasabi** 0.8.2, **watchdog** 2.1.3, **wcwidth** 0.2.5, **weasyprint** 53.0, **webencodings** 0.5.1, **websocket-client** 1.1.0, **websockets** 9.1, **wheel** 0.36.2, **widgetsnbextension** 3.5.1, **wrapt** 1.12.1, **XlsxWriter** 1.4.5, **yfinance** 0.1.63, **zopfli** 0.1.8

LaTeX environment

pdfTeX 3.141592653-2.6-1.40.23 (TeX Live 2021/W32TeX)

kpathsea version 6.3.3

Copyright 2021 Han The Thanh (pdfTeX) et al.

There is NO warranty. Redistribution of this software is

covered by the terms of both the pdfTeX copyright and

the Lesser GNU General Public License.

For more information about these matters, see the file

named COPYING and the pdfTeX source.

Primary author of pdfTeX: Han The Thanh (pdfTeX) et al.

Compiled with libpng 1.6.37; using libpng 1.6.37

Compiled with zlib 1.2.11; using zlib 1.2.11

Compiled with xpdf version 4.03

References

[ALG]   Dasgupta, S., Papadimitriou, C.H., & Vazirani, U.V. (2006). *Algorithms*. McGraw-Hill Higher Education. `https://books.google.com/books?id=DJSUCgAAQBAJ`

[APN]   Harris, Charles R., Millman, K. Jarrod, van der Walt, Stéfan J., Gommers, Ralf, Virtanen, Pauli, Cournapeau, David, Wieser, Eric, Taylor, Julian, Berg, Sebastian, Smith, Nathaniel J., Kern, Robert, Picus, Matti, Hoyer, Stephan, van Kerkwijk, Marten H., Brett, Matthew, Haldane, Allan, del Río, Jaime Fernández, Wiebe, Mark, Peterson, Pearu, Gérard-Marchant, Pierre, Sheppard, Kevin, Reddy, Tyler, Weckesser, Warren, Abbasi, Hameer, Gohlke, Christoph, & Oliphant, Travis E. (2020). Array programming with NumPy. *Nature, 585*(7825), 357-362. `https://www.nature.com/articles/s41586-020-2649-2`

[AST]   Amazon Web Services. Amazon Transcribe Streaming SDK. Retrieved on March 19, 2021. Retrieved from `https://pypi.org/project/amazon-transcribe/`

[AXM]   Jenks, Richard D., & Sutor, Robert S. (1992). *axiom: The Scientific Computation System*. New York, NY: Springer New York. `https://doi.org/10.1007/978-1-4612-2940-7`

[CAT]   City of Greater Dandenong. Registered Cats – Greater Dandenong. Retrieved on April 20, 2021. Retrieved from `https://data.gov.au/data/dataset/dandenong-registered-cats`

[CLD]   Mertz, David (2021). *Cleaning Data for Effective Data Science*. Packt Publishing. `https://subscription.packtpub.com/book/data/9781801071291`

[CNW]   Lamare, Amy. The Richest Person In The World Every Year, From 1987 to 2020. Retrieved on March 24, 2021. Retrieved from `https://www.celebritynetworth.com/articles/billionaire-news/richest-person-in-the-world-1987-to-2000/`

[CPA]   Knuth, Donald E. (1974). Computer Programming as an Art. *Commun. ACM, 17*(12), 667–673. `https://doi.org/10.1145/361604.361612`

[DAN]   The City of Greater Dandenong. Greater Dandenong: City of Opportunity. Retrieved on April 22, 2021. Retrieved from `https://www.greaterdandenong.vic.gov.au/`

[DIL]   McKerns, Mike. Dill. Retrieved on April 1, 2021. Retrieved from `https://pypi.org/project/dill/`

[DLG]   Goodfellow, Ian, Bengio, Yoshua, & Courville, Aaron (2016). *Deep Learning*. Adaptive Computation and Machine Learning series. MIT Press.

[DRQ]   The Quantum Daily. A Detailed Review of Qubit Implementations for Quantum Computing. Retrieved on May 18, 2021. Retrieved from `https://thequantumdaily.com/2020/05/21/tqd-exclusive-a-detailed-review-of-qubit-implementations-for-quantum-computing/`

[DWQ]   Sutor, Robert S. (2019). *Dancing with Qubits*. Packt Publishing. `https://www.packtpub.com/data/dancing-with-qubits`

[EAS]   Wikipedia. Extended ASCII. Retrieved on February 22, 2021. Retrieved from `https://en.wikipedia.org/wiki/Extended_ASCII`

[ECA]   Comer, Douglas (2017). *Essentials of Computer Architecture* (2nd ed.). Chapman & Hall/CRC.

[FIT]   Bennett, Liz. Five invaluable techniques to improve regex performance. *SolarWinds loggly*. Retrieved on March 18, 2021. Retrieved from `https://www.loggly.com/blog/five-invaluable-techniques-to-improve-regex-performance/`

[FPA]   IEEE (2019). *IEEE Std 754-2019 (Revision of IEEE 754-2008): IEEE Standard for Floating-Point Arithmetic*. IEEE.

[GSA]   Grover, Lov K.. A Fast Quantum Mechanical Algorithm for Database Search. *Proceedings of the Twenty-Eighth Annual ACM Symposium on the Theory of Computing, Philadelphia, Pennsylvania, USA, May 22-24, 1996*, 212–219.

[GSG]   Google. Google Python Style Guide. Retrieved on February 22, 2021. Retrieved from `https://google.github.io/styleguide/pyguide.html`

[GST]   Google. Using the Speech-to-Text API with Python. Retrieved on March 19, 2021. Retrieved from `https://codelabs.developers.google.com/codelabs/cloud-speech-text-python3/#0`

[HPY]   Wikipedia. History of Python. Retrieved on February 22, 2021. Retrieved from `https://en.wikipedia.org/wiki/History_of_Python`

[IDL]   Neary, Michael (2021). Getting Started With Python IDLE. *Real Python*. Retrieved on May 25, 2021. Retrieved from `https://realpython.com/python-idle/`

[INL]   IBM. Watson Natural Language Understanding. Retrieved on March 23, 2021. Retrieved from `https://www.ibm.com/cloud/watson-natural-language-understanding`

[IST]   IBM. The IBM Watson™ Speech to Text. Retrieved on March 19, 2021. Retrieved from `https://cloud.ibm.com/apidocs/speech-to-text?code=python`

[JSN]   IETF Trust (2017). The JavaScript Object Notation (JSON) Data Interchange Format, RFC 8259. Retrieved on February 22, 2021. Retrieved from `https://tools.ietf.org/html/rfc8259`

[JUP]   Project Jupyter (2021). Project Jupyter. Retrieved on May 26, 2021. Retrieved from `https://jupyter.org/`

[KER]   Keras Special Interest Group (2021). Keras: the Python deep learning API. Retrieved on May 20, 2021. Retrieved from `https://keras.io/`

[LQCQ]   Asfaw, Abraham, Corcoles, Antonio, Bello, Luciano, Ben-Haim, Yael, Bozzo-Rey, Mehdi, Bravyi, Sergey, Bronn, Nicholas, Capelluto, Lauren, Carrera Vazquez, Almudena, Ceroni, Jack, Chen, Richard, Frisch, Albert, Gambetta, Jay,

Garion, Shelly, Gil, Leron, De La Puente Gonzalez, Salvador, Harkins, Francis, Imamichi, Takashi, Kang, Hwajung, h. Karamlou, Amir, Loredo, Robert, McKay, David, Mezzacapo, Antonio, Minev, Zlatko, Movassagh, Ramis, Nannicini, Giacomo, Nation, Paul, Phan, Anna, Pistoia, Marco, Rattew, Arthur, Schaefer, Joachim, Shabani, Javad, Smolin, John, Stenger, John, Temme, Kristan, Tod, Madeleine, Wood, Stephen, & Wootton, James (2020). Learn Quantum Computation Using Qiskit. Retrieved on May 25, 2021. Retrieved from `http://community.qiskit.org/textbook`

[MAT]   The Matplotlib development team (2021). Matplotlib: Visualization with Python. Retrieved on February 24, 2021. Retrieved from `https://matplotlib.org/`

[MOL]   Quine, W. V. (1982). *Methods of logic* (4th ed.). Cambridge, MA: Harvard University Press.

[MPL]   The Matplotlib development team (2021). Matplotlib: matplotlib.pyplot.plot. Retrieved on February 28, 2021. Retrieved from `https://matplotlib.org/stable/api/_as_gen/matplotlib.pyplot.plot.html`

[NDG]   numpydoc Maintainers (2019). numpydoc Docstring Guide. Retrieved on February 22, 2021. Retrieved from `https://numpydoc.readthedocs.io/en/latest/format.html`

[NLTK]   NLTK Project (2020). Natural Language Toolkit. Retrieved on March 23, 2021. Retrieved from `https://www.nltk.org/`

[NPY]   NumPy (2020). NumPy: The fundamental package for scientific computing with Python. Retrieved on February 22, 2021. Retrieved from `https://numpy.org/`

[NUM]   Johansson, Robert (2019). *Numerical Python Scientific Computing and Data Science Applications with Numpy, SciPy and Matplotlib* (2nd ed.). Berkeley, CA: Apress. `https://doi.org/10.1007/978-1-4842-4246-9`

[PAN]   The pandas Development Team (2021). pandas. Retrieved on April 25, 2021. Retrieved from `https://pandas.pydata.org/`

[PCB]   Harrison, Matt, & Petrou, Theodore (2020). *Pandas 1. x Cookbook: Practical Recipes for Scientific Computing, Time Series Analysis, and Exploratory Data Analysis Using Python* (2nd ed.). Birmingham, England: Packt Publishing.

[PDS]   Bader, Dan (2021). Common Python Data Structures. *Real Python*. Retrieved on February 22, 2021. Retrieved from `https://realpython.com/python-data-structures/`

[PEP008]   Python Software Foundation (2013). PEP 8 – Style Guide for Python Code. Retrieved on February 22, 2021. Retrieved from `https://www.python.org/dev/peps/pep-0008/`

[PEP020]   Python Software Foundation (2004). PEP 20 – The Zen of Python. Retrieved on February 22, 2021. Retrieved from `https://legacy.python.org/dev/peps/pep-0020/`

[PEP247]   Python Software Foundation (2001). PEP 257 – Docstring Conventions. Retrieved on February 22, 2021. Retrieved from `https://www.python.org/dev/peps/pep-0257/`

[PEP485]   Python Software Foundation (2015). PEP 485 – A Function for testing approximate equality. Retrieved on February 22, 2021. Retrieved from `https://www.python.org/dev/peps/pep-0485/`

[PIL]   Lundh, Fredrik, Clark, Alex, & GitHub Contributors (2021). Pillow. Retrieved on February 22, 2021. Retrieved from `https://python-pillow.org/`

[PIW]   Skvortsov, Victor (2021). Python behind the scenes #8: how Python integers work. *Ten thousand meters*. Retrieved on February 22, 2021. Retrieved from `https://tenthousandmeters.com/blog/python-behind-the-scenes-8-how-python-integers-work/`

[PLR]   O'Grady, Stephen (2021). The RedMonk Programming Language Rankings: June 2021. Retrieved on August 11, 2021. Retrieved from `https://redmonk.com/sogrady/2021/08/05/language-rankings-6-21/`

[PLY]   Plotly (2021). Plotly Python Open Source Graphing Library. Retrieved on April 30, 2021. Retrieved from `https://plotly.com/python/`

[PML]   Raschka, Sebastian, & Mirjalili, Vahid (2019). *Python machine learning : machine learning and deep learning with python, scikit-learn, and tensorflow 2* (3rd ed.). Birmingham, England: Packt Publishing.

[PPM]   Sturz, John (2021). Python Modules and Packages – An Introduction. *Real Python*. Retrieved on February 22, 2021. Retrieved from `https://realpython.com/python-data-structures/`

[PTA]   Shor, Peter W. (1999). Polynomial-Time Algorithms for Prime Factorization and Discrete Logarithms on a Quantum Computer. *SIAM review, 41*(2), 303-332. `https://doi.org/10.1137/S0036144598347011`

[PVZ]   The pandas development team (2021). pandas Visualization. Retrieved on April 25, 2021. Retrieved from `https://pandas.pydata.org/pandas-docs/stable/user_guide/visualization.html`

[PYB]   Python Software Foundation. The Python Standard Library: Built-in Functions. Retrieved on February 22, 2021. Retrieved from `https://docs.python.org/3/library/functions.html`

[PYD]   Python Software Foundation. Python 3 Documentation. Retrieved on February 22, 2021. Retrieved from `https://docs.python.org/3/`

[PYF]   Python Software Foundation. The Python Standard Library: Format Specification Mini-Language. Retrieved on February 25, 2021. Retrieved from `https://docs.python.org/3/library/string.html#format-specification-mini-language`

[PYL]   Python Software Foundation. The Python Standard Library. Retrieved on February 22, 2021. Retrieved from `https://docs.python.org/3/library/`

[PYPI]   Python Software Foundation. Python Package Index. Retrieved on February 22, 2021. Retrieved from `https://pypi.org/`

[PYR1]   Python Software Foundation. Regular Expression HOWTO. Retrieved on March 12, 2021. Retrieved from `https://docs.python.org/3/howto/regex.html`

[PYR2]   Python Software Foundation. The Python Standard Library: re – Regular expression operations. Retrieved on March 18, 2021. Retrieved from `https://docs.python.org/3/library/re.html`

[PYT]   Python Software Foundation. The Python Tutorial. Retrieved on February 22, 2021. Retrieved from `https://docs.python.org/3/tutorial/`

[PYTST]   Krekel, Holger, & pytest-dev Team. pytest: helps you write better programs. Retrieved on February 22, 2021. Retrieved from `https://docs.pytest.org/en/latest/`

[PYU]   Python Software Foundation. Unicode HOWTO. Retrieved on February 22, 2021. Retrieved from `https://docs.python.org/3/howto/unicode.html`

[PYV]   Python Software Foundation. The Python Tutorial: Virtual Environments and Packages. Retrieved on July 15, 2021. Retrieved from `https://docs.python.org/3/tutorial/venv.html`

[Q40]   Preskill, John (2021). Quantum computing 40 years later. Retrieved on June 29, 2021. Retrieved from `https://arxiv.org/abs/2106.10522`

[QCP]   Loredo, Robert (2020). *Learn Quantum Computing with Python and IBM Quantum Experience*. Packt Publishing. `https://www.packtpub.com/product/learn-quantum-computing-with-python-and-ibm-q-experience/9781838981006`

[QEF]   Preskill, John (2012). Quantum computing and the entanglement frontier. Retrieved on June 29, 2021. Retrieved from `https://arxiv.org/abs/1203.5813v3`

[QIS]   Qiskit (2021). Qiskit: Open-Source Quantum Development. Retrieved on February 24, 2021. Retrieved from `https://qiskit.org/`

[QQP]   Qiskit Development Team. Qiskit IBM Quantum Provider. Retrieved on April 24, 2021. Retrieved from `https://pypi.org/project/qiskit-ibmq-provider/`

[RST]   Brandl, Georg, & the Sphinx team. reStructuredText. Retrieved on February 22, 2021. Retrieved from `https://www.sphinx-doc.org/en/master/usage/restructuredtext/`

[SCA]   Explosion (2021). spaCy: Annotation Specifications. Retrieved on March 22, 2021. Retrieved from `https://spacy.io/api/annotation`

[SCY]   Explosion (2021). spaCy: Industrial-Strength Natural Language Processing. Retrieved on March 22, 2021. Retrieved from `https://spacy.io/`

[SEA]   Waskom, Michael (2020). Seaborn: Statistical Data Visualization. Retrieved on April 29, 2021. Retrieved from `https://seaborn.pydata.org/`

[SKL]   scikit-learn Developers (2020). scikit-learn: Machine Learning in Python. Retrieved on May 20, 2021. Retrieved from `https://scikit-learn.org`

[SKLN]   scikit-learn Developers (2020). scikit-learn: Neural network models (supervised). Retrieved on May 20, 2021. Retrieved from `https://scikit-learn.org/stable/modules/neural_networks_supervised.html`

[SPH]   Brandl, Georg, & the Sphinx team. Sphinx Python Documentation Generator. Retrieved on August 14, 2021. Retrieved from `https://www.sphinx-doc.org/en/master/`

[SQU]   Laserson, Uri. squarify. Retrieved on April 29, 2021. Retrieved from `https://pypi.org/project/squarify/`

[TBL]   Loria, Steven. textblob. Retrieved on March 23, 2021. Retrieved from `https://pypi.org/project/textblob/`

[TCN]   Shi, Yaoyun (2002). Both Toffoli and controlled-Not need little help to do universal quantum computation. Retrieved on February 22, 2021. Retrieved from `https://arxiv.org/abs/quant-ph/0205115`

[TFL]   Google Inc. (2021). TensorFlow: Large-Scale Machine Learning on Heterogeneous Systems. Retrieved on May 20, 2021. Retrieved from `https://www.tensorflow.org/`

[TOR]   Facebook, Inc. (2020). PyTorch: An open source machine learning framework that accelerates the path from research prototyping to production deployment. Retrieved on March 12, 2021. Retrieved from `https://pytorch.org/`

[UNI]   The Unicode Consortium (2017). About the Unicode® Standard. Retrieved on February 22, 2021. Retrieved from `http://www.unicode.org/standard/standard.html`

[VEN]   Tretyakov, Konstantin (2020). matplotlib-venn. Retrieved on April 30, 2021. Retrieved from `https://pypi.org/project/matplotlib-venn/`

[VSC]   Microsoft, & open source developers (2021). Visual Studio Code. Retrieved on May 26, 2021. Retrieved from `https://code.visualstudio.com/`

[VSL]   Microsoft, & open source developers (2021). Linting Python in Visual Studio Code. Retrieved on May 27, 2021. Retrieved from `https://code.visualstudio.com/docs/python/linting`

[ZEN]   Janhangeer, Abdur-Rahmaan (2020). The Zen Of Python: A Most In Depth Article. *Python kitchen*. Retrieved on February 22, 2021. Retrieved from `https://www.pythonkitchen.com/zen-of-python-in-depth/`

www.packt.com

Subscribe to our online digital library for full access to over 7,000 books and videos, as well as industry-leading tools to help you plan your personal development and advance your career. For more information, please visit our website.

Why subscribe?

- Spend less time learning and more time coding with practical eBooks and Videos from over 4,000 industry professionals
- Learn better with Skill Plans built especially for you
- Get a free eBook or video every month
- Fully searchable for easy access to vital information
- Copy and paste, print, and bookmark content

Did you know that Packt offers eBook versions of every book published, with PDF and ePub files available? You can upgrade to the eBook version at www.Packt.com and as a print book customer, you are entitled to a discount on the eBook copy. Get in touch with us at customercare@packtpub.com for more details.

At www.Packt.com, you can also read a collection of free technical articles, sign up for a range of free newsletters, and receive exclusive discounts and offers on Packt books and eBooks.

Other Books You May Enjoy

If you enjoyed this book, you may be interested in these other books by Packt:

Learn Quantum Computing with Python and IBM Quantum Experience

Robert Loredo

ISBN: 978-1-83898-100-6

- Explore quantum computational principles such as superposition and quantum entanglement
- Become familiar with the contents and layout of the IBM Quantum Experience
- Understand quantum gates and how they operate on qubits
- Discover the quantum information science kit and its elements such as Terra and Aer
- Get to grips with quantum algorithms such as Bell State, Deutsch-Jozsa, Grover's algorithm, and Shor's algorithm
- How to create and visualize a quantum circuit

Quantum Computing in Practice with Qiskit® and IBM Quantum Experience®

Hassi Norlen

ISBN: 978-1-83882-844-8

- Visualize a qubit in Python and understand the concept of superposition
- Install a local Qiskit® simulator and connect to actual quantum hardware
- Compose quantum programs at the level of circuits using Qiskit® Terra
- Compare and contrast Noisy Intermediate-Scale Quantum computing (NISQ) and Universal Fault-Tolerant quantum computing using simulators and IBM Quantum® hardware
- Mitigate noise in quantum circuits and systems using Qiskit® Ignis
- Understand the difference between classical and quantum algorithms by implementing Grover's algorithm in Qiskit®

Packt is searching for authors like you

If you're interested in becoming an author for Packt, please visit `authors.packtpub.com` and apply today. We have worked with thousands of developers and tech professionals, just like you, to help them share their insight with the global tech community. You can make a general application, apply for a specific hot topic that we are recruiting an author for, or submit your own idea.

Share your thoughts

Now you've finished *Dancing with Python*, we'd love to hear your thoughts! Scan the QR code below to go straight to the Amazon review page for this book and share your feedback or leave a review on the site that you purchased it from.

https://packt.link/r/1-801-07785-1

Your review is important to us and the tech community and will help us make sure we're delivering excellent quality content.

Index

Index Formatting Examples

- Python function and method names: *insert, sort, set*
- Python keyword and reserved names: def, True, not in
- Python module names: **shutil, cmath, qiskit**
- Python class and type names: int, float, set

Function, method, and property index

Class index

Module and package index

General index